THEORY OF SIMPLE GLASSES

This pedagogical and self-contained text describes the modern mean field theory of simple structural glasses. The book begins with a thorough explanation of infinite-dimensional models in statistical physics, before reviewing the key elements of the thermodynamic theory of liquids and the dynamical properties of liquids and glasses. The central feature of the mean field theory of disordered systems, the existence of a large multiplicity of metastable states, is then introduced. The replica method is then covered, before the final chapters describe important, advanced topics such as Gardner transitions, complexity, packing spheres in large dimensions, the jamming transition and the rheology of glasses. Presenting the theory in a clear and pedagogical style, this is an excellent resource for researchers and graduate students working in condensed matter physics and statistical mechanics.

GIORGIO PARISI is a professor of physics at Sapienza University of Rome. His research is broadly focused on theoretical physics – from particle physics to glassy systems. He has been the recipient of numerous awards, including the Boltzmann Medal, the Enrico Fermi Prize, the Max Planck Medal, the Lars Onsager Prize and an ERC advanced grant. He is president of the Accademia dei Lincei and a member of the collaboration Cracking the Glass Problem, funded by the Simons Foundation.

PIERFRANCESCO URBANI is a CNRS researcher. His research activity focuses on statistical physics of disordered and glassy systems. After a joint PhD between Sapienza University of Rome and University of Paris-Sud, he joined the Institut de Physique Théorique of CEA, first as a post-doctoral researcher and then as a permanent researcher.

FRANCESCO ZAMPONI is a CNRS research director and an associate professor at Ecole Normale Supérieure. His research is broadly focused on complex systems, ranging from glasses to agent-based models for macroeconomy. He has been awarded an ERC consolidator research grant, and he is a member of the collaboration Cracking the Glass Problem, funded by the Simons Foundation.

THEORY OF SIMPLE GLASSES

Exact Solutions in Infinite Dimensions

GIORGIO PARISI

Sapienza University of Rome

PIERFRANCESCO URBANI

Institut de physique théorique, Université Paris Saclay, CNRS, CEA

FRANCESCO ZAMPONI

Ecole Normale Supérieure

CAMBRIDGE
UNIVERSITY PRESS

University Printing House, Cambridge CB2 8BS, United Kingdom

One Liberty Plaza, 20th Floor, New York, NY 10006, USA

477 Williamstown Road, Port Melbourne, VIC 3207, Australia

314–321, 3rd Floor, Plot 3, Splendor Forum, Jasola District Centre, New Delhi – 110025, India

79 Anson Road, #06–04/06, Singapore 079906

Cambridge University Press is part of the University of Cambridge.

It furthers the University's mission by disseminating knowledge in the pursuit of education, learning, and research at the highest international levels of excellence.

www.cambridge.org
Information on this title: www.cambridge.org/9781107191075
DOI: 10.1017/9781108120494

© Giorgio Parisi, Pierfrancesco Urbani and Francesco Zamponi 2020

First published 2020

A catalogue record for this publication is available from the British Library.

ISBN 978-1-107-19107-5 Hardback

Contents

Preface

State of the Art: In Search of a Fully Microscopic Theory of Glasses

Most classical solid state textbooks are almost entirely devoted to crystals, see, e.g., [21]. The main reason is that, while the theory of crystalline solids is fully developed, that of amorphous solids is still very incomplete. As a first approximation, crystals can be understood as perfectly symmetric periodic lattices, around which particles undergo small vibrations. A low-temperature harmonic expansion can then be constructed to obtain the thermodynamic properties in terms of harmonic excitations – i.e., phonons. Defects (mostly dislocations) can then be added to the theory to describe crystal flow (or plasticity) and melting [180, 212]. Crystals are well understood mainly because they can be thought of as small perturbations of a perfectly symmetric lattice, the small parameters being the amplitude of thermal vibrations and the density of defects.

Yet most of the solid matter in nature is not crystalline but amorphous: glasses, foams, pastes, granulars and plastics are but a few examples. These materials are not only ubiquitous but also extremely important for practical, everyday applications. For simplicity, in the rest of this book, we call these materials 'glasses'. Glasses display all kind of anomalies with respect to crystals: in particular, their vibrations cannot simply be understood in terms of plane waves, their flow is not mediated by well defined defects, and their dynamics is extremely complex. Unlike crystals, glasses offer no guiding symmetry principle to construct a microscopic theory, and no natural 'small' parameter can be used to organise a perturbative expansion.

Constructing a complete first-principle theory of glasses has then turned out to be an extremely difficult task. Yet a lot of progress has been made and, recently, several books on glasses written (or edited) by theoretical physicists [37, 53, 168, 357] have appeared. These books are largely devoted to the phenomenology of real (or realistic models of) materials, as known from experiments and numerical simulations, and the theoretical approaches they discuss mostly make use of approximate methods. This is, of course, an excellent idea given the complexity of the problem, and it is typical of theoretical physics.

The aim and style of this book is, however, quite different and, in our opinion, complementary to previous efforts. We discuss here the exact solution of a microscopically well-defined model which, we believe, can be taken as the simplest realistic model of a glass. By 'exact solution', we mean that one is able to compute in a mathematically exact way all the relevant observables of the model. Although the solution is not mathematically rigorous, we argue that it is exact from a theoretical physicist's perspective. We believe, as we discuss in the rest of this preface, that the material presented in this book constitutes a useful first step towards reaching the goal of constructing a complete and fully microscopic theory of glasses.

A Digression on the Structure of Scientific Theories

Let us make a philosophical digression to discuss what we can reasonably expect from a theory of glasses. A convenient definition of a 'scientific theory', given[1] by Lucio Russo in [312, section 1.3], is obtained by requiring the following three properties:

1. Its statements do not concern concrete objects pertaining to the real world but specific abstract mathematical objects.
2. It has a deductive structure: it is made by a few postulates concerning its objects and by a method to derive from them a potentially infinite number of consequences.
3. Its application to the real world is based on a series of 'correspondence rules' between the abstract objects of the theory and those of the real world.

According to Russo, a useful criterion for determining whether a theory has these properties is to check if one can compile a collection of exercises that can be solved within the theory. Solving a problem in the context of the theory is then nothing but an (arbitrarily difficult) 'exercise'.

As an example of this structure we can take Newtonian mechanics, where (1) the abstract objects are point particles interacting via forces, (2) the postulates are minimal but the theory is extremely powerful because from these postulates one can deduce an enormous variety of results, and (3) the point particles of the theory can be put in correspondence with many real-world objects, ranging from atoms to planets, depending on the context in which one wishes to use the theory.[2]

---

[1] Together with a nice discussion of its limits, that is not reproduced here.
[2] Note that we are nowadays used to this kind of logical structure, which is, however, the result of an extremely long historical process. Even the mathematical definition of 'point' has long been debated [312].

This example highlights that the choice of correspondence rules is extremely delicate. We know very well that atoms and planets are not point particles. They have a complex internal structure, which limits the applicability of the theory, giving rise to important physical phenomena.

'Scientific theories', as defined earlier, are powerful for two main reasons:

(i) Working on two parallel but distinct levels (the mathematical model and the real world) allows for a very flexible reasoning. In particular, one can guarantee the 'truth' of scientific statements by limiting them to the domain of the model.
(ii) The theory can be extended, by using the deductive method and introducing new correspondence rules, to treat situations that were not a priori included in the initial objectives for which the theory was developed.

At the same time, it is important to keep in mind that any 'scientific theory' has a limited utility. In general, it can only be used to model phenomena that are not too 'far' from those that motivated its elaboration. Theories that become inadequate to describe a new phenomenology must, for this reason, be substituted. They remain, however, according to our definition, 'scientific theories', and one can continue to use them in their domain of validity [312].[3] This last statement is particularly important because it reminds us that 'the' theory of a given class of phenomena – e.g., the glass transition – will never exist. Scientific theories are never unique or everlasting. They are models of reality, and there is no problem in using different models of the same phenomenon and in replacing current models with more powerful ones when they are found.

Towards a Scientific Theory of Glasses

The aim of this book is to make some steps in the application of the programme mentioned earlier to the problem of glasses. Let us state from the very beginning that we will not be able here to complete this programme. This book is mainly concerned with steps 1 and 2 – i.e., constructing an abstract mathematical theory that has a deductive structure and describes at least the basic phenomenology of glass formation and of the amorphous solid phase. The difficult problem of establishing a correspondence with the real world is left aside here. We give only hints and references to the literature so that the reader can form their own opinion on the proper correspondence rules and judge the quality of the theory according to their own taste. Another book will have to be written on the subject in the future.

---

[3] For instance, Newtonian mechanics did not become useless once quantum mechanics was developed, and the fact that it gives incorrect predictions (e.g., the instability of the hydrogen atom) does not mean that it is plain wrong.

We believe that the approach described above has important advantages, mostly based on points *(i)* and *(ii)*. Let us give two examples. Concerning point *(i)*, the problem of glasses is extremely complex and many different theories have been proposed. In the attempt to describe real-world materials, most of these theories make heavy use of approximations, to the point that it often becomes quite difficult to establish whether the statements made by the theory are true even within the logical structure of the theory itself. We instead introduce a simple and solvable mathematical model of glass: a system of Newtonian point particles in the limit of infinite spatial dimensions. Our aim is to discuss the mathematical solution of this model, which is already extremely rich and complex. But, although we believe the solution to be exact, a mathematically rigorous proof of its exactness is still lacking, and we hope that presenting the non-rigorous solution in a clear and aspirationally pedagogical way will help progress towards a rigorous proof. Our statements are limited to this mathematical model, and one is then able to decide whether they are true or false in a well-defined mathematical sense. Concerning point *(ii)*, we will see a spectacular example of its power in Chapter 9. The model, originally designed to describe the liquid and glass phases of atomic materials, also displays a phase transition that can be put in correspondence with the jamming transition of granular materials. Thus, the model shows a potential to unify phenomena (glass and jamming transitions) and materials (atomic glasses, colloidal glasses, granular glasses) that were thought to be somehow distinct.

The main, and very important, drawback of this approach is that the infinite-dimensional limit is quite abstract. Hence, the final step of establishing correspondence rules between the different phases and observables of the abstract model and their real-world counterparts (i.e., real liquids, real glasses, real granular materials) is non-trivial and remains largely open to debate. In granular materials, for example, the role of friction remains to be clarified. If successfully performed, this step would ultimately correspond to constructing a scientific theory of real-world glasses (i.e., to implement step 3), but it requires a lot of additional discussion which goes much beyond the scope of this book. Here, when discussing each specific aspect of the mathematical solution of the model, we limit ourselves to a few hints and references to direct the reader towards real-world phenomena that could potentially be described by this solution. This is done at the end of each chapter, in the Further Reading sections. We do not, however, specify completely the list of phenomena that could be accounted for by such a theory, nor do we try to establish precise correspondence rules between the mathematical model discussed here and real-world objects. Discussing this issue with all the needed details will be part of the follow-up publications to the present book.

Historical Note

The idea of using infinite-dimensional solvable models has been extremely successful in condensed matter, appearing in the context of atomic physics [337], liquids [154, 206], ferromagnetic systems [163] and strongly correlated electrons, where it has led to the celebrated dynamical mean field theory [164]. See also [356] for related ideas in high-energy physics.

In the context of the glass transition, the idea of solving the problem in the infinite dimensional limit was first proposed in [206]. A complete and exact solution of the infinite-dimensional problem was then obtained more recently, in a long series of research articles to which we contributed together with many other colleagues [3, 88, 89, 220, 221, 239, 291, 292, 299]. The methods used in this solution are deeply rooted in the theory of spin glasses [79, 254], using dynamical methods [109, 111] and replica methods [144, 260] specifically developed for the glass problem. These methods were first applied to approximate the behaviour of glasses in finite dimensions [76, 252, 253, 256, 292, 366, 368] before it was realised that they become exact in the infinite-dimensional limit [220, 291]. The phase diagram turns out to be similar to that of a class of spin glass models (Ising p-spin and Potts glasses [79, 160, 170, 171, 203, 204, 207]), which confirms the main assumption behind the random first-order transition (RFOT) theory of the glass transition [205, 208, 357]. The solution also reproduces the essential features of the mode-coupling theory of the glass transition [168], as discussed in [239].

The aim of this book is to collect these results and organise them in a pedagogically coherent way. We did not include here any new material (except for some polishing of the original work), and we do not wish to take any additional credit for the results. The original papers are thus carefully referenced along the book.

Target Audience

This book is written having in mind two distinct types of readers. The first are young students (at the level of the final year of undergraduate studies or at the beginning of their graduate studies). We expect these readers to have no or very little background knowledge of the physics of glasses. Reading this book only requires a basic background in statistical mechanics: the mean field (Landau) approach to phase transitions in magnetic materials and some basics in liquid state theory (as covered, e.g., in [175]). We ask these readers to believe our conjecture that a system of infinite-dimensional atoms can be a good mathematical model of glass and to follow us in the mathematical study of the model. Along the way, they will learn advanced statistical mechanics techniques (e.g., the replica method) as well as many deep concepts that pervade the physics of disordered systems (e.g., long-lived metastable states, complex free energy landscapes, ergodicity breaking). By working out all the calculations in this example, they will learn methods that can be used in many different contexts ranging from spin glasses [254] to optimisation problems [251] and neural networks [10]. However, they will not acquire sufficient background about the phenomenology of glasses and on the many different approaches that are used in their theoretical description. For this purpose, we refer to other existing excellent books and reviews [37, 40, 53, 80, 168, 357].

The second group of readers are experienced researchers working on the physics of glasses. We expect them to be already acquainted with the main physical concepts discussed in the book and to be familiar with the material contained in [37, 40, 53, 80, 168, 357]. Yet we hope that these readers will find here a way to put many different pieces of knowledge into a common perspective. Some of these more experienced readers might also be interested in learning the details of the methods mentioned earlier.

Structure of the Book

The book presents the main logical steps of the derivation of the exact solution of amorphous infinite-dimensional particle systems. All the non-trivial steps are presented, but leaving to the reader some trivial intermediate steps (for which we provide references to the original work). At the end of each chapter, a Summary section is provided to recapitulate the main points discussed in the chapter. The correspondence with real-world objects, as well as the comparison with approximate theories of glasses, is logically separated from the main stream of the book.[4] Some elements are presented in short Further Reading sections at the end of the relevant chapters.

The structure of the book chapters is the following.

- Chapter 1 reviews classical results in the statistical mechanics of infinite-dimensional systems, using the Ising model as a paradigmatic example. The notion of metastability is introduced and discussed.
- Chapter 2 provides a short review of classical results in liquid state theory [175] and then introduces the strategy to solve the thermodynamics of atomic liquids in infinite dimensions.
- Chapter 3 presents the solution of the equilibrium liquid dynamics in infinite dimensions by a series of simple arguments, following Szamel [339].

---

[4] Except in Chapter 9, where the problem is easier and some discussion of this correspondence is provided in the main text.

The existence of a sharp dynamical glass transition, at which the equilibrium diffusion constant vanishes, is discussed.

- Chapter 4 discusses the central feature of the mean field theory of disordered systems, namely the existence of a large multiplicity of metastable states. The replica method is introduced in this context, following Franz and Parisi [144]. The appearance of metastable states is directly connected to the dynamical arrest of the liquid. The exact expression of the replicated free energy of an atomic system is derived, following the ideas of [220, 299] and the detailed derivation of [59]. From this basic object, the glass phase diagram of two model systems is derived. This chapter also contains the core of the mathematical solution of the model.

- Chapter 5 provides a short compendium of basic notions of replica symmetry breaking (RSB) and the associated ultrametric distribution of states in phase space. This chapter is a review of classical results in spin glass theory [254].

- Chapter 6 discusses RSB effects in the Franz-Parisi construction introduced in Chapter 4. The phase diagram of two simple glass models is derived. The notion of a Gardner transition is introduced, and its physical meaning is discussed using the results of Chapter 5. This chapter reviews results originally presented in [88, 299, 315].

- Chapter 7 introduces another replica scheme due to Monasson [260], which allows one to study a different class of metastable states. The phase diagram of hard spheres is derived and compared with that of Chapter 6. The connection with the Edwards approach to the study of jammed packings is discussed. This chapter reviews results originally presented in [88, 93, 291, 292].

- Chapter 8 introduces the sphere-packing problem. A brief review of mathematical and physical results is provided. It is shown how the results obtained in Chapters 4, 6 and 7 provide insight on this problem. This chapter reviews results originally presented in [240, 291, 292, 325].

- Chapter 9 focuses on the jamming transition. It is explained why jamming is a critical point, and the associated critical exponents are described. The non-trivial scaling form of the RSB equations in the vicinity of jamming is discussed in detail. This chapter presents results originally derived in [88, 89, 91].

- Chapter 10 discusses the response of the glass to an applied shear strain. The elasticity and yielding of the glass are discussed. This chapter presents results originally derived in [58, 298, 299, 346].

The reader has certainly noticed that the title of this book makes explicit reference to the classic *Theory of Simple Liquids* by Hansen and McDonald [175]. Our intention is indeed to apply to glasses the same program that was applied to liquids in the 1960s. The word 'simple' refers, here and in [175], to the fact that we only consider the simplest model of liquids and glasses: a collection of classical point particles, modelling atoms. There are, of course, much more complex glass-forming systems (e.g., polymer glasses and network glasses, or anisotropic granular systems like hard ellipsoids), but their description falls beyond the scope of this book.

Acknowledgements

We would like to especially thank Patrick Charbonneau, who – besides participating actively to the research effort that has led to this book – has carefully reviewed the original manuscript and provided very useful comments.

The main results presented in this book are the outcome of a long and much broader research effort that involved many important ideas and tools in this and related fields. This has been possible thanks to the efforts of many people, colleagues and students. It is a pleasure to warmly thank all of them. We would like, in particular, to thank our most recent collaborators that have been directly involved in the results presented in this book: Elisabeth Agoritsas, Ada Altieri, Ludovic Berthier, Giulio Biroli, Jean-Philippe Bouchaud, Carolina Brito, Eric Corwin, Silvio Franz, Sungmin Hwang, Atsushi Ikeda, Harukuni Ikeda, Hugo Jacquin, Jorge Kurchan, Yuliang Jin, Thibaud Maimbourg, Marc Mézard, Corrado Rainone, Federico Ricci-Tersenghi, Camille Scalliet, Antonio Sclocchi, Mauro Sellitto, Guilhem Semerjian, Beatriz Seoane, Gilles Tarjus, Marco Tarzia, Matthieu Wyart and Hajime Yoshino. We would also like to thank all the members of the Simons Collaboration on Cracking the Glass Problem for many useful discussions and for providing an ideal environment for developing this project.

Part of the research described in this book has received funding from the European Research Council (ERC) under the European Union's Horizon 2020 research and innovation programme (grant agreement n. 694925 – Low Temperature Glassy System, Giorgio Parisi and n. 723955 – GlassUniversality, Francesco Zamponi), by a grant from the Simons Foundation (#454949, Giorgio Parisi and #454955, Francesco Zamponi) and by the Investissements d'Avenir program of the French government via the LabEx PALM (ANR-10-LABX-0039-PALM).

1

Infinite-Dimensional Models in Statistical Physics

Infinite-dimensional models are core to statistical physics. They can be used to understand liquids and glasses, as they are in this book [153, 206, 292, 362], but also strongly coupled electrons [164], atomic physics [337] and gauge field theory [132], to name a few. The reason is that infinite-dimensional models are exactly solvable using mean field methods. The aim of this chapter is to give an example of this construction in the context of the Ising model of magnetism.

It will be assumed that the reader is already familiar with the basic properties of the Ising model as presented, for example, in the first chapters of [69]: its main observables (magnetisation, magnetic susceptibility), its phase diagram and phase transitions, and its dynamics. The aim of this chapter is mostly to present these properties in the context of a large dimensional expansion and to introduce the concept of a thermodynamic (stable or metastable) state, identified with a local minimum of a suitable free energy function.

1.1 The Ising Model

1.1.1 Definitions

Although some of the concepts presented in this chapter are fairly general, it is instructive to focus on a specific setting. We thus consider a model of N Ising spins $\{S_i\}_{i=1,\ldots,N}$, with $S_i = \pm 1$. The energy of a spin configuration $\underline{S} = \{S_i\}$ is given by the Hamiltonian function

$$H[\underline{S}] = -\frac{1}{2} \sum_{ij} J_{ij} S_i S_j - \sum_i B_i S_i, \qquad (1.1)$$

where $J_{ij} = J_{ji} \in \mathbb{R}$ denotes the (symmetric) exchange coupling between spins i and j (with $J_{ii} = 0, \forall i$), and $B_i \in \mathbb{R}$ denotes an external magnetic field acting on spin i. The factor $1/2$ in front of the exchange energy ensures that each pair ij is counted only once. The Hamiltonian in Eq. (1.1) summarises many cases of interest,

1

such as each spin interacting with all others (the 'fully connected' model) [69], or with a random subset of them (the 'Bethe lattice' or 'random graph' model) [251]. In general, the set of non-zero couplings defines the 'interaction graph'; its nodes are the spins and its edges are the pairs $\langle ij \rangle$, such that $J_{ij} \neq 0$. In the following, the neighbourhood of spin i, denoted by

$$\partial i = \{j : J_{ij} \neq 0\}, \tag{1.2}$$

is the set of all spins j that interact with spin i.

The ferromagnetic Ising model in d dimensions corresponds to a particular choice of interaction graph: a d-dimensional cubic lattice of unit spacing and linear size L, containing $N = L^d$ lattice points. Each spin i sits at a point $\mathbf{x}(i)$ of the lattice and interacts only with its nearest neighbours, located at unit distance away in each principal direction. Because there is a bijective correspondence between labels $i = 1, \ldots, N$ and lattice points $\mathbf{x}$, both labels can be used equivalently. In the following, spins will thus be denoted either S_i or $S_{\mathbf{x}}$ depending on which labelling is more convenient. Each pair of nearest neighbour spins interact via a coupling $J_{ij} = J = \frac{1}{2d}$, and $J_{ij} = 0$ otherwise. In lattice notation, this corresponds to $J_{\mathbf{xy}} = \delta_{|\mathbf{x}-\mathbf{y}|,1}/(2d)$, where $|\mathbf{x} - \mathbf{y}|$ is the Euclidean distance between points $\mathbf{x}, \mathbf{y}$ and δ_{ab} is the Kronecker delta.

Because a cubic lattice has a boundary, one needs to specify boundary conditions. Three choices are commonly used [158].

- **Periodic boundary conditions** – Each face of the lattice is identified with its opposite face. In this case, the lattice is translationally invariant in all directions, and each spin has $2d$ nearest neighbours, corresponding to displacements in all possible directions on the lattice; hence, the size of ∂i is $2d$. In $d = 2$, the topology is that of a torus, generalised to a hypertorus in $d \geq 3$.
- **Open boundary conditions** – The lattice is considered isolated. In this case, the system is not translationally invariant. In particular, the spins on the boundary have fewer interactions than those in the bulk.
- **Frozen boundary conditions** – A layer of external spins is added to each face of the lattice. The external spins are frozen in prescribed positions: for example, they are all fixed to $+1$ or to -1. In this case, each spin has $2d$ nearest neighbours, but spins on the boundary interact with one frozen spin (or more, for those on cube edges), acting as an effective external magnetic field. Also in this case, the system is not translationally invariant.

Note that the overall magnitude of the couplings J only sets the energy and temperature scales and is therefore irrelevant for the properties of the model. The choice $J = \frac{1}{2d}$ guarantees that the exchange energy remains finite when $d \to \infty$, for any value of N. For instance, with periodic boundary conditions, the fully magnetised spin configuration $\underline{1} = \{S_i = 1, \forall i\}$ has

$$H[\underline{1}] = -\sum_{\langle ij \rangle} J_{ij} - \sum_i B_i = -NJd - \sum_i B_i = -\frac{N}{2} - \sum_i B_i, \qquad (1.3)$$

which explicitly shows that the exchange energy is finite for any N. Because the exchange energy remains of the same order as the entropy and the magnetic field energy, the model behaviour is interesting at finite temperature and magnetic field.[1]

1.1.2 Thermodynamic Free Energy and Observables

In equilibrium statistical mechanics, within the canonical ensemble, the probability of observing a spin configuration $\underline{S}$ is given by the Gibbs–Boltzmann distribution

$$P_{\mathrm{GB}}[\underline{S}] = \frac{e^{-\beta H[\underline{S}]}}{Z}, \qquad (1.4)$$

where $Z = \sum_{\underline{S}} e^{-\beta H[\underline{S}]}$ is the partition function and $\beta = 1/T$ is the inverse temperature.[2] Solving the model amounts to computing the free energy of the system at temperature T:

$$F = -T \log Z = -T \log \sum_{\underline{S}} e^{\frac{\beta}{2} \sum_{ij} J_{ij} S_i S_j + \beta \sum_i B_i S_i}. \qquad (1.5)$$

Note that F is extensive – i.e., proportional to N. In the following, capital letters are used for extensive thermodynamic quantities, while lowercase letters are used for intensive quantities that remain finite in the thermodynamic limit, $N \to \infty$ – e.g., $f = F/N$. From the free energy, one can derive the statistical average of any observable, which we denote by brackets $\langle \bullet \rangle$. For example, the average energy is

$$U = \langle H[\underline{S}] \rangle = \sum_{\underline{S}} H[\underline{S}] \frac{e^{-\beta H[\underline{S}]}}{Z} = \frac{\partial(\beta F)}{\partial \beta}, \qquad (1.6)$$

the entropy is

$$S = -\sum_{\underline{S}} P_{\mathrm{GB}}[\underline{S}] \log P_{\mathrm{GB}}[\underline{S}] = \beta(U - F), \qquad (1.7)$$

---

[1] All the thermodynamic quantities are functions of T/J and B/J for dimensional reasons. Hence, different choices of J require rescaling both T and B with d to avoid a trivial behaviour of the model dominated by entropy or energy. Once T and B are properly rescaled, the result is invariant. For instance, the choice $J = 1$ leads to an exchange energy of order d and thus requires $T, B \propto d$.

[2] The Boltzmann constant k_B is fixed to unity throughout this book. In other words, T stands for $k_B T$ in such a way that temperatures are measured in the same units as energies. Similarly, entropy S stands for S/k_B and is thus adimensional.

the local magnetisation is

$$m_i = \langle S_i \rangle = \sum_{\underline{S}} S_i \frac{e^{-\beta H[\underline{S}]}}{Z} = -\frac{\partial F}{\partial B_i}, \tag{1.8}$$

and the local magnetic susceptibility is

$$\chi_{ij} = \frac{\partial m_i}{\partial B_j} = -\frac{\partial^2 F}{\partial B_i \partial B_j} = \beta(\langle S_i S_j \rangle - \langle S_i \rangle \langle S_j \rangle). \tag{1.9}$$

By construction, χ_{ij} is a positive matrix; i.e., all its eigenvalues are greater than or equal to zero.[3] The global magnetic susceptibility is the variation of the global magnetisation $m = N^{-1} \sum_i m_i$ with respect to a uniform magnetic field $B_i = B, \forall i$. Under a variation of the global field, dB, all local fields change by $dB_i = dB$, hence $dB_i/dB = 1, \forall i$ and

$$\chi = \frac{dm}{dB} = \frac{1}{N} \sum_i \frac{dm_i}{dB} = \frac{1}{N} \sum_{ij} \frac{\partial m_i}{\partial B_j} \frac{dB_j}{dB} = \frac{1}{N} \sum_{ij} \chi_{ij}. \tag{1.10}$$

In general, any physical observable can be written as a linear combination of products of spin variables,[4]

$$\mathcal{O}[\underline{S}] = \sum_{n=0}^{N} \sum_{i_1 \cdots i_n} \mathcal{O}_{i_1 \cdots i_n} S_{i_1}, \ldots, S_{i_n}. \tag{1.11}$$

Therefore, in order to characterise all thermodynamic averages, it suffices to consider the average of spin products $\langle S_{i_1}, \ldots, S_{i_n} \rangle$, which can be computed as multiple derivatives of the free energy with respect to the relevant magnetic fields.

1.1.3 Free Energy as a Function of the Local Magnetisation

In order to investigate the high dimensional limit, it is convenient to introduce the free energy function $F[\underline{m}]$ which gives, roughly speaking, the free energy of a system constrained to have a prescribed set of local magnetisations $\underline{m} = \{m_i\}_{i=1,\ldots,N}$.

---

[3] Writing equivalently Eq. (1.9) as $\chi_{ij} = \beta \langle (S_i - m_i)(S_j - m_j) \rangle$, it follows, for any vector v, that

$$v^T \chi v = \sum_{ij} v_i \chi_{ij} v_j = \beta \sum_{ij} \langle v_i (S_i - m_i)(S_j - m_j) v_j \rangle = \beta \left\langle \left[\sum_i v_i (S_i - m_i) \right]^2 \right\rangle \geq 0.$$

This result holds in particular for the normalised eigenvectors of χ, for which $\chi v = \lambda v$, with λ the corresponding eigenvalue; hence, $v^T \chi v = \lambda v^T v = \lambda \geq 0$. This relationship proves that the eigenvalues are all positive.

[4] Equation (1.11) is easily justified by noting that $\mathcal{O}[\underline{S}]$ can take at most 2^N values, one for each spin configuration. The expression on the right-hand side of Eq. (1.11) contains $\binom{N}{n}$ coefficients for each n, which gives $\sum_{n=0}^{N} \binom{N}{n} = 2^N$ coefficients in total.

The local magnetisations are fixed by introducing a set of auxiliary local magnetic fields $\underline{b} = \{b_i\}_{i=1,...,N}$, and defining the corresponding auxiliary free energy:

$$\Omega[\underline{b}] = -T \log \sum_{\underline{S}} e^{-\beta H[\underline{S}] + \beta \sum_i b_i S_i}. \qquad (1.12)$$

Note that the 'physical' magnetic fields B_i are included in $H[\underline{S}]$, while the auxiliary fields are denoted explicitly as we will use them to constrain the local magnetisations. Obviously, Eqs. (1.8) and (1.9) hold equivalently if one takes derivatives of $\Omega[\underline{b}]$ with respect to the auxiliary fields b_i. Because the matrix of its second derivatives is $-\chi_{ij}$, which is negative, $\Omega[\underline{b}]$ is necessarily a concave function. The Legendre transform of $\Omega[\underline{b}]$ defines $F[\underline{m}]$ as

$$-\beta F[\underline{m}] = -\beta \max_{\underline{b}} \left[\Omega[\underline{b}] + \sum_i b_i m_i \right]$$

$$= \min_{\underline{b}} \left[\log \sum_{\underline{S}} e^{-\beta H[\underline{S}] + \beta \sum_i b_i (S_i - m_i)} \right], \qquad (1.13)$$

which can be justified as follows. For any finite N, $\Omega[\underline{b}]$ is everywhere differentiable, and its concavity ensures that the maximum over $\underline{b}$ exists and is unique. In this case, the value $\underline{b} = \underline{b}[\underline{m}]$ that corresponds to the maximum is the unique solution of $\partial\Omega[\underline{b}]/\partial b_i = -m_i$; i.e., it is the set of local fields $b_i[\underline{m}]$ needed to enforce the magnetisations m_i. Once these values are computed, the value of $\Omega[\underline{b}[\underline{m}]]$ gives the free energy of the system with field $\underline{b}[\underline{m}]$. By subtracting the additional magnetic energy due to the external field, $-\sum_i b_i m_i$ in Eq. (1.13), one obtains the free energy $F[\underline{m}]$ of the system constrained to have local magnetisation $\underline{m}$. However, in the thermodynamic limit, $\Omega[\underline{b}]$ can develop singularities (in the vicinity of a phase transition) and become non-differentiable. This complicates the discussion of the Legendre transform, as will be detailed in Section 1.4.

Note that the derivative of $F[\underline{m}]$ (when it exists) is the auxiliary field $\underline{b}[\underline{m}]$ that corresponds to the maximum in Eq. (1.13):

$$b_i = \frac{\partial}{\partial m_i} F[\underline{m}], \qquad \frac{\partial b_i}{\partial m_j} = \frac{\partial^2 F[\underline{m}]}{\partial m_i \partial m_j} = (\chi^{-1})_{ij}. \qquad (1.14)$$

The matrix χ^{-1} is positive, because χ is positive. $F[\underline{m}]$ is thus a convex function,[5] and the free energy $\Omega[\underline{b}]$ can be recovered as its inverse Legendre transform:

$$-\beta\Omega[\underline{b}] = -\beta \min_{\underline{m}} \left[F[\underline{m}] - \sum_i b_i m_i \right]. \qquad (1.15)$$

---

[5] Note that the convexity of $F[\underline{m}]$ also directly follows from its definition, Eq. (1.13), because the maximum (over $\underline{b}$) of linear functions (of $\underline{m}$) is a convex function.

The stationarity condition implies that $\underline{m}[\underline{b}]$ is a solution of Eq. (1.14), and it must be a minimum because the second derivative of $F[\underline{m}]$ is positive. This result leads to an important observation: if there are no auxiliary fields, $b_i = 0$, then $\Omega[\underline{b} = \underline{0}] = F$ is equal to the thermodynamic free energy, as defined in Eq. (1.5). We obtain from Eq. (1.15) that

$$F = \min_{\underline{m}} F[\underline{m}]. \tag{1.16}$$

The thermodynamic free energy thus corresponds to the minimum of $F[\underline{m}]$ over all possible sets of local magnetisations $\underline{m}$, and the set of local magnetisations that achieves the minimum in Eq. (1.16) corresponds to the equilibrium magnetisations in absence of any auxiliary field.

1.2 Large Dimension Expansion for the Ising Model

The function $F[\underline{m}]$ contains a lot of information but is unfortunately impossible to compute explicitly for the Ising model when d is finite. One can nonetheless try to obtain information by constructing a perturbative expansion. The simplest perturbative expansion of $F[\underline{m}]$ is the high-temperature expansion [296], which here also coincides with the large d expansion [163], as shown in this section.

1.2.1 Infinite Temperature

At infinite temperature ($\beta = 0$), the Gibbs–Boltzmann probability distribution $e^{-\beta H[S]}/Z$ is uniform over all spin configurations. In order to fix the magnetisations, we need strong magnetic fields; we thus rescale the fields b_i introducing $\lambda_i = \beta b_i$. Because the entropy S at infinite temperature is finite, while the energy U vanishes, the free energy $F = U - TS$ diverges proportionally to $T = 1/\beta$ when $T \to \infty$. It is thus better to consider $A = -\beta F = S - \beta U$ which remains finite even at $\beta = 0$. With these rescalings, Eq. (1.12) becomes

$$A_0[\underline{\lambda}] = -\beta\Omega_0[\underline{b}] = \log \sum_{\underline{S}} e^{\sum_i \lambda_i S_i} = \sum_i \log[2\cosh(\lambda_i)], \tag{1.17}$$

where the suffix 0 highlights that this is the zeroth order ($\beta = 0$) expression. The free energy is an analytic function of the auxiliary fields, as obviously there is no phase transition at infinite temperature. The Legendre transform can thus be computed by differentiation. The condition that determines λ_i is

$$m_i = -\frac{\partial\Omega_0[\underline{b}]}{\partial b_i} = \frac{\partial A_0[\underline{\lambda}]}{\partial\lambda_i} = \tanh(\lambda_i), \tag{1.18}$$

and, according to Eq. (1.13),

$$-\beta F_0[\underline{m}] = A_0[\underline{\lambda}[\underline{m}]] - \sum_i \lambda_i[\underline{m}]m_i$$

$$= \sum_i \{\log[2\cosh(\lambda_i[\underline{m}])] - \lambda_i[\underline{m}]m_i\} = \sum_i s_0(m_i), \qquad (1.19)$$

$$s_0(m_i) = -\left(\frac{1+m_i}{2}\log\frac{1+m_i}{2} + \frac{1-m_i}{2}\log\frac{1-m_i}{2}\right).$$

The function $s_0(m)$ is the entropy of a single spin constrained to have magnetisation[6] m; at infinite temperature, spins are independent, and the total entropy – which coincides with $-\beta F_0[\underline{m}]$ – is the sum of the single-spin entropies.

The first small β correction can be computed easily. We only sketch here the derivation. First, one expands $A[\underline{\lambda}]$ in powers of β as follows:

$$A[\underline{\lambda}] = \log\sum_{\underline{S}} e^{\sum_i \lambda_i S_i} e^{-\beta H[\underline{S}]} = \log\sum_{\underline{S}} e^{\sum_i \lambda_i S_i}(1 - \beta H[\underline{S}]) + O(\beta^2)$$

$$= A_0[\underline{\lambda}] - \beta\frac{\sum_{\underline{S}} e^{\sum_i \lambda_i S_i} H[\underline{S}]}{\sum_{\underline{S}} e^{\sum_i \lambda_i S_i}} + O(\beta^2). \qquad (1.20)$$

The first correction is then given by the average of $H[\underline{S}]$ over independent spins subjected to magnetic fields λ_i. The average of S_i is $\tanh(\lambda_i)$, and, therefore,[7] the average of $H[\underline{S}]$ is $H[\tanh(\underline{\lambda})]$, where $\tanh(\underline{\lambda}) = \{\tanh(\lambda_i)\}_{i=1,\dots,N}$. We thus have $A[\underline{\lambda}] = A_0[\underline{\lambda}] + \beta A_1[\underline{\lambda}]$, with $A_1[\underline{\lambda}] = -H[\tanh(\underline{\lambda})]$. The equation for $\underline{\lambda}$ then becomes $m_i = \tanh(\lambda_i) + \beta\partial A_1[\underline{\lambda}]/\partial\lambda_i$, and $\lambda_i = \lambda_i^0 + \beta\lambda_i^1$ with $\lambda_i^0 = \mathrm{atanh}(m_i)$. The Legendre transform is obtained by writing

$$-\beta F[\underline{m}] = A_0[\underline{\lambda}^0 + \beta\underline{\lambda}^1] + \beta A_1[\underline{\lambda}^0 + \beta\underline{\lambda}^1] - \sum_i (\lambda_i^0 + \beta\lambda_i^1)m_i$$

$$= A_0[\underline{\lambda}^0] + \beta A_1[\underline{\lambda}^0] - \sum_i \lambda_i^0 m_i = \sum_i s_0(m_i) - \beta H[\underline{m}]. \qquad (1.21)$$

---

[6] Because an Ising spin has only two states, the probability distribution $p(S)$ of a single spin is expressed by two real numbers $p(1)$ and $p(-1)$ satisfying $p(1) + p(-1) = 1$. It is thus specified by one real number, which can be conveniently chosen to be the average magnetisation of the spin, $m = \sum_S Sp(S)$; hence, $p(S) = (1 + mS)/2$. With this choice, the single spin entropy

$$s_0(m) = -p(1)\log p(1) - p(-1)\log p(-1)$$

coincides with Eq. (1.19).

[7] Note that this is true only if $H[\underline{S}]$ is explicitly written as a linear combination of products of spin variables, which is true for our reference expression Eq. (1.1). Any general spin Hamiltonian can also be written in this form, using Eq. (1.11).

Note that the contributions of $\underline{\lambda}^1$ have disappeared from Eq. (1.21) for the following reasons. In A_1, $\underline{\lambda}^1$ can be eliminated because it gives contributions of order β^2. In the remaining terms, $A_0[\underline{\lambda}] - \sum_i \lambda_i m_i$, the derivative with respect to $\underline{\lambda}$ vanishes identically due to the Legendre transform condition, and, therefore, at first order, the correction to $\underline{\lambda}$ disappears. Finally, the terms $A_0[\underline{\lambda}^0] - \sum_i \lambda_i^0 m_i$ give the infinite-temperature result, $\sum_i s_0(m_i)$, while the correction is $\beta A_1[\underline{\lambda}^0]$ $= -\beta H[\tanh(\underline{\lambda}^0)] = -\beta H[\underline{m}]$.

1.2.2 High-Temperature Expansion

At first order in β, the free energy function $F[\underline{m}] = H[\underline{m}] - T \sum_i s_0(m_i)$ is thus given by the free energy of independent spins with magnetisations $\underline{m}$, as shown by Eq. (1.21). The computation of higher-order corrections in β goes along the same lines as the first order correction, as described in [163]. At third order, for instance, the result is

$$
\begin{aligned}
-\beta F[\underline{m}] = &- \sum_i \left(\frac{1+m_i}{2} \log \frac{1+m_i}{2} + \frac{1-m_i}{2} \log \frac{1-m_i}{2} \right) \\
&+ \beta \frac{1}{2} \sum_{ij} J_{ij} m_i m_j + \beta \sum_i B_i m_i \\
&+ \frac{\beta^2}{4} \sum_{ij} J_{ij}^2 (1-m_i^2)(1-m_j^2) \\
&+ \frac{\beta^3}{6} \left[2 \sum_{ij} J_{ij}^3 m_i (1-m_i^2) m_j (1-m_j^2) \right. \\
&\left. + \sum_{ijk} J_{ij} J_{ik} J_{jk} (1-m_i^2)(1-m_j^2)(1-m_k^2) \right] + O(\beta^4),
\end{aligned}
\tag{1.22}
$$

and

$$
\begin{aligned}
\beta b_i[\underline{m}] = \frac{\partial \beta F[\underline{m}]}{\partial m_i} = \operatorname{atanh}(m_i) &- \beta \left[\sum_j J_{ij} m_j + B_i \right] \\
&+ \beta^2 m_i \sum_j J_{ij}^2 (1-m_j^2) + O(\beta^3).
\end{aligned}
\tag{1.23}
$$

Imposing the absence of auxiliary fields, $b_i = 0$, is equivalent to minimising $F[\underline{m}]$. In this case, one obtains the so-called Thouless–Anderson–Palmer (TAP) equations [342],

$$m_i = \tanh \beta h_i,$$

$$h_i = B_i + \sum_j J_{ij} m_j - \beta m_i \sum_j J_{ij}^2 (1 - m_j^2) + O(\beta^2), \qquad (1.24)$$

where h_i is the effective magnetic field provided by the spins neighbouring site i. Note that the TAP Eq. (1.24) are examples of a general class of mean field equations for disordered systems. These equations can be derived through high-temperature expansions, as discussed here, or alternatively via probabilistic methods called the 'cavity method' or 'belief propagation' [251, 254, 372].

1.2.3 Large Dimension Expansion

The high-temperature expansion in Eq. (1.22) can be specialised to the ferromagnetic d-dimensional Ising model, with periodic boundary conditions and with a uniform external magnetic field $B_i = B$. In this case, the system is translationally invariant. It is, therefore, reasonable to expect that the free energy minimum is realised by a uniform magnetisation $m_i = m$. In the following, the free energy per spin corresponding to a uniform magnetisation, also called 'potential', is denoted[8] by v(m). Equation (1.22) gives

$$v(m) = \frac{F[\{m_i = m, \forall i\}]}{N} = -T s_0(m) - \frac{1}{2} m^2 - Bm$$

$$- \frac{\beta}{8d} (1 - m^2)^2 - \frac{\beta^2}{12 d^2} m^2 (1 - m^2)^2 + O(\beta^3). \qquad (1.25)$$

With the choice of coupling scale $J = 1/(2d)$, the first three terms, which represent entropy, exchange energy and magnetic field energy, respectively, remain finite for $d \to \infty$, as discussed in Section 1.1.1. The correction terms form a series in powers of $1/d$: the high-temperature (small β) expansion can be used to construct the

---

[8] To clarify the rationale behind this notation, it is convenient to anticipate here briefly a few notions that will be clarified later in the chapter. In a coarse grained representation the spatial dependent magnetisation profile is $m(\mathbf{x})$. The thermodynamic free energy $F(m)$ corresponding to a global magnetisation m is the minimum of $F[\underline{m}] \sim F[m(\mathbf{x})]$ over all configurations $m(\mathbf{x})$ such that $m = N^{-1} \sum_{i=1}^{N} m_i \sim V^{-1} \int d\mathbf{x} m(\mathbf{x})$. Although the minimum is often realised by uniform configurations, this is not always the case. Examples are antiferromagnets or systems in the phase coexistence region. In a coarse grained representation, corresponding to Landau theory, one can approximate

$$F[m(\mathbf{x})] \sim \int d\mathbf{x} \{c[\nabla m(\mathbf{x})]^2 + v[m(\mathbf{x})]\},$$

where v(m) is the free energy of a uniform configuration, which then provides a 'potential' term in the total free energy. If v(m) has a unique minimum, then the profile is uniform; the gradient term disappears and $F(m) = Nv(m)$. Otherwise, phase coexistence can lead to $F(m) < Nv(m)$ for some values of m. See Section 1.4 for a more detailed discussion.

large d (small $1/d$) expansion.[9] The same result holds also in presence of disorder, although the proper scaling of the couplings might then differ.[10]

It is important to stress that while the true function $F[\underline{m}]$ is guaranteed to be convex, any truncation at a finite order of its high-temperature expansion in Eq. (1.22) does not necessarily share this property. For example, it is very easy to check that Eq. (1.25), truncated at any order in β, is not a convex function of m if the temperature is low enough. This point will be further discussed in Section 1.4.

1.3 Second-Order Phase Transition of the Ising Ferromagnet

The large d expansion in Eq. (1.25) can be used to investigate the ferromagnetic phase transition, a second-order phase transition that is observed in many magnetic materials and is well described by the Ising model [11, 69, 282]. In this section, we discuss the nature of this transition in the limit $d \to \infty$, the corrections in $1/d$ and the nature of spin–spin correlations around the transition.

1.3.1 Mean Field: The Curie–Weiss Model

For a uniform magnetisation profile $m_i = m$, keeping only the terms that remain finite for $d \to \infty$, the potential is

$$v(m) = -T s_0(m) - \frac{1}{2}m^2 - Bm, \tag{1.26}$$

and the TAP equations that determine m simplify to reproduce the mean field result,

$$m = \tanh[\beta(B + m)]. \tag{1.27}$$

The global magnetic susceptibility $\chi = dm/dB$ is then obtained by differentiating Eq. (1.27) with respect to B:

$$\frac{dm}{dB} = \{1 + \tanh[\beta(B + m)]^2\}\beta\left(1 + \frac{dm}{dB}\right) \Rightarrow \chi = \frac{\beta(1 + m^2)}{1 - \beta(1 + m^2)}. \tag{1.28}$$

Similar expressions have been derived by Curie and Weiss under a mean field approximation, and for this reason, the infinite-dimensional Ising model is also known as the Curie–Weiss model [69]. In fact, Eq. (1.27) can be equivalently obtained by assuming that the neighbours of a given spin i (whose set we denote ∂i, see Eq. (1.2)) are uncorrelated and replacing the spins by their average. The local field, $h_i = B_i + \frac{1}{2d}\sum_{j\in\partial i} S_j$, is then on average equal to $h_i = B + m$. The magnetisation of an isolated spin in a field h_i is $m = \tanh(\beta h_i)$, from which one

---

[9] See [163, figure 3] for additional correction terms.
[10] For instance, if J_{ij} are independent Gaussian variables with zero mean, the natural scaling is $J_{ij} \propto 1/\sqrt{d}$.

obtains Eq. (1.27). This construction is the origin of the name 'mean field'. The high-temperature expansion shows that this approximation actually becomes exact when $d \to \infty$. The local field h_i then becomes the sum of a large number of terms, and by the central limit theorem, it can be replaced by its mean. Neighbouring spins then also become independent, as will be shown in Section 1.3.3.

In absence of magnetic field, $B = 0$, the Curie–Weiss model can undergo a phase transition. According to Eq. (1.16), the absolute minimum of $F[m]$ corresponds to the thermodynamic equilibrium state of the system. Because of the translational invariance of the ferromagnetic Ising model, it is reasonable to expect that in equilibrium all spins have the same magnetisation and $m_i = m$. Under this assumption, and using Eq. (1.25), the thermodynamic equilibrium is given by the absolute minimum of $v(m)$. At high temperature, the potential has a unique minimum in $m = 0$, the 'paramagnetic' state. This state has zero energy and entropy density $s = S/N = \log 2$; hence, $v(m = 0) = -T \log 2$. One can study the stability of this state by expanding the free energy around $m = 0$. Expanding $s_0(m) = \log 2 - m^2/2 - m^4/12 + O(m^6)$ gives

$$\delta v(m) \equiv v(m) - v(m = 0) = -Bm + \frac{a_0}{2}m^2 + \frac{b_0}{4}m^4 + O(m^6), \qquad (1.29)$$

with $a_0 = T - 1$ and $b_0 = T/3$. From this expression one deduces that for $B = 0$, the minimum in $m = 0$ becomes unstable when $a_0 < 0$ – i.e., $T < 1$. At low temperatures $T < 1$, the function $v(m)$ has two degenerate minima at $m = \pm m_{\text{eq}}(T, B = 0)$, which correspond to 'ferromagnetic' states. Close to the critical temperature $T_c = 1$, the magnetisation behaves as

$$m_{\text{eq}}(T \to T_c^-, B = 0) \sim \sqrt{-a_0/b_0} \sim (T_c - T)^{1/2}. \qquad (1.30)$$

Moreover, from Eq. (1.28), the magnetic susceptibility in the paramagnetic phase is given by

$$\chi(T, B = 0) = \frac{1}{T - 1} \quad \Rightarrow \quad \chi(T \to T_c^+, B = 0) \sim (T - T_c)^{-1}, \qquad (1.31)$$

and diverges upon approaching the critical temperature. Finally, if one keeps the temperature fixed at the critical value $T = 1$, and adds a small magnetic field B, the magnetisation is then

$$m_{\text{eq}}(T = T_c, B \to 0) \sim (B/b_0)^{1/3} \sim B^{1/3}. \qquad (1.32)$$

The phase transition is said to be of second order, because the free energy and its first derivatives with respect to T and B are continuous at T_c; the first singular term in the free energy appears in its second derivatives.

Many physical observables scale as power laws in the vicinity of the critical temperature $T_c = 1$. We have seen that $m_{\text{eq}} \propto |T - T_c|^\beta$ with $\beta = 1/2$, the magnetic susceptibility $\chi \propto |T - T_c|^{-\gamma}$ with $\gamma = 1$, and $m_{eq} \propto B^{1/\delta}$ with $\delta = 3$. The power-law scaling of physical observables close to T_c is dubbed the critical behaviour, and β, γ, δ are the corresponding critical exponents. It is easy to see that adding other terms in the expansion of $s_0(m)$ in the vicinity of $m = 0$ (the next term is proportional to m^6) does not change the critical exponents, which are entirely determined by the first terms in the expansion that have been kept in Eq. (1.29), as recognised by Landau. For this reason, Eq. (1.29) is known as 'Landau free energy' and the resulting theory of the phase transition in the Ising model is the 'Landau theory' [69].

1.3.2 Corrections to Mean Field and Landau Theory

The next task is to investigate the corrections to Eq. (1.29) that come from higher-order terms in the $1/d$ expansion. Expanding the term of order $1/d$ for small m in Eq. (1.25), we obtain

$$\delta v(m) = \frac{a_1}{2}m^2 + \frac{b_1}{4}m^4 - Bm,$$
$$a_1 = T - 1 + \frac{\beta}{2d}, \quad b_1 = \frac{T}{3} - \frac{\beta}{2d}. \tag{1.33}$$

Because all orders of the $1/d$ expansion are analytic functions of m and preserve the inversion symmetry of the original model at $B = 0$, the form of the free energy remains a polynomial in m^2. The coefficients are modified by $1/d$ corrections, so their numerical values at a given d are modified, but they remain analytic functions of T at all orders. Hence, at any order in $1/d$, the free energy has the form of the Landau theory, and all the consequences of the Landau free energy remain valid. In particular, a second-order phase transition takes place with the same critical exponents β, γ, δ. The numerical values of the observables are, however, changed. For example, at first order in $1/d$ the critical temperature T_c, defined by $a_1 = 0$, is $T_c = 1 - \frac{1}{2d}$. One concludes that while the critical behaviour of the large d expansion is independent of the order in d as a consequence of analyticity and symmetry, as it was understood by Landau, its quantitative results strongly depend on the order at which the expansion is truncated.

How do these results compare with the true behaviour of the Ising model in finite d? In $d = 1$, there is no phase transition [69, 223], so the large d approximation fails badly, and this choice of d will not be further considered. In $d = 2$, the ferromagnetic Ising model has been solved exactly by Onsager for $B = 0$ [69]. In $d > 2$, while there is no exact solution, many properties of the model are known very accurately by a combination of Monte Carlo numerical simulation, renormalisation group and conformal bootstrap methods. In short:

- The critical properties of the model are correctly described by the $1/d$ expansion for all dimensions $d \geq d_u = 4$, where d_u is the upper critical dimension of the model. For $d = 3$ and $d = 2$, the phase transition remains of second order, but the critical exponents are different. Around the critical point, the potential has the 'scaling' form[11] [69, chapter 7]

$$\delta v(m) = -Bm + m^{1+\delta} \mathcal{F}[(T - T_c)m^{-1/\beta}], \qquad m \to 0, \quad T \to T_c, \quad (1.34)$$

where $\mathcal{F}(x)$ is a scaling function that is analytic in x. At the critical point $T = T_c$, in order to recover that $m_{eq}^{\delta} \propto B$ for small B, we must have $\mathcal{F}(0) > 0$. For $m \to 0$ at $T \neq T_c$ – i.e. slightly away from the critical point – the susceptibility $\chi \propto 1/v''(m = 0)$ is finite. All higher-order derivatives of $\delta v(m)$, being related to finite higher-order susceptibilities, remain finite too. Hence, $\delta v(m)$ must be an analytic function of m^2. This condition is satisfied if the function $\mathcal{F}(x)$ has the form, for $x \to \pm\infty$,

$$\mathcal{F}(x) = \sum_{n=1}^{\infty} c_n^{\pm} |x|^{\gamma - 2(n-1)\beta} = c_1^{\pm} |x|^{\gamma} + c_2^{\pm} |x|^{\gamma - 2\beta} + \cdots, \quad (1.35)$$

with[12] $\gamma/\beta = \delta - 1$. In fact, plugging this result in Eq. (1.34), one obtains for small m and $T \neq T_c$:

$$\delta v(m) = -Bm + c_1^{\pm} |T - T_c|^{\gamma} m^2 + c_2^{\pm} |T - T_c|^{\gamma - 2\beta} m^4 + \cdots, \quad (1.36)$$

where the coefficients c_n^+ correspond to $T > T_c$ and c_n^- to $T < T_c$. One concludes that $\delta v(m) \propto m^2$ with a finite coefficient $a \propto 1/\chi \propto |T - T_c|^{\gamma}$, and the susceptibility[13] is $\chi \sim |T - T_c|^{-\gamma}$. For $T < T_c$, assuming $c_1^- < 0$, this form implies that $m_{eq} \sim |T - T_c|^{\beta}$. Equations (1.34) and (1.35) thus encode correctly the critical behaviour around T_c. They also imply that for $d < 4$, in which case some critical exponents are not integers (Table 1.1), the free energy is not an analytic function of T and m around the critical point. Such a non-analytic behaviour cannot be derived from the $1/d$ expansion at any finite order. One must therefore use other techniques, such as the renormalisation group, to understand it in detail[14] [11, 69, 282].

- The quantitative properties of the model – that is, the numerical values of many of its observables of interest – can be obtained by truncating the $1/d$ expansion at a suitable order. As an illustration, we consider the value of the critical temperature

---

[11] For the free energy, Eq. (1.34) holds for $T \sim T_c$ and $m \sim 0$, but with the additional constraint that $|m| > m_{eq}$ for $T < T_c$. When $T < T_c$ and $|m| < m_{eq}$, the free energy is given by the Maxwell construction (Section 1.4). This constraint, however, does not affect the derivation of the scaling relations from Eq. (1.34). Note that for $d \to \infty$ one has $\mathcal{F}(x) = x/2 + T/12$, which satisfies all the requirements discussed below and reproduces Eq. (1.29).

[12] This result is an example of a scaling relation that is obeyed by the critical exponents in all dimensions.

[13] From Eq. (1.36) one can show that the nonlinear susceptibility χ_3, defined by $m_{eq} = \chi B + \chi_3 B^3 + \cdots$, diverges as $\chi_3 \sim |T - T_c|^{-\gamma_3}$ with $\gamma_3 = 3\gamma + 2\beta$, which is another scaling relation.

[14] For the ferromagnetic Ising model, appropriate resummations of the $1/d$ expansion can nonetheless be used to obtain reasonable estimates of the critical exponents. See [163] for a more complete discussion.

Table 1.1 *Critical properties of the Ising model as a function of spatial dimension d. The critical exponents are independent of d for all $d \geq 4$, and for this reason, they are not reported explicitly for $d > 4$. The values in $d = 2$ are from Onsager's exact solution. The critical exponents in $d = 3$ are obtained from conformal bootstrap [137].*

d	β	γ	δ	T_c(exact/MC)
2	1/8	7/4	15	0.567296...
3	0.326419(3)	1.237075(10)	4.78984(1)	0.75192(2) [54, eq. (3.1)]
4	1/2	1	3	0.835033(3) [237]
5				0.877844(2) [54, figure 5]
6				0.90290(5) [165]
7				0.91921(5) [165]
8				0.9311(6) [4]

[142, 163]. Fisher and Gaunt [142] have obtained the following result at the fifth order in $1/d$, and additional terms up to order twelve are given in [71]:

$$T_c = 1 - \frac{1}{q} - \frac{4}{3q^2} - \frac{13}{3q^3} - \frac{979}{45q^4} - \frac{2009}{15q^5} - \frac{176749}{189q^6}$$

$$- \frac{6648736}{945q^7} - \frac{765907148}{14175q^8} - \frac{5446232381}{14175q^9} - \frac{829271458256}{467775q^{10}} \quad (1.37)$$

$$+ \frac{164976684314}{22275q^{11}} + \frac{6495334834824112}{638512875q^{12}} \cdots, \qquad q = 2d.$$

This result is compared with the true value of T_c (obtained from the exact Onsager solution in $d = 2$ and from Monte Carlo simulations for $d > 2$) in Figure 1.1. The fourth-order result gives a remarkably good approximation to the true result in all $d \geq 2$, but adding more terms to the expansion actually worsens the agreement at the lowest dimensions. The convergence properties of the series are still poorly understood. See [71, 142, 163] for a more detailed discussion.[15]

---

[15] There are essentially two possible scenarios. One possibility is that the series converges for $d \geq 4$, the upper critical dimension. The other is that the series is asymptotic, as in the spherical model or in the $O(n)$ model in the infinite n limit [71, 142]. An asymptotic series in $1/d$ is a formal series expression of the form $T_c(d) = \sum_{k=0}^{\infty} a_k d^{-k}$ such that, truncating the series to a given order K, one has $|T_c(d) - \sum_{k=0}^{K} a_k d^{-k}| \leq C_K d^{-(K+1)}$ for some K-dependent constant C_K. In other words, by truncating the series to a given fixed order, the error decreases upon increasing d. If one instead fixes d and increases K, the error first decreases but then diverges for large K because the series is divergent for all finite d.

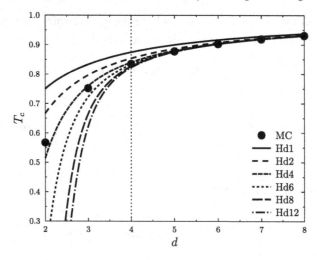

Figure 1.1 The critical temperature of the ferromagnetic Ising model as a function of d. Monte Carlo results are compared with different orders in the $1/d$ expansion, from $1/d$ to $1/d^{12}$ [71, 142, 163]. The vertical dotted line indicates the upper critical dimension $d_u = 4$.

1.3.3 Correlations in Large Dimensions

The spin–spin correlation function is a very important quantity for analysing the phase transition of the Ising model. According to Eq. (1.9), it is proportional to χ_{ij}, and its explicit expression in the large d limit can be obtained as follows. As shown in Section 1.3.1, only the first two lines of Eq. (1.22) remain finite for $d \to \infty$ in the ferromagnetic case. Then, from Eq. (1.14), we have

$$(\chi^{-1})_{ij} = \frac{\partial^2 F[\underline{m}]}{\partial m_i \partial m_j} = \frac{T}{1 - m_i^2}\delta_{ij} - J_{ij}. \tag{1.38}$$

Specialising to the paramagnetic phase, where $m_i = 0$, one has $(\chi^{-1})_{ij} = T\delta_{ij} - J_{ij}$, which must be inverted to obtain the correlation matrix.

To perform the inversion, one can use the periodicity of the lattice, working in Fourier space. This standard procedure is here only sketched; see [282] for details. Following the discussion of Section 1.1.1, we consider for simplicity a d-dimensional cubic lattice of side L with periodic boundary conditions in all directions. This choice conveniently guarantees that translational invariance is preserved, and other choices of boundary conditions would, in any case, lead to the same result in the $L \to \infty$ limit. We label lattice points by d-dimensional vectors – e.g., $\mathbf{x}, \mathbf{y}$. The normalised eigenvectors of $(\chi^{-1})_{\mathbf{xy}}$ are plane waves of the form $v_{\mathbf{x}}^{(\mathbf{q})} = L^{-d/2}e^{i\mathbf{q}\mathbf{x}}$. The periodic boundary conditions impose that the wavevectors

$\mathbf{q}$ have the form $\mathbf{q} = \frac{2\pi}{L}(n_1, \ldots, n_d)$ where the n_μ are integers that belong to the interval $(-L/2, L/2)$. The eigenvector equation is then

$$\sum_\mathbf{y} (\chi^{-1})_{\mathbf{xy}} v_\mathbf{y}^{(\mathbf{q})} = \sum_\mathbf{y} (T\delta_{\mathbf{x,y}} - J_{\mathbf{x,y}}) L^{-d/2} e^{i\mathbf{qy}}$$

$$= L^{-d/2} e^{i\mathbf{qx}} \left[T - \frac{1}{d} \sum_{\mu=1}^d \cos(q_\mu) \right] = \lambda_\mathbf{q} v_\mathbf{x}^{(\mathbf{q})}, \tag{1.39}$$

with associated eigenvalues $\lambda_\mathbf{q} = T - \frac{1}{d} \sum_{\mu=1}^d \cos(q_\mu)$. Denoting complex conjugation by an overline, one then obtains

$$\chi_{\mathbf{xy}} = \sum_\mathbf{q} \frac{1}{\lambda_\mathbf{q}} v_\mathbf{x}^{(\mathbf{q})} \overline{v_\mathbf{y}^{(\mathbf{q})}} = \frac{1}{L^d} \sum_\mathbf{q} \frac{e^{i\mathbf{q(x-y)}}}{T - \frac{1}{d} \sum_{\mu=1}^d \cos(q_\mu)}. \tag{1.40}$$

In the thermodynamic limit $L \to \infty$, the sum over $\mathbf{q}$ becomes an integral, and one obtains

$$\chi_{\mathbf{xy}} = \int_{[-\pi,\pi]^d} \frac{d\mathbf{q}}{(2\pi)^d} \frac{e^{i\mathbf{q(x-y)}}}{T - \frac{1}{d} \sum_{\mu=1}^d \cos(q_\mu)} = \beta \langle S_\mathbf{x} S_\mathbf{y} \rangle. \tag{1.41}$$

The spin–spin correlation function obviously respects the translational symmetry of the lattice and is therefore a function of $\mathbf{x} - \mathbf{y}$. Note that $\sum_\mathbf{x} e^{i\mathbf{qx}} = (2\pi)^d \delta(\mathbf{q})$, and thus, $\chi = L^{-d} \sum_{\mathbf{xy}} \chi_{\mathbf{xy}} = \sum_\mathbf{x} \chi_{\mathbf{x},0} = 1/(T-1)$. As expected, we recover the Curie–Weiss expression for the susceptibility given in Eq. (1.28). Note that the critical temperature is here $T_c = 1$ because only the leading terms in $d \to \infty$ have been retained. The expression of $\chi_{\mathbf{xy}}$ is thus only correct for $T > T_c = 1$.

To evaluate $\chi_{\mathbf{xy}}$, we choose $\mathbf{y} = 0$ for convenience. To facilitate the integration over $\mathbf{q}$, we also choose to evaluate the correlation along one of the principal directions of the lattice, say 1, without loss of generality. Calling $x = x_1$, while $x_\mu = 0, \forall \mu = 2, \ldots, d$, then

$$\langle S_x S_0 \rangle = \int_{[-\pi,\pi]^d} \frac{d\mathbf{q}}{(2\pi)^d} \frac{e^{iq_1 x}}{1 - \frac{\beta}{d} \sum_{\mu=1}^d \cos(q_\mu)}$$

$$= \int_{[-\pi,\pi]^d} \frac{d\mathbf{q}}{(2\pi)^d} e^{iq_1 x} \int_0^\infty d\lambda \, e^{-\lambda[1 - \frac{\beta}{d} \sum_{\mu=1}^d \cos(q_\mu)]} \tag{1.42}$$

$$= \int_0^\infty d\lambda \, e^{-\lambda} I_x(\beta\lambda/d) I_0(\beta\lambda/d)^{d-1},$$

where $I_n(z)$ are the modified Bessel functions of the first kind. This last expression can easily evaluated to visualise the shape of the correlation function; see Figure 1.2. A detailed analysis of its asymptotic properties can be found in [282]. At a given $T > 1$, there are two distinct regimes of correlation decay.

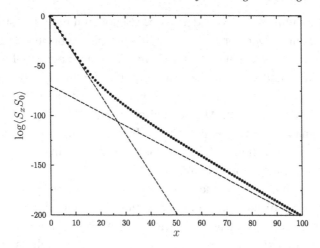

Figure 1.2 Spin–spin correlation function of the ferromagnetic Ising model, evaluated using Eq. (1.42), in dimension $d = 25$ at temperature $\beta = 0.96$. The short-distance exponential decay corresponding to Eq. (1.43) is well visible, while at large distances, the correlation slowly converges to the critical exponential decay given by Eq. (1.44).

- If the distance x is kept fixed while $d \to \infty$, one can use the small-argument expansion of the Bessel functions to show that[16]

$$\langle S_x S_0 \rangle \sim \left(\frac{\beta}{2d} \right)^x = e^{-x/\xi_1}, \qquad \xi_1 = -\frac{1}{\log(\beta/(2d))}. \qquad (1.43)$$

In this regime, the correlation decays exponentially on a length scale ξ_1. When $d \to \infty$, $\xi_1 \to 0$, and hence, for any $x \neq 0$, the correlation vanishes. One concludes that in the $d \to \infty$ limit, each spin is uncorrelated from all the other spins, consistently with the mean field construction. Remarkably, ξ_1 does not diverge at the critical point $\beta = 1$. In other words, correlations decay quickly even at the critical point.

- If instead d is kept fixed, while $x \to \infty$, one can use the large-n asymptotics of $I_n(z)$ to show that

$$\langle S_x S_0 \rangle \sim e^{-x/\xi_2}, \qquad \xi_2 = \frac{1}{\operatorname{arccosh}(d/\beta - d + 1)} \propto (1 - \beta)^{-1/2}. \qquad (1.44)$$

In this large-distance regime, the correlation decays exponentially with a length scale ξ_2 that diverges at the critical point, with critical exponent $1/2$.

---

[16] This result can also be derived either at the first order in the high-temperature expansion or by considering a model defined on a Bethe lattice in the large connectivity limit [251].

Therefore, the limits $d \to \infty$ and $x \to \infty$ do not commute. For strictly infinite d, correlations vanish even for nearest-neighbour spins. For very large but finite d, there are two regimes: at small distances, the decay is exponential with a length ξ_1, which remains small even at the critical point; for much larger distances (that diverge upon increasing d), the decay is characterised by a second length ξ_2, which diverges at the critical point. At a second-order phase transition, even in the mean field approximation, which keeps only the leading terms in $d \to \infty$, a correlation length diverges in any finite d. Including additional terms in the $1/d$ expansion does not qualitatively change this picture.

Note that when d is large, the total magnetic susceptibility is dominated by the short range decay of the spin–spin correlation. The reason is the following. The correlation between a spin and its nearest neighbours ($x = 1$) is $\beta/(2d)$ and, thus, vanishes quickly. However, there are $2d$ nearest neighbours, so their total contribution to the susceptibility is $\beta/(2d) \times 2d = \beta$, which remains finite for $d \to \infty$. The Curie–Weiss expression of the total susceptibility, $\chi = 1/(T - 1)$, which is exact for $d \to \infty$, can be recovered from the short-distance regime in Eq. (1.43) by a careful computation, in which the correlations are also evaluated off the principal axes. The large-distance regime gives negligible contributions in large d. Hence, while correlations between individual spins vanish, the number of neighbouring spins diverges. The two effects compensate one another, giving rise to an overall finite susceptibility. In mean field, one does not need a diverging length scale to obtain a diverging χ, because χ diverges while ξ_1 remains finite. Remember, however, that ξ_2 diverges even if its associated correlations do not contribute to χ.

1.4 Low-Temperature Ferromagnetic Phase

In Section 1.3, we discussed the simplest possible situation, that in which $v(m)$ has a single minimum, corresponding to the paramagnetic state at $m = 0$. We studied how this minimum becomes unstable and bifurcates into two minima, thus giving rise to a second-order phase transition at T_c. In this section, we study in more detail what happens in the ferromagnetic phase, for $T < T_c$, in which $v(m)$ has more than one minimum.

Recall that in Section 1.1, it was shown that the thermodynamic free energy is the absolute minimum of $F[\underline{m}]$, as expressed by Eq. (1.16). In Section 1.2.3, the special case of uniform magnetisation was considered, defining the potential

$$v(m) = \frac{1}{N} F[\{m_i = m, \forall i\}], \tag{1.45}$$

which is the free energy of a system constrained to have uniform magnetisation m and is given by Eq. (1.25). It is also useful to introduce

$$f(m) = \frac{1}{N} \min_{\underline{m}:m=N^{-1}\sum_i m_i} F[\underline{m}], \tag{1.46}$$

which is the true thermodynamic free energy of a system constrained to have a given global magnetisation $m = N^{-1} \sum_i m_i$ and otherwise free to choose the set of local magnetisations that minimises $F[\underline{m}]$. From these definitions, it follows that $f(m) \leq v(m)$ and that the thermodynamic free energy per spin is $f = \min_m f(m)$. Note that there is no guarantee that the absolute minimum of $v(m)$ gives the thermodynamic free energy because the absolute minimum of $F[\underline{m}]$ could be realised by a non-uniform configuration, as it happens in antiferromagnets.

1.4.1 Multiple Equilibrium States and Phase Coexistence

The structure of the equilibrium states of the ferromagnetic d-dimensional Ising model is very well understood and is rigorously proven in most cases [158, 297, 310, 328]. We thus know for certain that the following properties hold.

- An 'equilibrium state' of an infinite system is defined as follows. To construct the infinite system size limit, one considers a sequence of systems of N spins with some prescribed boundary conditions $\mathcal{B}_N$ for each N (examples have been given in Section 1.1.1). If the limit

$$\langle S_{i_1} \cdots S_{i_n} \rangle = \lim_{N \to \infty} \langle S_{i_1} \cdots S_{i_n} \rangle_{N, \mathcal{B}_N} \tag{1.47}$$

exists for all possible sets of n spins,[17] the resulting set of correlation functions defines an equilibrium state of the infinite system.

- A 'translationally invariant equilibrium state' is an equilibrium state such that the expectation value of any observable is translationally invariant – i.e., $\langle S_{\mathbf{x}_1} \cdots S_{\mathbf{x}_n} \rangle = \langle S_{\mathbf{x}_1+\mathbf{y}} \cdots S_{\mathbf{x}_n+\mathbf{y}} \rangle, \forall \mathbf{y}$.

- If $B \neq 0$, or if $B = 0$ and $T > T_c$, there is only one equilibrium state, and it is translationally invariant. In other words, in the thermodynamic limit, all possible sequences of boundary conditions give the same correlation functions and, in particular, the same uniform magnetisation $m = \langle S_i \rangle$.

- For $B = 0$ and $T < T_c$, let us denote by $\langle S_{i_1} \cdots S_{i_n} \rangle_+$ and $\langle S_{i_1} \cdots S_{i_n} \rangle_-$ the equilibrium states obtained by fixing frozen boundary conditions (Section 1.1.1), with external spins all fixed to $S = 1$ and $S = -1$, respectively. The thermodynamic limit with these boundary conditions exists and defines equilibrium states that are translationally invariant and are obviously related by inversion symmetry: $\langle S_{i_1} \cdots S_{i_n} \rangle_+ = (-1)^n \langle S_{i_1} \cdots S_{i_n} \rangle_-$. In particular, $m_{\text{eq}} = \langle S_i \rangle_+ = -\langle S_i \rangle_-$. These

---

[17] According to the discussion of Section 1.1.2, this implies the existence of the thermodynamic limit for all observables.

two states thus correspond to uniform states with positive and negative magnetisation $\pm m_{eq}$. By symmetry, they have the same free energy $v(m_{eq}) = f(m_{eq})$, which is the lowest possible free energy. It therefore corresponds to the thermodynamic free energy – i.e., $F = Nf(m_{eq}) = Nf(-m_{eq})$.

- Both the unique equilibrium state for $B \neq 0$ or $B = 0, T > T_c$, and the two states $+$ and $-$ for $B = 0$ and $T < T_c$, enjoy the so-called 'clustering property'. That is, the spin–spin correlation function $\langle (S_i - m)(S_j - m) \rangle = \langle S_i S_j \rangle - m^2$ that appears in Eq. (1.9) vanishes when the distance between spins diverges.[18] Well-separated spins are thus uncorrelated – i.e., $\langle S_i S_j \rangle \sim \langle S_i \rangle \langle S_j \rangle = m^2$. This property is necessary, for example, to ensure that the susceptibility $\chi = N^{-1} \sum_{ij} \chi_{ij}$ is finite. Moreover, recalling that $\chi_{ij} = \partial m_i / \partial B_j$, the clustering property ensures that the response of spin i to a variation of the local field B_j acting on a faraway spin vanishes. This condition is a minimal stability requirement for a physical system. Equilibrium states that satisfy the clustering property are said to be 'pure states'.

- For all $T < T_c$, the two pure states $+$ and $-$ can also be constructed by taking the limit of zero magnetic field of the unique equilibrium state that exist for $B \neq 0$ – i.e.,

$$\langle \bullet \rangle_{\pm} = \lim_{B \to 0^{\pm}} \langle \bullet \rangle_B . \tag{1.48}$$

This relation is very important. It implies that pure states can be constructed equivalently by a sequence of appropriate boundary conditions or by adding a weak external magnetic field that is sent to zero after taking the thermodynamic limit.

- For $B = 0$ and $T < T_c$, any other translationally invariant equilibrium state can be written as a convex combination of the two pure states:

$$\left\langle S_{i_1} \cdots S_{i_n} \right\rangle_{\alpha} = \alpha \left\langle S_{i_1} \cdots S_{i_n} \right\rangle_{+} + (1 - \alpha) \left\langle S_{i_1} \cdots S_{i_n} \right\rangle_{-}, \tag{1.49}$$

with $\alpha \in [0, 1]$. In particular, the magnetisation $m_{\alpha} = \alpha m_{eq} - (1 - \alpha) m_{eq} = (1 - 2\alpha) m_{eq}$ can take any possible value in $[-m_{eq}, m_{eq}]$. Note that according to Eq. (1.49), the system can be found with probability α in state $+$ and with probability $1 - \alpha$ in state $-$, uniformly in space. This behaviour is unphysical; a classical spin system cannot be in two states simultaneously. Another, perhaps more transparent, reason why these states are unphysical is that they do not enjoy the clustering property. In fact, $\left\langle S_i S_j \right\rangle_{\alpha} = \alpha \left\langle S_i S_j \right\rangle_{+} + (1 - \alpha) \left\langle S_i S_j \right\rangle_{-} = \left\langle S_i S_j \right\rangle_{+}$ by symmetry. Then, at large distance, $\left\langle S_i S_j \right\rangle_{\alpha} \sim \left\langle S_i S_j \right\rangle_{+} \sim (m_{eq})^2 \neq (m_{\alpha})^2$ unless $\alpha = 0$ or $\alpha = 1$. This result implies that the response of spin i to a weak field B_j remains large even when i and j are very distant. Equivalently, a

---

[18] A similar property is true for all n-spin 'connected' correlation functions. For a detailed discussion, see, e.g., [158, 282].

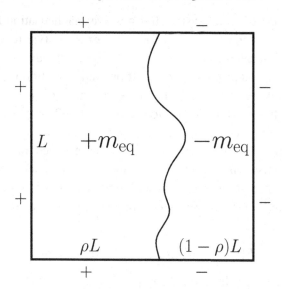

Figure 1.3 Schematic illustration of phase coexistence in the ferromagnetic Ising model in $d = 2$. Here $x \in [0, L]$ is the coordinate of the horizontal direction. Frozen boundary conditions are used, with $S = 1$ ($S = -1$) on the left (right) vertical faces, while on the horizontal faces $S = 1$ for $x < \rho L$ and $S = -1$ for $x > \rho L$. A fluctuating interface separates the two coexisting states with equilibrium magnetisation $\pm m_{\mathrm{eq}}$. Here it is assumed that L is large enough that the interface looks smooth at this scale. Note that the interface is not sharp for $d = 2$ because its fluctuations diverge for $L \rightarrow \infty$, contrary to what happens for $d \geq 3$ and low enough temperature.

weak field applied on a single spin can perturb the entire system, which has a catastrophically destabilising effect. Therefore, although translationally invariant states defined in Eq. (1.49) formally exist, they are not physical. The only stable and translationally invariant states are the two pure states[19] + and −.

- For $B = 0$ and $T < T_c$, non-translationally invariant equilibrium states may exist. This happens if one can separate the system in homogeneous regions separated by sharp interfaces. Such sharp interfaces can only exist for $d \geq 3$ and sufficiently low temperatures, $T < T_r < T_c$ [158], and the interfacial phase transition that happens at T_r is called 'roughening transition'. A particularly interesting example of such states is obtained by choosing one particular direction, say $x \in [0, L]$, and setting frozen boundary conditions with $S = 1$ on one side, $x < \rho L$, and $S = -1$ on the opposite side, $x > \rho L$, with $\rho \in [0, 1]$, as illustrated in Figure 1.3. In

---

[19] In general, by considering the class of translationally invariant states, one can prove that clustering states cannot be written as a convex linear combination of other states, and that states which cannot be written as a convex linear combination of other states are clustering [158].

this case, for large L, the system has positive magnetisation m_{eq} for $x \ll \rho L$ and negative magnetisation $-m_{eq}$ for $x \gg \rho L$. The two regions are separated by an interface located around $x \sim \rho L$. In the thermodynamic limit, a fraction ρ of the volume has positive magnetisation, and a fraction $(1 - \rho)$ has negative magnetisation. We denote $\underline{m}_\rho$ the local magnetisation of such a configuration, and the corresponding total magnetisation is $m_\rho = (1 - 2\rho)m_{eq} + O(1/L)$. Far from the interface, the two regions of positive and negative magnetisation have the same free energy per spin, $v(m_{eq})$. Around the interface, the mismatch between the two magnetisations introduces an additional free energy term. Because the couplings are short range, this term only receives contributions from the region of space around the interface and is thus proportional to the size L^{d-1} of the interface,[20]

$$F[\underline{m}_\rho] = L^d v(m_{eq}) + L^{d-1}\sigma(\beta) + o(L^{d-1}). \tag{1.50}$$

The quantity $\sigma(\beta) = \lim_{L\to\infty}\{F[\underline{m}_\rho] - L^d v(m_{eq})\}/L^{d-1}$ is called the 'surface tension', and it is independent of ρ in the thermodynamic limit.[21] From Eq. (1.50), one deduces that

$$v(m_{eq}) \leq f(m_\rho) \leq v(m_{eq}) + \sigma(\beta)/L. \tag{1.51}$$

The right inequality follows from Eq. (1.46), which implies $f(m_\rho) \leq F[\underline{m}_\rho]/L^d$, with $F[\underline{m}_\rho]$ given by Eq. (1.50). The left inequality follows from the fact that $v(m_{eq}) = f(m_{eq})$ is the absolute minimum of $F[\underline{m}]/N$, as stated above, hence $f(m) \geq v(m_{eq}), \forall m$. One concludes that in the thermodynamic limit $L \to \infty$, $f(m_\rho) \to v(m_{eq}), \forall \rho \in [0, 1]$. In other words, the function $f(m)$ is constant and equal to $v(m_{eq})$ for $m \in [-m_{eq}, m_{eq}]$. This is a particular case of the more general 'Maxwell construction' that will be discussed in Section 1.5. We have constructed an example of non-translationally invariant states that have any possible magnetisation $m \in [-m_{eq}, m_{eq}]$ and have the same free energy as the pure states in the thermodynamic limit. This mechanism, by which one can construct equilibrium states with the two phases ($+$ and $-$) occupying different regions of the system, is called 'phase coexistence'.

In summary, for $T < T_c$ and $B = 0$, the function $f(m)$ has the form depicted in Figure 1.4. The two homogeneous pure states minimise the free energy at $m = \pm m_{eq}$. In the interval $m \in (-m_{eq}, m_{eq})$, phase coexistence ensures that $f(m)$ is

---

[20] For example, suppose that temperature is very low, such that $m_{eq} \sim 1$. In this limit, the interface is extremely sharp, and $m(\mathbf{x}) = \text{sgn}(\rho L - x)$. The energy excess over the ground state energy, given by Eq. (1.3), due to the interface is $2JL^{d-1}$. One deduces that Eq. (1.50) holds with $\sigma(T \to 0) = 2J = 1/d$.

[21] To be more precise, one should take into account in Eq. (1.50) the presence of the interface between the bulk of the system and the boundary, which also gives a contribution of the order of the surface. The first term in Eq. (1.50) should then be replaced by the free energy of a system with all $+$ boundary condition [158].

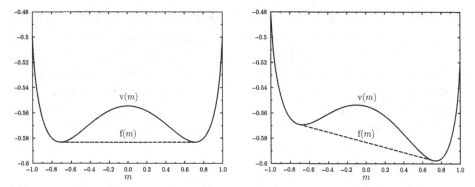

Figure 1.4 The functions $v(m)$, Eq. (1.26), and $f(m)$ for the Ising ferromagnet at $T = 0.8 < T_c$, in the thermodynamic limit $N \to \infty$ and for $d \to \infty$. Left: $B = 0$. Right: $B = 0.02$.

constant, $f(m) = v(m_{eq})$. The main point of this section is that the presence of a flat region in the free energy $f(m)$ generally signals the presence of multiple equilibria and phase coexistence. At the same time, the Legendre transform of $f(m)$, which is the free energy $f(B)$ in presence of a uniform magnetic field B, has a discontinuous derivative in $B = 0$ – i.e., $m(B)$ jumps at $B = 0$.

1.4.2 Phase Coexistence in Large Dimension

We now examine the behaviour of the free energy in the phase coexistence region in the large d limit. First of all, we note that in this limit, connected correlations of distinct spins vanish in pure states. The discussion in Section 1.3.3 can indeed be easily extended to the low-temperature phase and even to non-translationally invariant pure states that appear in disordered systems. In other words, $\langle S_i S_j \rangle = \langle S_i \rangle \langle S_j \rangle = m_i m_j$, $\forall i \neq j$. Therefore, in the infinite-dimensional limit, a pure state is completely determined by the set of local magnetisations,[22] m_i, $i = 1, \dots, N$, which in the ferromagnetic Ising model are all equal by translational invariance. For pure states in $d \to \infty$, spins are thus independent, and one can write

$$P(S_1, \dots, S_N) \sim \prod_{i=1}^{N} p_i(S_i), \qquad p_i(S) = \frac{1 + m_i S}{2}, \tag{1.52}$$

where $p_i(S)$ is a single-spin probability distribution[6] with $m_i = \langle S_i \rangle$.

We now specialise to the translationally invariant case with $m_i = m$. The $1/d$ expansion starts from a probability distribution of independent spins with fixed magnetisation m for $d \to \infty$ and perturbs systematically around it by introducing

---

[22] Note that this is not true for non-pure states. For example, for the states $\langle \bullet \rangle_\alpha$ discussed in Section 1.4.1, one has $\langle S_i S_j \rangle = (m_{eq})^2 \neq m^2$.

correlations in a controlled way. This approach makes sense for the translationally invariant pure states at $m = \pm m_{eq}$ because the magnetisation then remains homogeneous at all orders in $1/d$. The same reasoning holds for $|m| > m_{eq}$, when the free energy $v(m) = f(m)$ is also dominated by translationally invariant states.

But what happens when $m \in (-m_{eq}, m_{eq})$? To illustrate the situation, one can focus on $m = 0$. In this case, the starting point of the perturbation expansion in $1/d$ is the completely random distribution $P(S_1, \ldots, S_N) = 2^{-N}$, in which each spin can equally likely be found in the state 1 or -1, with a local spin distribution $p(S) = 1/2$. However, we know from the analysis of Section 1.4.1 that the free energy $f(m = 0)$ is dominated by completely different spin configurations, in which half of the system has $m = m_{eq}$, while the other half has $m = -m_{eq}$, as illustrated in Figure 1.3. This phase-separated configuration cannot be constructed perturbatively from the homogeneous $m = 0$ state. The perturbative free energy $v(m)$ given in Eq. (1.25), truncated at any given order in $1/d$, therefore represents the free energy of the the the homogeneous $m = 0$ state – i.e., the paramagnetic state. This state has a higher free energy than phase-separated inhomogeneous states, and for this reason, $v(m = 0) > f(m = 0)$ in the phase coexistence region, as represented in Figure 1.4. Moreover, $v(m)$ is not convex in this region and thus cannot be the correct free energy. The large-d expansion must therefore be handled with some care in the phase coexistence region.

It is also possible to consider an inhomogeneous density profile as the starting point of the $1/d$ expansion. For instance, we can consider Eq. (1.22) with ferromagnetic couplings J_{ij}, divide the system in two by an interface and set $m_i = m_{eq}$ for half of the system and $m_i = -m_{eq}$ for the other half. In that case, we correctly get the thermodynamic free energy $f = F[m]/N = v(m_{eq})$ in the thermodynamic limit. We thus see that the convexity of $f(m)$ holds (weakly) in the phase coexistence region, as seen in Figure 1.4, but is lost when the perturbative expansion in $1/d$ around the homogeneous state is truncated at any finite order.

1.4.3 Dynamics in the Phase Coexistence Region

To conclude the discussion of phase coexistence, and in preparation for the discussion of metastability in Section 1.5, it is important to mention the dynamical properties of the Ising model in the phase coexistence region – i.e., at $B = 0$ and $T < T_c$.

The dynamics in this region depends strongly on the boundary conditions. As an illustration, consider a system of finite size L, with periodic boundary conditions, undergoing a local dynamics[23] that satisfies detailed balance (e.g., Metropolis or

---

[23] A local dynamics is such that spins are updated according to a rule that involves a finite number of neighbouring spins.

Glauber dynamics, which will be reviewed in more detail in Chapter 3), starting at time $t = 0$ from the configuration with all spins $S_i(t = 0) = -1$, hence with magnetisation $m(t = 0) = -1$. The following statements can then be proven rigorously, at least in $d = 2$ [246].

- In a time that remains finite for $L \to \infty$, the magnetisation reaches a value $m(t) \sim -m_{eq}$, characteristic of the negatively magnetised pure state.
- At larger times, the magnetisation can diffuse randomly to values $m > -m_{eq}$. In particular, the time needed to reach any positive value of the magnetisation, $m(t) > 0$ (e.g., $m(t) \sim m_{eq}$), is

$$\tau \sim e^{2\beta\sigma(\beta)L^{d-1}}, \qquad (1.53)$$

where $\sigma(\beta)$ is the surface tension introduced in Section 1.4.1.

This result shows that the time needed to explore all the thermodynamically relevant values of the magnetisation, $m \in [-m_{eq}, m_{eq}]$ (these are the values that minimise the free energy), actually diverges exponentially in L^{d-1} for $L \to \infty$. In this limit, the system is said to be non-ergodic, because it takes an infinite amount of time to explore all the thermodynamically relevant part of the configuration space.

The scaling in Eq. (1.53) can be understood by the following simple argument. In order to reach positive values of the magnetisation, at some point the system needs to visit configurations with $m = 0$. We have already seen that the minimal possible free energy at $m = 0$ is realised by phase coexistence. With periodic boundary

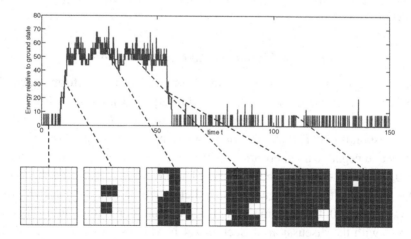

Figure 1.5 Illustration of a transition between negative (white) and positive (black) magnetisation, for a ferromagnetic Ising model in $d = 2$, in a square lattice with $L = 10$ and $N = 100$, at temperature $T = 0.25$ under periodic boundary conditions. Intermediate configurations have $m \sim 0$; in these configurations, two black and white regions are separated by two interfaces.

conditions, the best possible arrangement is to have half of the system in the m_{eq} state and the other half in the $-m_{\text{eq}}$ state, the two parts being separated by two interfaces, as illustrated in Figure 1.5. The free energy is then $F = \text{v}(m_{\text{eq}})L^d + 2\sigma(\beta)L^{d-1}$, and the free energy excess over equilibrium is $\delta F = 2\sigma(\beta)L^{d-1}$. Because the system relaxes to a partially equilibrated state around magnetisation $m(t) \sim -m_{\text{eq}}$ in a finite time, one can use the theory of equilibrium fluctuations [223] to estimate the probability of observing such a fluctuation starting from $m = -m_{\text{eq}}$. The resulting escape probability is $p_{\text{e}} \sim \exp(-\beta\delta F)$, from which it follows that the time needed to observe such fluctuation is $\tau \sim 1/p_{\text{e}}$ and, hence, Eq. (1.53).

1.5 Metastable States

The aim of this section is to introduce metastable states, which are crucial in the study of glasses. In this section, we present the essential ideas while still focusing on the ferromagnetic state, in which the system is translationally invariant, and metastable states respect this symmetry. The equivalent concept for disordered systems will be discussed in Chapter 4. Detailed reviews on metastability can be found in [52, 120], and a short pedagogical introduction can be found in [55].

In Section 1.4, we first discussed the structure of the equilibrium states in finite d (Section 1.4.1) and then the large d approximation (Section 1.4.2). This was possible because phase coexistence in finite d is mathematically well understood. The theory of metastability in finite d is, by contrast, not well established. It is therefore convenient to discuss first the large d behaviour and then its finite d extension.

1.5.1 Local Minima of the Free Energy in Large d

Consider the free energy of the Ising ferromagnet for a uniform magnetisation $m_i = m, \forall i$ and in the limit $d \to \infty$, given in Eq. (1.26). Its shape for $T < T_c$ and small enough B is shown in Figure 1.4. As in the case $B = 0$, the potential $\text{v}(m)$ that corresponds to uniform magnetisation is non-convex. It has two local minima at positive and negative magnetisation, separated by a local maximum situated around $m = 0$. The term $-Bm$ in the free energy has the effect of lowering the free energy of the minimum that is magnetised parallel to the field, i.e., such that $Bm > 0$. Hence, for $B \neq 0$, the function $\text{v}(m)$ has a unique global minimum. This result is coherent with the discussion of Section 1.4.1, which stated that for $B \neq 0$, at all temperatures, there is a single translationally invariant pure state. This state is the global minimum of $\text{v}(m)$.

As in the case $B = 0$, in the region where $\text{v}(m)$ is not convex, one can examine what happens if the magnetisation is allowed to be non-uniform. Consider a global magnetisation m. One can always choose two arbitrary values of magnetisation $m_\pm$, such that $m \in [m_-, m_+]$, and separate the system in two regions, in such a way that a fraction ρ of the system has magnetisation m_- and a fraction $1 - \rho$ has m_+, with ρ fixed by $m = \rho m_- + (1 - \rho)m_+ + O(1/L)$. If the two regions are separated by a flat interface of size L^{d-1}, Section 1.4.1, the free energy per spin is $f(m) = \rho v(m_-) + (1-\rho)v(m_+) + \sigma(\beta, B)/L$, where $\sigma(\beta, B)$ is the surface tension. In the thermodynamic limit, one then obtains

$$
\begin{aligned}
m &= \rho m_- + (1 - \rho)m_+, \qquad m \in [m_-, m_+], \\
f(m) &= \rho v(m_-) + (1 - \rho)v(m_+).
\end{aligned}
\tag{1.54}
$$

Therefore, the plot of $f(m)$ is a straight line connecting the points $\{m_-, v(m_-)\}$ and $\{m_+, v(m_+)\}$. Geometrically it is clear that for all m, $f(m)$ is minimised by choosing m_- and m_+ such that $f(m)$ is the convex envelope of $v(m)$, as illustrated in Figure 1.4. From this geometrical construction, which is nothing but the Maxwell construction [52], one can see that for any $B \neq 0$, there is a compact interval $[m_-(B), m_+(B)]$ (the phase coexistence region) over which $v(m)$ is not convex. For $m \notin [m_-(B), m_+(B)]$, one has $f(m) = v(m)$, while for $m \in [m_-(B), m_+(B)]$, $f(m)$ is the convex envelope of $v(m)$, according to Eq. (1.54).

Without loss of generality, one can restrict the discussion to the case $B > 0$; the other case is symmetric under $m \to -m$. The equilibrium magnetisation $m_{eq}(B) > 0$, which corresponds to the global minimum of $v(m)$, falls outside of the phase coexistence region, consistently with the equilibrium state being translationally invariant. Instead, the local minimum with $m_m(B) < 0$ is very close to $m_-(B)$ but falls within the phase coexistence region. The aim of the rest of this section is to show that the local minimum can be interpreted as a metastable state. A metastable state is similar to a thermodynamic state, around which the system, if it is initialised with a sufficiently negative m, spends a lot of time before eventually reaching the equilibrium state. This discussion therefore requires a careful investigation of the dynamics. Beforehand, it is useful to stress three points.

1. For large enough B, the secondary minimum disappears. The value of B at which this minimum disappears defines a 'spinodal point' [52]. Beyond the spinodal, no metastability exists.
2. The picture described in this section is derived explicitly in $d \to \infty$ and remains valid in any order in the $1/d$ expansion. However, the dynamics in the strict limit $d \to \infty$ is very different from that in any finite d, even when d is very large. This will be discussed in Section 1.5.3.
3. Because m_m is a local minimum of $v(m)$, all the thermodynamic relations, such as Eqs. (1.8) and (1.9), remain valid. The usual thermodynamic relations hold in a metastable state during the time the system remains confined within it.

1.5.2 Metastable Dynamics in Infinite Dimensions

In this section, in order to discuss metastable dynamics in infinite dimensions, we consider a simple example for which the dynamics and the role of the local minima of $v(m)$ can be discussed exactly. This is a microscopic realisation of the ferromagnetic Curie–Weiss model, already introduced in Section 1.3.1. Note that we here follow closely the discussion of [311].

Dynamics of an Ising Spin System

Consider a system of N spins, interacting with each other via the Ising Hamiltonian in Eq. (1.1), and undergoing a continuous time Markovian dynamics entirely characterised by the rates $w_{\underline{S};\underline{S}'}$ of jumping from configuration $\underline{S}'$ to $\underline{S}$. The evolution of the probability distribution of spin configurations $p_t(\underline{S})$ can be described by a master equation [159]:

$$\partial_t p_t(\underline{S}) = \sum_{\underline{S}'} \left[w_{\underline{S};\underline{S}'} p_t(\underline{S}') - w_{\underline{S}';\underline{S}} p_t(\underline{S}) \right]. \tag{1.55}$$

The first term describes jumps from any $\underline{S}'$ into $\underline{S}$, while the second term described jumps from $\underline{S}$ to any $\underline{S}'$. We denote by $\underline{S}_{\updownarrow i} = \{S_1, \ldots, -S_i, \ldots, S_N\}$ the configuration that differs from $\underline{S}$ by flipping spin i, and specialise to dynamics that can only flip spins sequentially. In other words, we assume that $w_{\underline{S};\underline{S}'}$ vanishes unless $\underline{S}' = \underline{S}_{\updownarrow i}$ for some i. The energy change under one spin flip is then[24]

$$\Delta E = H(\underline{S}) - H(\underline{S}_{\updownarrow i}) = -2\hat{h}_i S_i, \qquad \hat{h}_i = B_i + \sum_{j(\neq i)} J_{ij} S_j, \tag{1.56}$$

and transition rates are assumed to depend only on ΔE. In this case, one has $w_{\underline{S};\underline{S}_{\updownarrow i}} = w(\Delta E) = w(-2\hat{h}_i S_i)$; hence, we can write the master equation as

$$\partial_t p_t(\underline{S}) = \sum_i \left[w(-2\hat{h}_i S_i) p_t(\underline{S}_{\updownarrow i}) - w(2\hat{h}_i S_i) p_t(\underline{S}) \right]. \tag{1.57}$$

The first term describes the probability of flipping spin i so that the system goes into state $\underline{S}$ from $\underline{S}_{\updownarrow i}$. The second term describes all the events by which the system leaves $\underline{S}$ by a single-spin flip. For simplicity, we choose

$$w(\Delta E) = e^{-\beta \Delta E/2}, \tag{1.58}$$

which is consistent with the detailed balance condition

$$w(\Delta E) = e^{-\beta \Delta E} w(-\Delta E), \tag{1.59}$$

---

[24] The 'microscopic' field $\hat{h}_i$ bears a hat to emphasise that it depends on the configuration $\underline{S}$ and, thus, to distinguish it from the average field h_i given by Eq. (1.24).

but any other choice of rate that satisfies Eq. (1.59) guarantees that for $t \to \infty$, $p_t(\underline{S})$ converges to the Gibbs–Boltzmann equilibrium distribution with Hamiltonian given by Eq. (1.1). This point will be discussed in more detail in Chapter 3.

The Curie–Weiss Model

The study of Ising spin dynamics of d-dimensional systems, even in the limit $d \to \infty$, is fairly involved. The mean field Curie–Weiss model, which corresponds to Eq. (1.1) with $J_{ij} = 1/(2N)$ for any pair ij and constant magnetic field $B_i = B$, provides instead an easily tractable case. The Hamiltonian then depends only on the global magnetisation,

$$H = -\frac{1}{2N}M^2 - BM, \qquad M = \sum_i S_i. \tag{1.60}$$

By observing that the number of configurations with magnetisation M is $\binom{N}{(M+N)/2}$, one can define a free energy at constant magnetisation:

$$F(M) = -T \log \left[\sum_{\underline{S} | \sum_i S_i = M} e^{-\beta H(\underline{S})} \right] \tag{1.61}$$

$$= -T \log \left[\binom{N}{(M+N)/2} e^{\beta M^2/2 + \beta BM} \right] \underset{N \to \infty}{\approx} N v(m),$$

where $m = M/N$, $v(m)$ is given in Eq. (1.26) (and illustrated in Figure 1.4), and the limit $N \to \infty$ is easily obtained by Stirling's formula. The free energy per spin of the Curie–Weiss model in the thermodynamic limit thus coincides with Eq. (1.26), which is the leading order for $d \to \infty$ of the d-dimensional Ising ferromagnet with an appropriate choice of the scale of couplings, as discussed in Section 1.2.3. It is interesting to note, however, that the Curie–Weiss model defined by Eq. (1.60) is well defined for any finite N. It essentially describes the behaviour of a system of $N = L^d$ spins when L is finite and d is very large or, more precisely, the limit of the d-dimensional model when $d \to \infty$ before L does. Note that even $L = 2$, the minimal linear system size, is enough to obtain a thermodynamically large system of $N = 2^d$ spins when $d \to \infty$. For $N \to \infty$, minimising $v(m)$ with respect to m gives the thermodynamic free energy of the Curie–Weiss model and, as discussed earlier, two states with positive and negative magnetisation are present at low enough temperatures.

Because the Hamiltonian depends only on M, it follows that at any time t, $p_t(\underline{S})$ also only depends on M (provided that this is true at $t = 0$). We define $p_t(M)$ as

$$p_t(\underline{S}) = p_t(M) \binom{N}{(N+M)/2}^{-1}. \tag{1.62}$$

The master Eq. (1.57) acting on the 2^N spin configurations can thus be reduced to a simpler master equation, defined on the much smaller space of the $N + 1$ possible values of the magnetisation $M \in \{-N, -N + 2 \ldots N - 2, N\}$. For simplicity, we focus on the case $B = 0$, although the discussion can easily be generalised to $B \neq 0$. Injecting Eq. (1.62) into Eq. (1.57), we get

$$\partial_t p_t(M) = w_+(M - 2)p_t(M - 2) + w_-(M + 2)p_t(M + 2)$$
$$- [w_-(M) + w_+(M)]p_t(M),$$
(1.63)

with

$$w_+(M) = \frac{N - M}{2} w\left[-2\frac{M + 1}{N}\right] = \frac{N - M}{2} e^{\beta(M+1)/N},$$

$$w_-(M) = \frac{N + M}{2} w\left[2\frac{M - 1}{N}\right] = \frac{N + M}{2} e^{-\beta(M-1)/N},$$
(1.64)

which has the form of a one-dimensional birth-death process [159, section 7.1]. By construction, at long times, $p_t(M)$ converges to its equilibrium form

$$p_\infty(M) \propto \binom{N}{(M + N)/2} e^{\beta M^2/2} \sim e^{-\beta N v(m)}.$$
(1.65)

Therefore, in the high-temperature phase, $T > 1$, the magnetisation converges with very high probability to $m = 0$, while in the low-temperature phase, $T < 1$, it converges with very high probability to one of the values corresponding to the minima of $v(m)$ – i.e., $\pm m_{eq}$. More precisely, in the low-temperature phase, the magnetisation first relaxes on a finite[25] time scale τ_{rel} to the value $\pm m_{eq}$ that is closest to the initial condition, as it can be checked explicitly by analysing Eq. (1.63) (see [311, section 6] for details). Without loss of generality, suppose that the system is initialised in $M_{start} < 0$. The magnetisation then converges in a finite time to $-Nm_{eq}$ so that if $N \to \infty$ and times are kept finite, the system remains stuck forever in the negatively magnetised equilibrium state.

Metastability and Transition Rate

However, we know that the system must eventually sample the two states $\pm Nm_{eq}$ with equal frequency. We are thus interested in the transition rate between these two states. Transition rates for Eq. (1.63) can be calculated using standard techniques for estimating mean first-passage times in discrete, one-dimensional birth-death processes [159, section 7.4]. Consider a system starting in $M_{start} < 0$ in $t = 0$ and call τ the time at which the system first reaches a state $M_{end} > M_{start}$. We

---

[25] The time is finite if the initial condition has $m \neq 0$. If the system instead starts at $m = 0$, then it takes a time $\sim \log N$ to reach one of the two values $\pm m_{eq}$.

define the mean first passage time $\tau(M_{\text{start}} \to M_{\text{end}}) = \langle \tau \rangle$, where the average is over the dynamics described by Eq. (1.63). The system is confined by a reflecting barrier in $M = -N$, because the magnetisation cannot be smaller than this value. Using this boundary condition and [159, eq. (7.4.12)], we get (K, K', M denote magnetisations and therefore increase in steps of two)

$$\tau(M_{\text{start}} \to M_{\text{end}}) = \sum_{K=M_{\text{start}}}^{M_{\text{end}}} \phi(K) \sum_{K'=-N}^{K} \frac{1}{\phi(K')w_+(K')}, \tag{1.66}$$

with

$$\phi(M) = \prod_{K=-N+2}^{M} \frac{w_-(K)}{w_+(K)}. \tag{1.67}$$

Eq. (1.66) can be computed numerically for any finite N, in a time that grows only polynomially in N. One can thus compute the mean first passage time in $M_{\text{end}}/N = m_{\text{end}} > 0$ of a system that starts in $M_{\text{start}}/N = m_{\text{start}} < 0$ at time $t = 0$, for $B = 0$ and $T < T_c$. An asymptotic analysis shows that the result does not depend on the start and end points in the thermodynamic limit. It then reads

$$\tau(m_{\text{start}} \to m_{\text{end}}) = \frac{\pi}{\beta} \sqrt{\frac{1}{[1 - \beta(1 - m_{\text{eq}}^2)](\beta - 1)}} e^{\beta N[v(0) - v(m_{\text{eq}})]}. \tag{1.68}$$

In this expression, we recognise the Arrhenius law, which states that the leading order in the scaling of the characteristic time τ with N is given by

$$\tau \sim e^{\beta N \Delta}, \qquad \Delta = v(0) - v(m_{\text{eq}}), \tag{1.69}$$

where the prefactor Δ is the free energy barrier that separates the two equilibrium states. This result is easily generalised to the case $B \neq 0$, where the two minima are non-degenerate,

$$\tau(-m_{\text{eq}} \to m_{\text{eq}}) \sim e^{\beta N[v(m_{\text{max}}) - v(-m_{\text{eq}})]},$$
$$\tau(m_{\text{eq}} \to -m_{\text{eq}}) \sim e^{\beta N[v(m_{\text{max}}) - v(m_{\text{eq}})]}, \tag{1.70}$$

where m_{max} denotes the magnetisation at which $v(m)$ has a local maximum in $[-m_{\text{eq}}, m_{\text{eq}}]$.

Note that this barrier scaling can also be obtained by the following simple argument, along the lines of that presented at the end of Section 1.4.3. By continuity, the probability to jump from positive to negative magnetisation is bounded from above by the probability of reaching any intermediate value of m. Assuming that the system first equilibrates in the metastable state, this probability is given, according to the theory of equilibrium fluctuations [223], by $e^{-\beta N[v(m) - v(m_{\text{eq}})]}$. The best upper bound is obtained for $m = m_{\text{max}}$, and its inverse gives a lower bound on the mean first passage time.

1.5.3 Escaping the Metastable State through Nucleation

In Section 1.5.2, we studied the metastable dynamics of a system of N spins with fully connected interactions, which essentially amounts to considering the case in which the dimension of space goes to infinity before N does. We now study what happens in the opposite situation, when $N \rightarrow \infty$ at fixed d. In this case, the analysis of the previous section cannot be applied and, in particular, the magnetisation profile does not remain uniform. There exists a more efficient way to escape from the metastable state – namely, the inhomogeneous nucleation of a critical droplet.

Consider a system in a cubic box of size L, which is prepared at $T < T_c$ with negative magnetisation and in presence of a positive magnetic field $B > 0$. In this case, most spins take a negative value, with some rare positive exceptions. Due to fluctuations, an island of positive spins can nevertheless spontaneously form in the sea of negative spins. On a cubic lattice, the most favourable configuration for this process is a 'droplet' made of a sub-cube of side R of the whole lattice, filled with mostly positive spins (i.e., the equilibrium state) surrounded on all sides by negative spins (i.e., the metastable state). Let us denote $\delta v(B, T) > 0$ the difference of potential between the homogeneous metastable state and the homogeneous equilibrium state, which is positive by construction. Because of the presence of the more stable phase within the bulk of the sub-cube, the droplet gains a free energy $\delta v(B, T)R^d$. However, the surface of the droplet (with area $2dR^{d-1}$) then becomes an interface between the two states and therefore increases the free energy by $\sigma(B, T)2dR^{d-1}$, which is the associated surface tension, as discussed in Section 1.4.1. The total free energy difference between the droplet and the homogeneous initial metastable state is therefore

$$\delta F(R) = -\delta v(B, T)R^d + \sigma(B, T)2dR^{d-1}. \tag{1.71}$$

For small R, $\delta F(R)$ is positive; hence, the cost of forming an interface dominates over the bulk free energy gain. For large R, however, $\delta F(R)$ becomes negative as the bulk dominates. The two regimes are separated by a point at which $d\delta F/dR = 0$, with corresponding

$$R_c(B, T) = \frac{2(d-1)\sigma(B, T)}{\delta v(B, T)}. \tag{1.72}$$

A droplet of this size is said to be critical. Once such a critical droplet forms via spontaneous fluctuations, the system can lower its free energy by increasing the

droplet size; R grows and the equilibrium phase invades quickly all the volume.[26]
The associated free energy cost is

$$\delta F^\dagger(B,T) = \delta F(R_c) = 2\sigma(B,T) R_c(B,T)^{d-1}$$
$$= \frac{[2\sigma(B,T)]^d}{\delta v(B,T)^{d-1}}(d-1)^{d-1}; \tag{1.73}$$

hence, the time needed to observe the spontaneous formation of such a droplet is

$$\tau \sim e^{\beta \delta F^\dagger(B,T)}. \tag{1.74}$$

From this argument, one concludes that the time to form a critical droplet and escape from the metastable state remains finite when $L \to \infty$, for all spatial dimensions d and for all $B \neq 0$, at finite T. It only diverges when $T \to 0$ or $B \to 0$. These results can be confirmed rigorously [246]. Note that this case is very different from phase coexistence ($B = 0$), discussed in Section 1.4.3, in which the time to jump from negative to positive m diverges exponentially in L, and from the infinite dimensional case of Section 1.5.2 in which it diverges exponentially in N.

A rough estimate of the free energy cost of forming the critical droplet can be obtained at very low temperatures and small B. For $T \to 0$, the surface tension is simply the energy difference between up and down spins, $\sigma(B, T \to 0) \sim 2J$ $= 1/d$ for $J = 1/(2d)$, and the two states have magnetisation $m \approx \pm 1$. For small B, their potentials are thus $v_+ = v(m = 0) - Bm = v(m = 0) - B$ and $v_- = v(m = 0) + B$; hence, $\delta v(B \to 0, T \to 0) \sim 2B$, and

$$\delta F^\dagger(B \to 0, T \to 0) = \frac{(2/d)^d}{(2|B|)^{d-1}}(d-1)^{d-1} \underset{d \to \infty}{\approx} \frac{2e}{d} \frac{1}{|B|^{d-1}}. \tag{1.75}$$

When $d \to \infty$, from Eq. (1.72) with $\sigma(B,T) \sim 1/d$ and $\delta v(B,T)$ remaining finite, the size R_c of the critical droplet reaches a finite limit (which can be made arbitrarily large upon decreasing B). If the system has a finite size $L < R_c$, the critical droplet cannot form. In this regime, nucleation is impossible, and the only way to escape the metastable state is through a homogeneous jump of the magnetisation, following the mechanism discussed in Section 1.5.2. In that case, the time is exponential in $N = L^d$. We conclude that, roughly,

$$\tau \sim \begin{cases} e^{\beta L^d \Delta}, & \text{for } L \ll R_c(B,T), \\ e^{\beta \delta F^\dagger(B,T)}, & \text{for } L \gg R_c(B,T). \end{cases} \tag{1.76}$$

The logarithm of the escape time thus increases proportionally to L^d for small L, and then saturates at a finite value when $L \sim R_c(B,T)$. Note that $\delta F^\dagger(B,T) \propto$

---

[26] In an infinite volume, the nucleation argument is more complicated because many critical droplets can form concurrently at different locations in the system, but the conclusion is very similar.

$R_c(B,T)^d$ from Eq. (1.73), which ensures the continuity of Eq. (1.76) when $R_c(B,T) \sim L$.

The limits $L \to \infty$ and $d \to \infty$, therefore, do not always commute. If d goes to infinity at fixed $L \ll R_c(B,T)$, one is then asymptotically in the first case of Eq. (1.76) and the infinite-d dynamics of Section 1.5.2 is recovered. Note that, as discussed in Section 1.5.2, this is the typical situation in large d because a lattice of size $L = 2$ is enough to obtain a thermodynamically large system, while the critical droplet size $R_c(B,T)$ is large at small B. The metastable state then has a lifetime exponential in N. If instead $L \to \infty$ at fixed d, one is then always in the second case of Eq. (1.76), and nucleation dominates. The metastable state then has a finite lifetime, which becomes very long when d is large. In both cases, however, the lifetime of the metastable state becomes extremely long as d grows.

1.6 Wrap-Up

1.6.1 Summary

In this chapter, we have seen that

- Large-d expansions can be constructed conveniently by defining the free energy as a function of the appropriate order parameter and then using a high-temperature expansion (Section 1.2).
- The leading ($d = \infty$) term of this expansion, which corresponds to the exact solution of the infinite-dimensional model, coincides with the mean field result and has the analytical structure of the Landau theory (Section 1.3.1).
- Upon adding more terms to the $1/d$ expansion, the free energy keeps the form of a Landau theory, and, therefore, the critical exponents remain unchanged from the mean field ones. The truncated $1/d$ expansion thus describes the critical behaviour only for $d \geq d_u$, the upper critical dimension (Section 1.3.2). Estimating the critical exponents for $d < d_u$ requires a resummation of the series or the use of different techniques, such as the renormalisation group.
- The coefficients of the Landau free energy are themselves power series in $1/d$. Thus, by carefully truncating the $1/d$ expansion or performing appropriate resummations, one can obtain accurate estimates of some of the physical observables (Section 1.3.2).
- Correlations, even between nearest neighbouring spins, vanish in the limit $d \to \infty$. For example, $\langle S_i S_j \rangle \propto 1/(2d)$ for nearest neighbours in the ferromagnetic Ising model. This explains why in the limit $d \to \infty$, the mean field approximation becomes exact (Section 1.3.3).
- Yet the number of neighbours grows with d. For example, each site of a cubic lattice has $2d$ nearest neighbours; hence, their total contribution to the susceptibility, $2d \times 1/(2d)$, remains finite. The same happens for spins at larger distances.

For this reason, the susceptibility generally remains finite when $d \to \infty$ and diverges at the critical point. In mean field, one can thus have both uncorrelated spins and a divergent susceptibility (Section 1.3.3).

- Within mean field, one can define a correlation length ξ_2 that diverges at the critical point, by considering the leading $d \to \infty$ contributions to $\langle S_i S_j \rangle$ and taking the large-distance limit before the limit $d \to \infty$ (Section 1.3.3). This reproduces the result of Landau theory for the associated critical exponent.

- In the low-temperature phase, $T < T_c$, and for $B = 0$, there are two equivalent equilibrium states. These states are homogeneous and have magnetisation $\pm m_{eq}$. One can form non-homogeneous states characterised by phase coexistence of the two homogeneous states. These states have $m \in [-m_{eq}, m_{eq}]$ and the same free energy per spin as the homogeneous states (Section 1.4.1).

- In $d \to \infty$, the homogeneous states are local minima of $v(m)$, the free energy corresponding to uniform magnetisation $m_i = m$, which can be computed in a $1/d$ expansion. Non-homogeneous states make the function $f(m)$, which is the free energy for a global magnetisation m, constant over the interval $m \in [-m_{eq}, m_{eq}]$ (Section 1.4.2).

- Dynamically, below T_c, the system is not ergodic; it takes a time $\tau \sim \exp(L^{d-1})$ to reach positive m for a system that starts at negative m (Section 1.4.3).

- For $T < T_c$ and small enough $B \neq 0$, the function $v(m)$ in the limit $d \to \infty$ displays a single absolute minimum, which is the unique equilibrium state, and a secondary local minimum. The local minimum is a metastable state. It behaves thermodynamically as an equilibrium state but on a restricted time scale (Section 1.5.1).

- In the limit $d \to \infty$, if the system is initialised close to the metastable state, it remains there for a time $\tau \sim \exp(N)$. The metastable state therefore has an infinite lifetime in the $N \to \infty$ limit and is a real thermodynamic state (Section 1.5.2).

- For any finite d (even if large), there is a critical droplet size R_c. For a system size $L \ll R_c$, the droplet cannot form, and $\tau \sim \exp(L^d)$. If instead $L \gg R_c$, the system can escape from the metastable state in a finite time through nucleation and τ saturates to a finite value. In this case, the metastable state can be considered as a thermodynamic state only if its lifetime is large enough (Section 1.5.3).

Metastable states will play a central role in the analysis of the glass state. These ideas will be generalised to the glass starting from Chapter 4.

1.6.2 Further Reading

We provide here a list of references that can be consulted to further explore the subjects discussed in this chapter. These references have been chosen using the following criteria. (1) Whenever available, we privileged books and review articles, because they are usually more pedagogical. (2) Among research articles, we selected those that are, in our opinion, the most accessible to the reader. We also privileged more recent articles, because they usually contain more references to the previous literature. (3) Whenever possible, we selected articles presenting results that are directly related to those discussed in the chapter. We would like to stress that this bibliography is not meant to be exhaustive, and, in particular, it does not represent the historical development of the field. Hence, the fact that we cite an article should not be interpreted as an attribution of credit to its authors.

In the rest of this book, we focus only on the infinite-dimensional limit. Most of the concepts that have been introduced in this chapter are, therefore, only needed to situate this limit in the general setting of statistical physics. Yet further reading on how to go beyond the limit $d \to \infty$ and include spatial fluctuations beyond mean field is certainly useful. Concerning critical fluctuations around second-order phase transitions, the most important method is the renormalisation group, which is covered in many introductory textbooks and reviews, such as

- Amit and Martin-Mayor, *Field theory, the renormalization group, and critical phenomena* [11]
- Brézin, *Introduction to statistical field theory* [69]
- Parisi, *Statistical field theory* [282]
- Delamotte, *An introduction to the nonperturbative renormalization group* [124]

The last reference [124] focuses specifically on the non-perturbative methods that are needed in the context of disordered systems and glasses, where perturbative renormalisation group usually fails.

Good introductions to the theory of phase coexistence in equilibrium, metastability and nucleation can be found in

- Binder, *Theory of first-order phase transitions* [52]
- Gallavotti, *Statistical mechanics: A short treatise* [158]
- Martinelli, *Lectures on Glauber dynamics for discrete spin models* [246]

A detailed discussion of the dynamics of infinite-dimensional spin models such as the Curie–Weiss model discussed in Section 1.5.2, with a focus on critical dynamics and metastability, can be found in

- Ruijgrok and Tjon, *Critical slowing down and nonlinear response in an exactly solvable stochastic model* [311]
- Mora, Walczak and Zamponi, *Transition path sampling algorithm for discrete many-body systems* [265]

2

Atomic Liquids in Infinite Dimensions
Thermodynamics

The aim of this chapter is to review the key elements of the thermodynamic theory of liquids for understanding the subsequent chapters. The reader is assumed to be familiar with the classic book by Hansen and MacDonald [175], but the initial discussion nonetheless offers a short summary of basic notions borrowed from that book in order to define the main observables and notations (that are sometimes slightly different from [175]), to introduce the virial expansion and to show its equivalence to a large-dimensional expansion. The formal analogy with the discussion of Chapter 1 for magnetic systems will be highlighted. Then, the large-dimensional limit will be explicitly constructed for typical liquid potentials, and the results for the gas and liquid phases will be discussed.

2.1 Thermodynamics of Atomic Systems

Many gases, liquids and solids can be modelled as collections of classical[1] point particles, representing the atoms, interacting via spherically symmetric potentials. Simple models that contain identical atoms, such as Argon atoms [175, chapter 1], are said to be 'monodisperse'. Most amorphous solids, however, are not composed of identical atoms. Monodisperse systems typically crystallise very easily and cannot be kept in an amorphous solid state for a macroscopic time. Amorphous solids are thus usually composed either of mixtures of different atomic species (binary or polydisperse mixtures), of small molecules or of even more complex molecules such as long chains – i.e., polymers. Non-spherical interactions are sometimes present, especially in granular matter.

One could then be worried that in order to study glasses, it is necessary to avoid simple monodisperse and spherical systems and to study instead more complicated

---

[1] Quantum effects can certainly be important in some regimes, especially at very low temperatures, but they will not be discussed in this book.

systems. It turns out, however, that this complication is only necessary in $d = 2$ and $d = 3$. These, of course, are the relevant physical dimensions, but as soon as $d > 3$, monodisperse spherical systems remain disordered for extremely long times, and it is very hard to observe them crystallise spontaneously [330, 350] (we come back to this point in Chapter 8). Because this book is mostly concerned with the large-dimensional limit, it is thus sufficient for us to focus on monodisperse spherical systems. In infinite dimensions, polydispersity is not essential for glass formation, and a small enough deviation from monodispersity [184] or from sphericity [367] does not affect the qualitative shape of the phase diagram. The rest of this book is thus focused on monodisperse spherical systems, although the discussion could be quite easily generalised to binary, polydisperse or non-spherical systems [49, 102, 184, 185, 367]. As discussed in the Preface, establishing to what extent an infinite-dimensional monodisperse spherical system is a good model of real glassy materials goes beyond the scope of this book.

Note that for monodisperse spherical systems, even if crystallisation is kinetically heavily suppressed in large dimensions (in other words, if the system is prepared in the liquid phase it will remain there forever), it is still highly probable that the true equilibrium thermodynamic state at large densities might be a crystal (see Chapter 8). If this is true, then the liquid phase would be but metastable with respect to the crystal. However, this is not a problem because as we have seen in Chapter 1, in infinite dimensions, metastable states are minima of a suitable free energy function. In this chapter, we define the free energy as a functional of the density profile. In this framework, the liquid state is a minimum of the free energy with uniform density while the crystal is a minimum with a periodically modulated density. From a theoretical standpoint, the two situations can then be distinguished without any ambiguity. In Chapter 4, we thus describe how this approach can be generalised to the glass phase.

2.1.1 Definitions

Consider a system of N identical atoms, enclosed in a compact region of $\mathbb{R}^d$ with volume V. Atoms are modelled as point particles. Their positions are specified by a set of d-dimensional vectors $\underline{X} = \{\mathbf{x}_i\}_{i=1,\dots,N}$, each $\mathbf{x}_i$ having components $x_{i\mu}$ for $\mu = 1, \dots, d$. In most atomic systems, the interaction energy is dominated by pairwise interactions, and the total interaction energy is

$$V(\underline{X}) = \sum_{i<j} v(|\mathbf{x}_i - \mathbf{x}_j|). \qquad (2.1)$$

Interactions involving more than two particles could also be considered, but they are not essential in most cases, and for simplicity, they will be neglected [175, chapter 1]. Because the particles are pointlike, the interaction between a given pair ij of them must be spherically symmetric – i.e., it can only depend on their distance $r_{ij} = |\mathbf{x}_i - \mathbf{x}_j|$. Examples of typical interaction potentials $v(r)$ are given in [175, chapter 1] and will be discussed in Section 2.3.2.

The main thermodynamic quantities are defined following [175, chapter 2], to which the reader is referred for a detailed discussion. The configurational integral at temperature[2] $T = 1/\beta$,

$$Z_N = \int d\underline{X}\, e^{-\beta V(\underline{X})}, \qquad d\underline{X} = \prod_{i=1}^{N} d\mathbf{x}_i, \qquad (2.2)$$

plays a central role in the thermodynamics of particle systems. In fact, for Hamiltonian systems, the total energy, kinetic plus potential, is the Hamiltonian function $H(\underline{P}, \underline{X}) = \sum_i \frac{\mathbf{p}_i^2}{2m} + V(\underline{X})$, and the canonical partition function is

$$Q_N = \frac{1}{h^{dN} N!} \int d\underline{P}d\underline{X}\, e^{-\beta H(\underline{P}, \underline{X})} = \frac{Z_N}{\Lambda^{dN} N!}. \qquad (2.3)$$

The factor $N!$ in the denominator takes into account the indistinguishability of the particles, while the constant h (which must be present for dimensional reasons, because Q_N must be adimensional) can be identified with the Planck constant. Both factors can be deduced from a purely classical treatment [158], or they can be obtained by taking the semiclassical limit of a quantum mechanical treatment [223]. With $\hbar = h/(2\pi)$, the 'De Broglie wavelength' $\Lambda = \sqrt{2\pi\beta\hbar^2/m}$ appears after the Gaussian integration over the momenta has been performed explicitly. The average of an observable $\mathcal{O}[\underline{X}]$ in the canonical ensemble is

$$\langle \mathcal{O}[\underline{X}] \rangle = \frac{1}{Q_N} \int \frac{d\underline{P}d\underline{X}}{h^{dN} N!}\, e^{-\beta H(\underline{P}, \underline{X})}\, \mathcal{O}[\underline{X}] = \frac{1}{Z_N} \int d\underline{X}\, e^{-\beta V(\underline{X})}\, \mathcal{O}[\underline{X}], \qquad (2.4)$$

and the Helmholtz free energy of a system with fixed N, V, T is

$$F(N, V, T) = -T \log Q_N = dT N \log \Lambda - T \log \left(\frac{Z_N}{N!}\right). \qquad (2.5)$$

The thermodynamic limit is taken by sending $N, V \to \infty$ with constant number density $\rho = N/V$ and temperature T. For this limit to exist, two conditions on the pair potential $v(r)$ are sufficient [158, 310].

1. 'Temperedness': for $r \to \infty$, $|v(r)| \leq A r^{-d-\delta}$ with constants $A \geq 0$ and $\delta > 0$. This condition ensures that the interaction energy decays sufficiently rapidly at large distance. In particular, finite range potentials, such that $v(r) = 0$ for $r \geq r_0$, trivially satisfy this condition.

---

[2] Recall that $k_B = 1$ throughout this book.

2. 'Stability': there exists an N-independent constant $B \geq 0$, such that $V(\underline{X}) \geq -BN$ for all particle configurations $\underline{X} \in \mathbb{R}^{dN}$ and for all N. This condition ensures that the collapse of infinitely many particles in a bounded region of space is impossible. There are many possible ways of imposing conditions on the pair potential to ensure stability; some of them are discussed in [310, chapter 3]. In particular, potentials with a hard core satisfy this condition.

Potentials that do not satisfy the temperedness condition are deemed to be 'long ranged', while potentials that do not satisfy the stability condition are said to be 'catastrophic'. In both cases, anomalies can appear in the thermodynamic limit. Such potentials – e.g., the gravitational potential – are not considered in this book.

In the thermodynamic limit, the equilibrium Helmholtz free energy is a function of ρ and T, which are the only state variables of the system. Moreover, because F is extensive (proportional to N), it is convenient to consider the free energy per particle, which remains finite:

$$
\begin{aligned}
\mathrm{f}(\rho, T) &= \lim_{N, V \to \infty, \, \rho=N/V} \frac{F(N, V, T)}{N} \\
&= dT \log \Lambda - T \lim_{N, V \to \infty, \, \rho=N/V} \frac{1}{N} \log\left(\frac{Z_N}{N!}\right).
\end{aligned}
\tag{2.6}
$$

From the Helmholtz free energy per particle, one can derive most of the interesting thermodynamic quantities – such as the average energy, the entropy, the pressure and the specific heat. One can consider different ensembles – for example, with pressure,

$$
P(\rho, T) = - \lim_{N, V \to \infty, \, \rho=N/V} \frac{\partial F(N, V, T)}{\partial V} = \rho^2 \frac{\partial \mathrm{f}(\rho, T)}{\partial \rho},
\tag{2.7}
$$

fixed instead of volume V (or chemical potential μ fixed instead of particle number N) – but because all ensembles are equivalent in the thermodynamic limit [158, 175], the (ρ, T) ensemble will most often be considered in this book. The main objective of the following discussion will thus be to compute $\mathrm{f}(\rho, T)$.

For an ideal gas of non-interacting particles, $v(r) = 0$, one has $Z_N = V^N$ and

$$
\begin{aligned}
\mathrm{f}^{\mathrm{id}}(\rho, T) &= dT \log \Lambda - T \lim_{N, V \to \infty, \, \rho=N/V} \frac{1}{N} \log\left(\frac{V^N}{N!}\right) \\
&= dT \log \Lambda - T(1 - \log \rho).
\end{aligned}
\tag{2.8}
$$

It is customary to separate the Helmholtz free energy between the ideal gas and the excess contributions:

$$
\mathrm{f}(\rho, T) = \mathrm{f}^{\mathrm{id}}(\rho, T) + \mathrm{f}^{\mathrm{ex}}(\rho, T),
$$

$$
\mathrm{f}^{\mathrm{ex}}(\rho, T) = -T \lim_{N, V \to \infty, \, \rho=N/V} \frac{1}{N} \log\left(\frac{Z_N}{V^N}\right).
\tag{2.9}
$$

The term $f^{ex}(\rho, T)$ contains all the non-trivial dependence on the interactions and depends only on the normalised configurational integral Z_N / V^N. Note that the term $dT \log \Lambda$ in $f^{id}(\rho, T)$ depends only on constants and on temperature T. Because its contribution to thermodynamic quantities is trivial, it will often be omitted in the following, although it can be reinserted at any time.

2.1.2 Local Density and Thermodynamic Observables

The discussion of Section 1.1.2 for magnetic systems can be adapted with only minor modifications to particle systems. The role of the local magnetisation is then played by the local density, defined as

$$\rho[\mathbf{x}; \underline{X}] = \sum_{i=1}^{N} \delta(\mathbf{x}_i - \mathbf{x}), \qquad \rho(\mathbf{x}) = \langle \rho[\mathbf{x}; \underline{X}] \rangle. \qquad (2.10)$$

Here the function $\rho[\mathbf{x}; \underline{X}]$ is the local density at point $\mathbf{x}$ for a given configuration $\underline{X}$, which is a sum of delta functions because classical particles are at well-defined positions. The function $\rho(\mathbf{x})$ is the thermodynamic average over $\underline{X}$ in the canonical ensemble defined by Eq. (2.4) or, equivalently, in any other ensemble. This quantity is usually a smooth function because particles move under the action of thermal fluctuations. Note that $\int d\mathbf{x}\rho(\mathbf{x}) = N$, as follows from Eq. (2.10).

Most of the interesting observables of particle systems can be written as linear combinations of products of local densities. In general, taking into account that particles are identical, a thermodynamic observable has the form

$$\mathcal{O}[\underline{X}] = \sum_{i} \mathcal{O}_1(\mathbf{x}_i) + \sum_{i \neq j} \mathcal{O}_2(\mathbf{x}_i, \mathbf{x}_j) + \sum_{i \neq j \neq k} \mathcal{O}_3(\mathbf{x}_i, \mathbf{x}_j, \mathbf{x}_k) + \cdots, \qquad (2.11)$$

where $\mathcal{O}_1$ is a one-body observable, $\mathcal{O}_2$ a two-body observable and so on. The potential energy in Eq. (2.1) is a typical example of a two-body observable, and so is pressure [175]. Eq. (2.11) can equivalently be written in terms of the local density,

$$\mathcal{O}[\underline{X}] = \int d\mathbf{x}\mathcal{O}_1(\mathbf{x})\rho[\mathbf{x}; \underline{X}]$$
$$+ \int d\mathbf{x}d\mathbf{y}\mathcal{O}_2(\mathbf{x}, \mathbf{y})\{\rho[\mathbf{x}; \underline{X}]\rho[\mathbf{y}; \underline{X}] - \rho[\mathbf{x}; \underline{X}]\delta(\mathbf{x} - \mathbf{y})\} + \cdots, \qquad (2.12)$$

which is the analog of Eq. (1.11) for particle systems.

The thermodynamic average then has the form

$$\langle \mathcal{O}[\underline{X}] \rangle = \int d\mathbf{x}\mathcal{O}_1(\mathbf{x})\rho(\mathbf{x}) + \int d\mathbf{x}d\mathbf{y}\mathcal{O}_2(\mathbf{x}, \mathbf{y})\rho^{(2)}(\mathbf{x}, \mathbf{y}) + \cdots, \qquad (2.13)$$

which can be decomposed as a sum of integrals of density-density correlations, where the one-body density is $\rho(\mathbf{x})$ and the two-body density is

$$\rho^{(2)}(\mathbf{x}, \mathbf{y}) = \langle \rho[\mathbf{x}; \underline{X}]\rho[\mathbf{y}; \underline{X}]\rangle - \rho(\mathbf{x})\delta(\mathbf{x} - \mathbf{y}). \tag{2.14}$$

Three-body and higher-order correlations are defined similarly [175, chapter 2]. The quantity $\rho^{(2)}(\mathbf{x}, \mathbf{y})\mathbf{dxdy}$ yields the probability of finding two distinct particles with coordinates in the volume element $\mathbf{dxdy}$, irrespective of the positions of the remaining particles and irrespective of momenta. From it, a normalised pair correlation function can be defined,

$$g(\mathbf{x}, \mathbf{y}) = \frac{\rho^{(2)}(\mathbf{x}, \mathbf{y})}{\rho(\mathbf{x})\rho(\mathbf{y})}. \tag{2.15}$$

This quantity is central in the theory of atomic systems. For an ideal gas, correlations between particles are absent, and one has $\rho^{(2)}(\mathbf{x}, \mathbf{y}) = \rho(\mathbf{x})\rho(\mathbf{y})$ and $g(\mathbf{x}, \mathbf{y}) = 1$. Deviations of $g(\mathbf{x}, \mathbf{y})$ from unity thus characterise the local structuring of the system due to interactions. In the gas and liquid phases, due to translational and rotational invariance, the local density $\rho(\mathbf{x}) = \rho$ is independent of $\mathbf{x}$, and the density-density correlation is only a function of distance, $r = |\mathbf{x} - \mathbf{y}|$. In this case $g(r) = \rho^{(2)}(r)/\rho^2$ is known as the 'radial distribution function'.

2.1.3 Free Energy as a Functional of the Local Density

As in magnetic systems, the thermodynamic phases of the system can be identified with the minima of the free energy $F[\rho(\mathbf{x})]$, expressed as a functional of the local density $\rho(\mathbf{x})$. However, computing this quantity directly in the canonical ensemble, where the total density is fixed, is cumbersome. Following the treatment of Chapter 1 for magnetic systems, it is convenient to introduce a field conjugated to the local density – i.e., the chemical potential – and compute the free energy as a functional of the chemical potential in the grand canonical ensemble. One can then perform a Legendre transformation to recover $F[\rho(\mathbf{x})]$.

Definition of the Free Energy Functional

In the grand canonical ensemble [175, chapter 2], the particle number can fluctuate, but its average is fixed by a chemical potential μ. To fix the local density, one can introduce an external local potential $\phi(\mathbf{x})$. Defining

$$z(\mathbf{x}) = \frac{e^{\beta\mu - \beta\phi(\mathbf{x})}}{\Lambda^d}, \tag{2.16}$$

the grand canonical partition function is

$$\Xi[z] = \sum_{N=0}^{\infty} \frac{e^{\beta \mu N}}{h^{dN} N!} \int d\underline{P} d\underline{X} \, e^{-\beta H(\underline{P}, \underline{X}) - \beta \sum_{i=1}^{N} \phi(\mathbf{x}_i)}$$

$$= \sum_{N=0}^{\infty} \frac{1}{N!} \int d\underline{X} \, e^{-\beta V(\underline{X})} \prod_{i=1}^{N} z(\mathbf{x}_i). \tag{2.17}$$

The free energy of the grand canonical ensemble is the 'grand potential'

$$\Omega[z] = -T \log \Xi[z] = -T \log \sum_{N=0}^{\infty} \frac{1}{N!} \int d\underline{X} \, e^{-\beta V(\underline{X}) + \int d\mathbf{x} \rho[\mathbf{x}; \underline{X}] \log z(\mathbf{x})}, \tag{2.18}$$

where $\prod_i z(\mathbf{x}_i) = \exp[\int d\mathbf{x} \rho[\mathbf{x}; \underline{X}] \log z(\mathbf{x})]$ is written in terms of the local density. From Eq. (2.18), it follows that the average local density and its correlations can be obtained by differentiating the grand potential with respect to $\log z(\mathbf{x}) = \beta \mu - \beta \phi(\mathbf{x}) - d \log \Lambda$. This treatment follows exactly how spin–spin correlations are obtained as derivatives[3] of the free energy with respect to the magnetic field (Section 1.1.2):

$$-\beta \frac{\partial \Omega}{\partial \log z(\mathbf{x})} = \langle \rho[\mathbf{x}; \underline{X}] \rangle = \rho(\mathbf{x}),$$

$$-\beta \frac{\partial^2 \Omega}{\partial \log z(\mathbf{x}) \partial \log z(\mathbf{y})} = \langle \rho[\mathbf{x}; \underline{X}] \rho[\mathbf{y}; \underline{X}] \rangle - \rho(\mathbf{x}) \rho(\mathbf{y}) \tag{2.19}$$

$$= \rho^{(2)}(\mathbf{x}, \mathbf{y}) + \rho(\mathbf{x}) \delta(\mathbf{x} - \mathbf{y}) - \rho(\mathbf{x}) \rho(\mathbf{y}) = \rho S(\mathbf{x}, \mathbf{y}).$$

The second derivative of $-\beta \Omega[z]$ is a positive operator; hence, $\Omega[z]$ is a concave function of $z(\mathbf{x})$.

Then, as it was done in Section 1.1.3 for magnetic systems, it is possible to define the free energy as a functional of $\rho(\mathbf{x})$ by a Legendre transform,

$$F[\rho] = \max_{z(\mathbf{x})} \left[\Omega[z] + T \int d\mathbf{x} \rho(\mathbf{x}) \log[\Lambda^d z(\mathbf{x})] \right], \tag{2.20}$$

from which it follows that

$$-\beta \frac{\partial F}{\partial \rho(\mathbf{x})} = -\log[\Lambda^d z(\mathbf{x})], \qquad -\beta \frac{\partial^2 F}{\partial \rho(\mathbf{x}) \partial \rho(\mathbf{y})} = -(\rho S)^{-1}(\mathbf{x}, \mathbf{y}), \tag{2.21}$$

where $(\rho S)^{-1}(\mathbf{x}, \mathbf{y})$ is the operator inverse of $\rho S(\mathbf{x}, \mathbf{y})$.

---

[3] Here and in the following, we use the same notation for regular and functional derivatives in order to emphasise this similarity.

Uniform Chemical Potential

The external potential $\phi(\mathbf{x})$ has been introduced with the only purpose of constructing the free energy functional of $\rho(\mathbf{x})$. In this book, we are mostly interested in the homogeneous case where $\phi(\mathbf{x}) = 0$ and $z(\mathbf{x}) = z$. The equation for $\rho(\mathbf{x})$ is then

$$\frac{\partial F}{\partial \rho(\mathbf{x})} = T \log(\Lambda^d z) = \mu. \tag{2.22}$$

This equation must be solved to determine the average local density profile that corresponds to a given μ. Let us denote the solution $\rho^*(\mathbf{x})$, and note that the global density is $\rho = \langle N \rangle / V = V^{-1} \int d\mathbf{x} \rho^*(\mathbf{x})$. In the gas and liquid phases, $\rho^*(\mathbf{x}) = \rho$ is homogeneous, but in a solid phase, $\rho^*(\mathbf{x})$ depends on $\mathbf{x}$. Note that in a solid phase, translational and rotational symmetries are broken.[4] There are then many pure states, one specified by $\rho^*(\mathbf{x})$ and a continuous set of other equivalent states corresponding to all the independent translations or rotations of $\rho^*(\mathbf{x})$. In this case, phase coexistence of different solids[5] is possible, leading to the same phenomenology as discussed in Section 1.4.

According to Eq. (2.20), the thermodynamic free energy $F[\rho^*]$ is[6]

$$F[\rho^*] = \Omega(z[\rho^*]) + T \int d\mathbf{x} \rho^*(\mathbf{x}) \log(\Lambda^d z[\rho^*]) = \Omega(\mu) + N\mu, \tag{2.23}$$

where we have used the fact that $z[\rho^*]$ does not depend on $\mathbf{x}$ because $\rho^*(\mathbf{x})$ is the solution of Eq. (2.22). Then, $F[\rho^*]$ is identified with $\Omega(\mu) + N\mu$, which is the Helmholtz free energy. Note that Eq. (2.22) for $\rho(\mathbf{x})$, in which the chemical potential is such that the average density is ρ, can also be obtained by adding a Lagrange multiplier μ to enforce the value of the density – i.e., determine $\rho(\mathbf{x})$ by minimising

$$F_\mu[\rho] = F[\rho] - \mu \left(\int d\mathbf{x} \rho(\mathbf{x}) - N \right). \tag{2.24}$$

We conclude that $F[\rho]$ gives the Helmholtz free energy associated to the local density $\rho(\mathbf{x})$ and that, given an expression for the functional $F[\rho]$, one has to find $\rho(\mathbf{x})$ by minimising it under the constraint that $V^{-1} \int d\mathbf{x} \rho(\mathbf{x}) = \rho$ is fixed. The minima define the equilibrium phases of the system.

---

[4] In crystals, these are only partially broken as a set of discrete symmetries remain unbroken. In amorphous solid phases, both symmetries are fully broken.

[5] For crystals, a state characterised by phase coexistence of different solids is called 'polycrystal', and the interfaces between coexisting regions are called 'grain boundaries'.

[6] Independently for each pure state if a symmetry is broken, but in this case, all pure states have the same free energy.

In infinite dimensions, as we have seen in Chapter 1, the local minima define metastable phases. The gas or liquid phase, which are translationally and rotationally invariant, are minima of $F[\rho]$ with uniform density profile – i.e., $\rho(\mathbf{x}) = \rho$ – and, thus, $\rho^{(2)}(\mathbf{x}, \mathbf{y}) = \rho^2 g(|\mathbf{x} - \mathbf{y}|)$. A crystal phase, instead, corresponds to a periodically modulated $\rho(\mathbf{x})$. By imposing translational invariance, we thus select the liquid phase and rule out any other phase (crystals or separated phases – e.g., gas–liquid coexistence).

We now focus on this choice. Introducing $h(\mathbf{r}) = g(\mathbf{r}) - 1$, and using Eq. (2.19), the second derivative of $\Omega[z]$ is proportional to

$$S(\mathbf{r}) = \delta(\mathbf{r}) + \rho h(\mathbf{r}), \qquad S(\mathbf{q}) = \int d\mathbf{r}\, e^{i\mathbf{q}\cdot\mathbf{r}} S(\mathbf{r}) = 1 + \rho h(\mathbf{q}), \qquad (2.25)$$

where the 'static structure factor' $S(\mathbf{q})$ is the Fourier transform[7] of $S(\mathbf{r})$. This quantity is also very important in liquid theory. It contains a lot of information about liquid structure and is routinely measured by neutron and radiation scattering [175, chapter 3]. From Eq. (2.21), the second derivative of $F[\rho]$ is proportional to $S^{-1}(\mathbf{x} - \mathbf{y})$, the operator inverse of $S(\mathbf{x} - \mathbf{y})$. In the translationally invariant case, it can be obtained in Fourier space as the numerical inverse of $S(\mathbf{q})$:

$$(S^{-1})(\mathbf{q}) = \int d\mathbf{r}\, e^{i\mathbf{q}\cdot\mathbf{r}} S^{-1}(\mathbf{r}) = \frac{1}{S(\mathbf{q})} = \frac{1}{1 + \rho h(\mathbf{q})} = 1 - \rho c(\mathbf{q}), \qquad (2.26)$$

which implies

$$S^{-1}(\mathbf{r}) = \delta(\mathbf{r}) - \rho c(\mathbf{r}). \qquad (2.27)$$

The function $c(\mathbf{q})$ and its inverse Fourier transform $c(\mathbf{r})$ are 'direct correlation functions' [175, chapter 3]. In rotationally invariant states, all these functions depend only on the modulus of their argument.

From the functions $g(r)$ and $S(q)$, one can deduce most of the thermodynamic observables. For example, using Eq. (2.13), and for pairwise additive interactions $V(\underline{X}) = \frac{1}{2} \sum_{i \neq j} v(|\mathbf{x}_i - \mathbf{x}_j|)$, the average potential energy is

$$\langle V(\underline{X}) \rangle = \frac{\rho^2}{2} \int d\mathbf{x} d\mathbf{y}\, g(|\mathbf{x} - \mathbf{y}|) v(|\mathbf{x} - \mathbf{y}|) = \frac{\rho N}{2} \int d\mathbf{r}\, g(r) v(r)$$
$$= \frac{\rho N \Omega_d}{2} \int_0^\infty dr\, r^{d-1} g(r) v(r), \qquad (2.28)$$

where $\Omega_d = 2\pi^{d/2} / \Gamma(d/2)$ is the solid angle in d dimensions, with $\Gamma(x)$ being the Euler gamma function. Similar expressions can be obtained for other observables

---

[7] In order to lighten the notation, we use the same symbol for a function and its Fourier transform when there is no ambiguity.

[175]. Finally, it is worth noting that, in translationally invariant phases, the radial distribution function can be written as a derivative of the free energy with respect to the pair potential, both in the canonical and grand canonical ensembles [175, section 3.4]:

$$\frac{1}{N}\frac{\partial F}{\partial v(\mathbf{r})} = \frac{1}{N}\frac{\partial \Omega}{\partial v(\mathbf{r})} = \frac{\rho}{2}g(\mathbf{r}). \tag{2.29}$$

Writing $V(\underline{X}) = \frac{1}{2}\sum_{i\neq j} v(\mathbf{x}_i - \mathbf{x}_j) = \frac{1}{2}\int d\mathbf{r} v(\mathbf{r}) \sum_{i\neq j} \delta(\mathbf{r} - \mathbf{x}_i + \mathbf{x}_j)$, from Eq. (2.29), one then also obtains

$$g(\mathbf{r}) = \frac{1}{\rho N}\left\langle \sum_{i\neq j} \delta(\mathbf{r} - \mathbf{x}_i + \mathbf{x}_j)\right\rangle. \tag{2.30}$$

2.2 The Virial Expansion

For particle systems, as for spin systems (Chapter 1), it is impossible to obtain a closed analytical expression of the free energy functional $F[\rho]$. The only way to make concrete calculations is to set up perturbative expansions for $F[\rho]$, the most famous being the virial expansion. It expands around the ideal gas limit, and as such is only formally reliable at high temperatures and low densities. In the high-temperature regime, one is above the critical point, and, hence, there is no distinction between gas and liquid. This is the (supercritical) 'fluid' phase. Interestingly, appropriate resummations of the virial expansion can also be used to describe the dense liquid phase [175].

2.2.1 The Ideal Gas

In the ideal gas case, $V(\underline{X}) = 0$, the grand potential can be easily computed:

$$\Xi^{id}[z] = \sum_{N=0}^{\infty} \frac{1}{N!}\int d\underline{X} \prod_{i=1}^{N} z(\mathbf{x}_i) = e^{\int d\mathbf{x} z(\mathbf{x})},$$
$$\Omega^{id}[z] = -T\int d\mathbf{x}\, z(\mathbf{x}). \tag{2.31}$$

It follows that the density profile is

$$\rho(\mathbf{x}) = -\beta\frac{\partial \Omega^{id}}{\partial \log z(\mathbf{x})} = z(\mathbf{x}). \tag{2.32}$$

Plugging these results in Eq. (2.20), one obtains

$$
\begin{aligned}
F^{\text{id}}[\rho] &= \Omega^{\text{id}}[z] + T \int d\mathbf{x} \rho(\mathbf{x}) \log[\Lambda^d z(\mathbf{x})] \\
&= dTN \log \Lambda - T \int d\mathbf{x} \rho(\mathbf{x})[1 - \log \rho(\mathbf{x})],
\end{aligned}
\tag{2.33}
$$

which is minimised by a uniform profile $\rho(\mathbf{x}) = \rho$. The Helmholtz free energy of the ideal gas, Eq. (2.8), is thus recovered.

2.2.2 Second Virial Diagram

The virial expansion is based on the following idea. Considering pairwise interactions with pair potential $v(r)$ as in Eq. (2.1), and introducing the Mayer function

$$
f(r) = e^{-\beta v(r)} - 1,
\tag{2.34}
$$

the potential energy term in the partition function can be expanded in powers of $f(r)$:

$$
\begin{aligned}
e^{-\beta V(\underline{X})} &= \prod_{i<j} e^{-\beta v(|\mathbf{x}_i - \mathbf{x}_j|)} = \prod_{i<j}[1 + f(|\mathbf{x}_i - \mathbf{x}_j|)] \\
&= 1 + \sum_{i<j} f(|\mathbf{x}_i - \mathbf{x}_j|) + \cdots
\end{aligned}
\tag{2.35}
$$

The motivation for this expansion is that, contrarily to the Gibbs–Boltzmann factor, the Mayer function decays to zero at large distances, $r \to \infty$. However, such an expansion only makes sense if $f(r)$ can be considered 'small' in the full range of r – which is correct, for instance, if β is small, or if the region where $f(r)$ is nonzero is small. The virial expansion can thus be considered as a partially resummed high-temperature expansion. Keeping only the first correction, one has[8]:

$$
\begin{aligned}
\Xi[z] &= \sum_{N=0}^{\infty} \frac{1}{N!} \int d\underline{X} \prod_{i=1}^{N} z(\mathbf{x}_i) \left[1 + \sum_{i<j} f(|\mathbf{x}_i - \mathbf{x}_j|) \right] \\
&= e^{\int d\mathbf{x} z(\mathbf{x})} + \sum_{N=2}^{\infty} \frac{1}{N!} \frac{N(N-1)}{2} \left(\int d\mathbf{x}\, z(\mathbf{x}) \right)^{N-2} \\
&\quad \times \int d\mathbf{x} d\mathbf{y}\, z(\mathbf{x}) z(\mathbf{y}) f(|\mathbf{x} - \mathbf{y}|) \\
&= e^{\int d\mathbf{x} z(\mathbf{x})} \left[1 + \frac{1}{2} \int d\mathbf{x} d\mathbf{y} z(\mathbf{x}) z(\mathbf{y}) f(|\mathbf{x} - \mathbf{y}|) \right].
\end{aligned}
\tag{2.36}
$$

---

[8] Note that terms containing the Mayer function can be present only if $N \geq 2$.

Therefore, at the first order in $f(r)$,

$$-\beta\Omega[z] = \int \mathrm{d}\mathbf{x} z(\mathbf{x}) + \frac{1}{2} \int \mathrm{d}\mathbf{x}\mathrm{d}\mathbf{y} z(\mathbf{x}) z(\mathbf{y}) f(|\mathbf{x} - \mathbf{y}|). \qquad (2.37)$$

This correction to $\Omega[z]$ also impacts the ideal gas result $\rho(\mathbf{x}) = z(\mathbf{x})$, which becomes $\rho(\mathbf{x}) = z(\mathbf{x}) + O(z^2)$. The $O(z^2)$ correction to $\rho(\mathbf{x})$, however, does not enter in $F[\rho]$ at the lowest order, for the same reason as in Section 1.2.1. We can then simply substitute $z(\mathbf{x}) = \rho(\mathbf{x})$ in the correction term of Eq. (2.37) to obtain the second virial correction to $F[\rho]$:

$$-\beta F[\rho] = -dN \log \Lambda + \int \mathrm{d}\mathbf{x} \rho(\mathbf{x})[1 - \log \rho(\mathbf{x})]$$
$$+ \frac{1}{2} \int \mathrm{d}\mathbf{x}\mathrm{d}\mathbf{y} \rho(\mathbf{x})\rho(\mathbf{y}) f(|\mathbf{x} - \mathbf{y}|). \qquad (2.38)$$

2.2.3 Full Virial Expansion

A convenient way to work out the full virial expansion is to represent its different terms as diagrams. The general formalism behind this approach is discussed in [175, chapter 3], and detailed derivations can be found in [118, 266]. Here, we summarise some of the key ideas. The second virial correction can be represented by a diagram of the form

$$\bullet\!\!-\!\!\bullet = \frac{1}{2} \int \mathrm{d}\mathbf{x}\mathrm{d}\mathbf{y} \rho(\mathbf{x})\rho(\mathbf{y}) f(|\mathbf{x} - \mathbf{y}|). \qquad (2.39)$$

In this diagram, each black dot (or vertex) represents a factor of $\rho(\mathbf{x})$ with a corresponding integration over $\mathbf{x}$, and each black line represents a factor $f(|\mathbf{x} - \mathbf{y}|)$. The diagram is also multiplied by a factor of $1/S$, where S is the 'symmetry number' of that diagram, which is calculated as follows. One first chooses an arbitrary labelling of the vertices; in the example of Eq. (2.39), one might call $\mathbf{x}$ the left vertex and $\mathbf{y}$ the right vertex. Then, S is the number of permutations of the labels that give rise to exactly the same labelling. In Eq. (2.39), exchanging $\mathbf{x}$ and $\mathbf{y}$ leads to the exact same result. Hence, $S = 2$, which explains the factor $1/2$ in front of the integral. Another example of diagram is

$$\triangle = \frac{1}{6} \int \mathrm{d}\mathbf{x}\mathrm{d}\mathbf{y}\mathrm{d}\mathbf{z} \rho(\mathbf{x})\rho(\mathbf{y})\rho(\mathbf{z}) f(|\mathbf{x} - \mathbf{y}|) f(|\mathbf{x} - \mathbf{z}|) f(|\mathbf{y} - \mathbf{z}|), \qquad (2.40)$$

where $S = 6$ because once an arbitrary labelling is chosen (e.g., $\mathbf{x}$ for the top vertex, $\mathbf{y}$ for the bottom right and $\mathbf{z}$ for the bottom left), there are three rotations of the labels and, for each rotation, one exchange of two labels that all give the same result.

The virial series for the Helmholtz free energy then has the form

$$-\beta F[\rho] = -\beta F^{\text{id}}[\rho] + \bullet\!\!-\!\!\bullet + \triangle + \square + \boxtimes\!\!\!\!\diagdown + \boxtimes + \cdots$$

(2.41)

The diagrams that appear in Eq. (2.41) are all the possible 'one-particle irreducible diagrams', which are those that cannot be disconnected in two or more parts by removing one vertex (i.e., one particle). Note that although the series in Eq. (2.41) has been derived through an expansion in powers of the Mayer function (the number of lines in each diagram), the diagrams in Eq. (2.41) are organised according to the number of vertices they contain, which is an easier and more natural scheme. Because the Mayer function is small at high temperatures, classifying the diagrams by the number of lines they contain essentially leads to a (resummed) high-temperature expansion, while classifying the diagrams according to the number of vertices amounts to an expansion in powers of $\rho(\mathbf{x})$ – i.e., a low-density expansion. In any case, one should keep in mind that a rigorous proof of convergence of the virial expansion requires both high temperatures and low densities, in addition to some regularity conditions on the interaction potential [158].

2.2.4 Virial Expansion in the Fluid Phase

In the fluid, gas or liquid phases, the homogeneous density $\rho(\mathbf{x}) = \rho$ is a solution of Eq. (2.22). From Eq. (2.41), the Helmholtz free energy per particle can be written as a power series of ρ as

$$-\beta f(\rho, T) = \frac{-\beta F[\rho(\mathbf{x}) = \rho]}{N} = -\beta f^{\text{id}}(\rho, T) - \sum_{n=2}^{\infty} \frac{B_n}{n-1} \rho^{n-1}, \qquad (2.42)$$

$$\beta P(\rho, T) = \rho^2 \frac{\partial \beta f(\rho, T)}{\partial \rho} = \rho + \sum_{n=2}^{\infty} B_n \rho^n = \rho + B_2 \rho^2 + B_3 \rho^3 + \cdots,$$

where the pressure $P(\rho, T)$ is obtained from Eq. (2.7), and B_n are the 'virial coefficients'.

Each virial coefficient is a sum of contributions coming from all one-particle irreducible diagrams $\mathcal{G}_n$ of n vertices. The algorithm to obtain the contribution of each diagram to B_n is best understood by examples. Consider the second virial coefficient, to which a single diagram contributes, Eq. (2.39) with $\rho(\mathbf{x}) = \rho$. Comparing Eq. (2.42) with Eq. (2.41), one obtains

$$B_2 = -\frac{1}{\rho N} \times \bullet\!\!-\!\!\bullet = -\frac{\rho^2}{2\rho N} \int d\mathbf{x}_1 d\mathbf{x}_2 f(|\mathbf{x}_1 - \mathbf{x}_2|) = -\frac{1}{2} \int d\mathbf{r} f(|\mathbf{r}|).$$

(2.43)

In the last expression, translational invariance has been used to change variables from $\{\mathbf{x}_1, \mathbf{x}_2\}$ to $\{\mathbf{r} = \mathbf{x}_1 - \mathbf{x}_2, \mathbf{x}_2\}$, in such a way that the integrand is independent of $\mathbf{x}_2$. The integration over $\mathbf{x}_2$ then gives a factor V that cancels the factors of ρ in front of the virial coefficient. Similarly, the third virial coefficient is given by

$$B_3 = -\frac{2}{\rho^2 N} \times \triangle = -\frac{1}{3} \int d\mathbf{r}_1 d\mathbf{r}_2 \, f(|\mathbf{r}_1|) f(|\mathbf{r}_2|) f(|\mathbf{r}_1 - \mathbf{r}_2|), \qquad (2.44)$$

where $\mathbf{r}_1 = \mathbf{x}_1 - \mathbf{x}_3, \mathbf{r}_2 = \mathbf{x}_2 - \mathbf{x}_3$, and the integration over $\mathbf{x}_3$ gives a factor V by translational invariance. The procedure can be iterated to obtain the general relation

$$B_n = -(n-1) \sum_{\mathcal{G}_n} \mathcal{B}(\mathcal{G}_n), \qquad (2.45)$$

where the sum is over all diagrams $\mathcal{G}_n$ having n vertices and a set of edges $E = \langle ij \rangle \in \mathcal{G}_n$. For a given diagram, the value of $\mathcal{B}(\mathcal{G}_n)$ is obtained by fixing one arbitrary vertex in $\mathbf{x}_n = \mathbf{r}_n = 0$, and the other vertices correspond to the points $\mathbf{r}_1 \ldots \mathbf{r}_{n-1}$. The diagram is multiplied by its symmetry factor $1/S(\mathcal{G}_n)$, with an integrand given by a product of Mayer functions $f(E) = f(|\mathbf{r}_i - \mathbf{r}_j|)$ assigned to each edge,

$$\mathcal{B}(\mathcal{G}_n) = \frac{1}{S(\mathcal{G}_n)} \int d\mathbf{r}_1 \ldots d\mathbf{r}_{n-1} \prod_{E \in \mathcal{G}_n} f(E), \qquad \text{with } \mathbf{r}_n = 0. \qquad (2.46)$$

Explicit expressions for the first few virial coefficients are known for several choices of interaction potential.

2.3 Liquids in Large Dimensions

In this section, we investigate the behaviour of the virial expansion in the limit of large spatial dimensions, $d \to \infty$. This problem was first considered by Frisch, Klein, Percus, Rivier and Wyler [154, 155, 210, 362]. As we will see, in large d, the virial expansion remains valid also in the dense fluid regime. For this reason, in the following, we generically refer to the homogeneous phase as the 'liquid' phase.

2.3.1 Virial Expansion for Hard Spheres

One of the most studied pair interactions is the hard-sphere potential, which models a collection of identical balls that cannot overlap and otherwise do not interact. For spheres of diameter ℓ, the potential is

$$v(r) = \begin{cases} 0 & \text{if } r > \ell, \\ \infty & \text{if } r \le \ell, \end{cases} \qquad (2.47)$$

which satisfies the temperedness and stability conditions discussed in Section 2.1.1, and, thus, the existence of the thermodynamic limit is guaranteed. The corresponding Mayer function is $f(r) = -\theta(\ell - r)$, where $\theta(x)$ is the Heaviside step function. Note that because $f(r)$ does not depend on temperature, the temperature scaling of free energy and pressure is trivial: βf^{ex} and βP are independent of T and only depend on density ρ. If temperature T is used as the unit of energy (which amounts to setting $T = 1$), all thermodynamic quantities are solely functions of ρ, which is thus the unique state variable [175]. The volume and surface of a ball of unit radius in d dimensions are, respectively,

$$V_d = \frac{\pi^{d/2}}{\Gamma(1 + d/2)}, \qquad \Omega_d = dV_d = \frac{2\pi^{d/2}}{\Gamma(d/2)}. \tag{2.48}$$

Note that Ω_d is also the solid angle in d dimensions, previously introduced in Eq. (2.28). The volume and surface of a sphere of radius r are then $V_d(r) = V_d r^d$ and $\Omega_d(r) = \Omega_d r^{d-1}$. For hard spheres, it is customary to define the 'packing fraction',

$$\varphi = \rho V_d \ell^d / 2^d, \tag{2.49}$$

which is the fraction of volume occupied by the spheres. For convenience, in this section, we also define $\overline{\varphi} = 2^d \varphi = \rho V_d \ell^d$. Without loss of generality, one can also choose ℓ as unit of length, which amounts to setting $\ell = 1$; this is the choice we make (unless otherwise specified) in this section to lighten the notation.

For hard spheres, many virial coefficients have been computed to high precision. The second virial coefficient is

$$B_2 = \frac{1}{2} V_d \ell^d. \tag{2.50}$$

Explicit expressions of B_3 and B_4 in terms of special functions for arbitrary dimension d are given in [196, 236]. Numerical values of B_n for $n \leq 11$ and $d \leq 8$ are given in [96, 319]. Even higher dimensions have been studied in [373]. In this section, we analyse the asymptotic behaviour of the virial series for hard spheres in the large d limit [154, 210, 362]. We will follow closely the treatment of [153]. The discussion is a bit long and technical, but the essential steps are quite simple. The reader can skip the proofs and jump straight to the conclusions, if desired.

Lower Bound on the Convergence Radius of the Virial Series

Denoting $\overline{\varphi}_{\text{conv}}$ the convergence radius of the virial series, Lebowitz and Penrose [226] proved that

$$\overline{\varphi}_{\text{conv}} \geq \frac{(W(e/2) - 1)^2}{W(e/2)} = 0.144767 \cdots = \overline{\varphi}_{\text{conv}}^l, \tag{2.51}$$

where $W(x)$ is the Lambert function defined by $W(x)e^{W(x)} = x$. The virial series is thus surely convergent for $\overline{\varphi} \leq \overline{\varphi}^l_{\text{conv}}$. In this region, the hard-sphere system has a unique pure state, the liquid phase, and there can be no phase transitions.

Ring Diagrams Dominate at Each Order

A special role in the series analysis is played by 'ring diagrams'. At any order in $n \geq 3$, the ring diagram $\mathcal{R}_n$ is the diagram with the fewest edges; it has exactly n edges arranged in a ring. For example, $\mathcal{R}_3$ is the triangle drawn in Eq. (2.40), $\mathcal{R}_4$ is the square in Eq. (2.41), $\mathcal{R}_5$ is a pentagon and so on. At any given finite order n, adding an edge to a diagram makes the resulting diagram exponentially smaller in d than the original one. Therefore, at any given order n, the ring diagram dominates the virial coefficient, and $B_n \sim -(n-1)\mathcal{B}(\mathcal{R}_n)$.

Proof A proof of this statement can be found in [153]. Here we follow a similar but slightly different route. By rotational and translational invariance, the Mayer functions in Eq. (2.46) depend only on $|\mathbf{r}_i - \mathbf{r}_j| = \sqrt{\mathbf{r}_i^2 + \mathbf{r}_j^2 - 2\mathbf{r}_i \cdot \mathbf{r}_j}$ – i.e., they depend only on the scalar products $q_{ij} = \mathbf{r}_i \cdot \mathbf{r}_j$. One can thus change the integration variables from $\mathbf{r}_i$ to q_{ij}. The details of this change of variable are given in Section 2.5, but recall nonetheless that $\mathbf{r}_n = \mathbf{0}$; hence, $\hat{q}$ is a $(n-1) \times (n-1)$ symmetric matrix. Following this scheme, Eq. (2.46) then becomes

$$\mathcal{B}(\mathcal{G}_n) = \frac{C_{n,d}}{S(\mathcal{G}_n)} \int_{\mathcal{Q}} d\hat{q} \, (\det \hat{q})^{\frac{d-n}{2}} \prod_{E \in \mathcal{G}_n} f(E),$$

$$C_{n,d} = 2^{1-n} \Omega_d \Omega_{d-1} \cdots \Omega_{d-n+2}.$$

(2.52)

Here $d\hat{q} = \prod_{i \leq j} dq_{ij}$, and the integration domain is

$$\mathcal{Q} = \{\hat{q} : q_{ii} \geq 0 \text{ and } |q_{ij}| \leq \sqrt{q_{ii}q_{jj}}\}.$$

(2.53)

In Eq. (2.52), the arguments of the Mayer functions are $\sqrt{q_{ii} + q_{jj} - 2q_{ij}}$ for all edges except those involving point n, which has $\mathbf{r}_n = \mathbf{0}$, and, therefore, edges $E = \langle in \rangle$ have $\sqrt{q_{ii}}$.

For $d \to \infty$ and finite n, Eq. (2.52) can be evaluated by the saddle point method. Neglecting constants and polynomial terms in d, the leading exponential order in d is given by maximising $\det \hat{q}$ over all matrices $\hat{q} \in \mathcal{Q}$ that also satisfy the constraints imposed by the requirement that the Mayer functions must be non-zero for all E:

$$|\mathcal{B}(\mathcal{G}_n)| \sim (\Omega_d)^{n-1} \mathcal{M}(\mathcal{G}_n)^{\frac{d}{2}},$$

$$\mathcal{M}(\mathcal{G}_n) = \max_{\hat{q} \in \mathcal{Q}: \, \prod_{E \in \mathcal{G}_n} f(E) > 0} |\det \hat{q}|.$$

(2.54)

The absolute value of $\mathcal{B}(\mathcal{G}_n)$ is taken because the Mayer functions are negative, and, therefore, each diagram has a sign $(-1)^{|\mathcal{G}_n|}$ where $|\mathcal{G}_n|$ is the number of edges of $\mathcal{G}_n$.

Adding edges to a diagram amounts to introducing more constraints, which, from Eq. (2.54), reduces the maximum value of $|\det \hat{q}|$ and thus suppresses exponentially the diagram. To show this explicitly, one can compute the leading order of the first diagrams from Eq. (2.54). The second virial is known exactly. The triangle diagram has a 2×2 matrix $\hat{q}$ with $\det \hat{q} = q_{11}q_{22} - q_{12}^2$, which has to be maximised under the constraints given by the three Mayer functions and Eq. (2.53). This results in $0 \leq q_{11} \leq 1, 0 \leq q_{22} \leq 1, 0 \leq q_{11}+q_{22}-2q_{12} \leq 1$. The maximising configuration is $q_{11} = q_{22} = 1$ and $q_{12} = 1/2$, which corresponds to a triangle of spheres all at unit distance from each other. Then

$$B\left(\triangle \right) \sim (\Omega_d)^2 \left(\max[q_{11}q_{22} - q_{12}^2] \right)^{d/2} = (\Omega_d)^2 \left(\frac{3}{4} \right)^{d/2}. \qquad (2.55)$$

At fourth order, there are three diagrams. The maximisation is over a 3×3 matrix, and it can be done easily with a numerical manipulation software. The ring diagram has

$$B\left(\square \right) \sim (\Omega_d)^3 \left(\frac{16}{27} \right)^{d/2}, \qquad \hat{q} = \begin{pmatrix} 1 & 2/3 & 1/3 \\ 2/3 & 4/3 & 2/3 \\ 1/3 & 2/3 & 1 \end{pmatrix}. \qquad (2.56)$$

The other two diagrams have an additional constraint, and, as expected, each additional edge makes the diagram exponentially smaller:

$$B\left(\boxslash \right) \sim (\Omega_d)^3 \left(\frac{9}{16} \right)^{d/2}, \qquad \hat{q} = \begin{pmatrix} 1 & 1/2 & 1/4 \\ 1/2 & 1 & 1/2 \\ 1/4 & 1/2 & 1 \end{pmatrix}, \qquad (2.57)$$

and

$$B\left(\boxtimes \right) \sim (\Omega_d)^3 \left(\frac{1}{2} \right)^{d/2}, \qquad \hat{q} = \begin{pmatrix} 1 & 1/2 & 1/2 \\ 1/2 & 1 & 1/2 \\ 1/2 & 1/2 & 1 \end{pmatrix}. \qquad (2.58)$$

The same reasoning can be applied to any higher-order n with similar results.

Asymptotic Behaviour of the Ring Diagrams

The previous result suggests that in large d, the nth virial coefficient is dominated by the ring diagram, except for B_2, which is given in Eq. (2.50). Therefore, the nth-order contribution to the free energy in Eq. (2.42) is

$$f_n = -\rho^{n-1} \frac{B_n}{n-1} = \begin{cases} -\frac{2^d \varphi}{2} = -\frac{\overline{\varphi}}{2} & \text{for } n = 2, \\ \rho^{n-1} B(\mathcal{R}_n) & \text{for } n \geq 3. \end{cases} \qquad (2.59)$$

At fixed n and $d \to \infty$, at leading exponential order in d, one can show that

$$|f_n| \sim \overline{\varphi}^{n-1} \left[\frac{n^{n-2}}{(n-1)^{n-1}} \right]^{d/2}. \tag{2.60}$$

For example, $|f_3| \sim \overline{\varphi}^2 (3/4)^{d/2}$, as in Eq. (2.55). Because $n^{n-2}/(n-1)^{n-1} < 1$ for all $n \geq 3$, if $\overline{\varphi} = 2^d \varphi$ is at most polynomial in d when $d \to \infty$, then all the f_n are exponentially smaller than f_2. In this case, one obtains

$$-\beta f(\rho, T) = -\beta f^{\mathrm{id}}(\rho, T) - \frac{\overline{\varphi}}{2},$$

$$p(\rho) = \frac{\beta P(\rho)}{\rho} = 1 + B_2 \rho = 1 + \frac{\overline{\varphi}}{2}, \tag{2.61}$$

where the reduced pressure $p(\rho)$ is derived from Eq. (2.7). If instead the scaled packing fraction $\overline{\varphi} = e^{\gamma d}$ (with $\gamma > 0$) grows exponentially with d, then the condition $|f_2| \gg |f_n|$ is equivalent to

$$e^\gamma < \left[\frac{(n-1)^{n-1}}{n^{n-2}} \right]^{\frac{1}{2(n-1)}} = \sqrt{1 - \frac{1}{n}} \, e^{\frac{\log n}{2(n-1)}}. \tag{2.62}$$

For $n \to \infty$, the right-hand side tends to 1, and Eq. (2.62) cannot be satisfied for any $\gamma > 0$. In this case, the virial series is divergent; the largest orders in n then dominate the series. Hence, the convergence radius $\overline{\varphi}_{\mathrm{conv}}$ of the virial series cannot grow faster than polynomially in d. The numerical results of [373] further suggest that the convergence radius $\overline{\varphi}_{\mathrm{conv}} \to 1$ for $d \to \infty$.

Proof   The proof of Eq. (2.60) can be obtained either from Eq. (2.54), or from a Fourier representation. We follow this second route because it will be useful for the next step. The Fourier transform of the Mayer function is

$$f(q) = \int d\mathbf{r} e^{i\mathbf{q} \cdot \mathbf{r}} f(r) = -V_d \mathcal{J}_d(q),$$

$$\mathcal{J}_d(q) = \Gamma\left(\frac{d}{2} + 1\right) \left(\frac{2}{q}\right)^{\frac{d}{2}} J_{\frac{d}{2}}(q), \tag{2.63}$$

where $J_\nu(x)$ is the Bessel function of order ν. An example of the behaviour of $\mathcal{J}_d(q)$ is given in Figure 2.1. Recalling that $S(\mathcal{R}_n) = 2n$ and using the Fourier transform, f_n can be written (for $n \geq 3$) as

$$f_n = \rho^{n-1} \mathcal{B}(\mathcal{R}_n) = \frac{\rho^{n-1}}{2nV} \int d\mathbf{r}_1 \dots d\mathbf{r}_n f(\mathbf{r}_1 - \mathbf{r}_2) f(\mathbf{r}_2 - \mathbf{r}_3) \dots f(\mathbf{r}_n - \mathbf{r}_1)$$

$$= \frac{\rho^{n-1}}{2n} \int \frac{d\mathbf{q}}{(2\pi)^d} f(q)^n = (-1)^n \frac{\overline{\varphi}^{n-1}}{2n} \frac{V_d \Omega_d}{(2\pi)^d} \int_0^\infty dq \, q^{d-1} \mathcal{J}_d(q)^n. \tag{2.64}$$

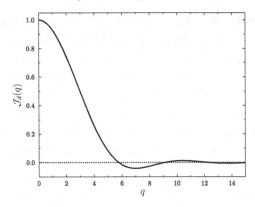

Figure 2.1 An example of the normalised Fourier transform of the Mayer function, the function $\mathcal{J}_d(q) = -f(q)/V_d$ defined in Eq. (2.63), for $d = 5$. The values of q corresponding to the first zero of $\mathcal{J}_d(q)$ and its first negative maximum both scale as $q \sim d/2$.

For large d, the integrand in q in Eq. (2.64) has an absolute maximum before $\mathcal{J}_d(q)$ changes sign for the first time, in the region $q \propto d/2$. This maximum dominates the integral for large d. One can use the asymptotic relation $J_\nu(\nu/\cosh\alpha)$ $\sim \exp[\nu(\tanh\alpha - \alpha)]$ (with $\alpha \geq 0$), choosing $\nu = d/2$ and $q = d/(2\cosh\alpha)$ $\in (0, d/2]$. Keeping only the leading exponential terms in d, the integral becomes

$$f_n \sim \frac{\overline{\varphi}^{n-1} V_d^2}{(2\pi)^d} \Gamma\left(\frac{d}{2} + 1\right)^n 2^{dn/2} \int_0^\infty dq \, q^{d(1-n/2)} J_{\frac{d}{2}}(q)^n$$

$$\sim \overline{\varphi}^{n-1} (2/e)^{d(n/2-1)} e^{d \max_\alpha \left[\left(\frac{n}{2}-1\right)\log(\cosh\alpha) + \frac{n}{2}(\tanh\alpha - \alpha)\right]}. \tag{2.65}$$

The maximisation over α gives $e^{2\alpha} = n - 1$, from which $\tanh\alpha = 1 - 2/n$ and $\cosh\alpha = n/(2\sqrt{n-1})$. Plugging these results in Eq. (2.65) gives Eq. (2.60).

Resummation of the Ring Diagrams

The Fourier expression in Eq. (2.64) can then be used to compute the sum of all f_n. The resulting free energy is

$$-\beta f(\rho, T) = -\beta f^{id}(\rho, T) + f_2 + \sum_{n=3}^\infty f_n$$

$$= -\beta f^{id}(\rho, T) - \frac{\overline{\varphi}}{2} - \frac{1}{2\overline{\varphi}} \frac{V_d \Omega_d}{(2\pi)^d} \int_0^\infty dq \, q^{d-1} L_3[\overline{\varphi}\mathcal{J}_d(q)], \tag{2.66}$$

$$L_3(x) = \sum_{n=3}^\infty (-1)^{n+1} \frac{x^n}{n} = \log(1 + x) - x + \frac{x^2}{2}.$$

The series expansion of $L_3(x)$ is convergent if $|x| < 1$; hence, $|\overline{\varphi}\mathcal{J}_d(q)| < 1$. Because $|\mathcal{J}_d(q)| \leq 1$, as illustrated in Figure 2.1, this condition is satisfied if $|\overline{\varphi}| < 1$. In this region, the series of f_n is convergent, supporting the asymptotic analysis of the virial coefficients in Eq. (2.60), and the numerical results of [373]. In the region $\overline{\varphi} = e^{\gamma d}$ with $\gamma < (1 - \log 2)/2 = 0.153426\ldots$, Eq. (2.66) is well defined and, in fact, coincides with Eq. (2.61). Hence, even if Eq. (2.60) indicates that the contribution of the ring diagrams increases exponentially upon increasing n for $\gamma > 0$, ring diagrams have alternating signs and cancel each other.

Proof The singularity that makes the series divergent corresponds to $q = 0$, where $\mathcal{J}_d(q) = 1$, and $\overline{\varphi} = -1$, so $\overline{\varphi}\mathcal{J}_d(q) = -1$ and the term $\log(1 + x)$ in $L_3(x)$ is singular. This singularity happens in the unphysical region of negative densities. By restricting the expression to positive densities, one can analytically continue Eq. (2.66) to much larger densities. In fact, Eq. (2.66) remains well defined if $\overline{\varphi}\min_q \mathcal{J}_d(q) > -1$, or, equivalently,

$$\overline{\varphi} < \overline{\varphi}_0 = -\frac{1}{\min_q \mathcal{J}_d(q)} \sim \left(\frac{e}{2}\right)^{\frac{d}{2}} = e^{d\frac{1-\log 2}{2}}. \tag{2.67}$$

The result in Eq. (2.67) is obtained by noting that, as illustrated in Figure 2.1, the absolute minimum of $\mathcal{J}_d(q)$ is its first minimum. The equation for the stationary points of $\mathcal{J}_d(q)$ is

$$0 = \frac{d}{dq}\mathcal{J}_d(q) \propto \frac{d}{dq}J_{\frac{d}{2}}(q) - \frac{d}{2q}J_{\frac{d}{2}}(q) = -J_{\frac{d}{2}+1}(q). \tag{2.68}$$

The first minimum corresponds to the first zero of $J_{\frac{d}{2}+1}(q)$, which is asymptotically found at $q_0 \sim d/2$. Using $J_{\frac{d}{2}}(d/2) \sim 1$, one has asymptotically $\mathcal{J}_d(d/2) \sim (4/d)^{d/2}\Gamma(d/2+1) \sim (2/e)^{d/2}$.

A complete proof that Eq. (2.66) coincides with Eq. (2.61) for $\overline{\varphi} < \overline{\varphi}_0$ is given in [153]; see also [284] for an alternative strategy. Here we give a simpler proof restricted to the interval $\gamma < \frac{1}{2}\log(4/3) = 0.143831\cdots$. First, one writes $L_3(x) = x^3\mathcal{L}_3(x)$ and notes that $\mathcal{L}_3(x) = L_3(x)/x^3 > 0$ is a positive and decreasing function of x for all $x > -1$, with $\mathcal{L}_3(0) = 1/3$. The last term in the free energy in Eq. (2.66) is then bounded by

$$\frac{1}{2\overline{\varphi}}\frac{V_d\Omega_d}{(2\pi)^d}\int_0^\infty dq\, q^{d-1}|L_3[\overline{\varphi}\mathcal{J}_d(q)]|$$

$$\leq \mathcal{L}_3[\overline{\varphi}\min_q \mathcal{J}_d(q)]\frac{\overline{\varphi}^2}{2}\frac{V_d\Omega_d}{(2\pi)^d}\int_0^\infty dq\, q^{d-1}|\mathcal{J}_d(q)|^3 \tag{2.69}$$

$$\sim \mathcal{L}_3[-\overline{\varphi}/\overline{\varphi}_0]\overline{\varphi}^2\left(\frac{3}{4}\right)^{d/2} \ll \overline{\varphi},$$

where the last line is obtained as Eq. (2.60) for $n = 3$, keeping only the leading exponential order in d. For $\gamma < \frac{1}{2}\log(4/3)$ and $d \to \infty$, one has $\overline{\varphi}/\overline{\varphi}_0 \to 0$ and the last inequality holds, which implies that Eq. (2.66) asymptotically coincides with Eq. (2.61).

Summary

In short, we have identified different density regimes of the virial series in $d \to \infty$, in terms of $\overline{\varphi} = 2^d \varphi$.

- $\overline{\varphi} < \overline{\varphi}_{\mathrm{conv}}^l = 0.144767\cdots$: the series is proven to be convergent, and it converges to Eq. (2.61). The excess free energy is then given by the second virial term alone.
- $\overline{\varphi}_{\mathrm{conv}}^l \leq \overline{\varphi} < 1$: the series is conjectured to be convergent to Eq. (2.61). The conjecture is supported by the asymptotic behaviour of the virial coefficients in Eq. (2.60), by the resummation in Eq. (2.66) and by the numerical results of [373].
- $1 < \overline{\varphi} < e^{\gamma d}$ with $\gamma < (1 - \log 2)/2 = 0.153426\cdots$: the series is formally divergent, but it can be resummed[9] to obtain Eq. (2.66), which at leading order also coincides with Eq. (2.61).

We conclude that if either $\overline{\varphi}$ is not exponentially large in d, or $\overline{\varphi} = e^{\gamma d}$ with $\gamma < (1 - \log 2)/2$, then the virial series in large d is dominated by the ideal gas term plus the second virial coefficient, corresponding to a direct two-particle interaction. All the other terms can be discarded. When $\overline{\varphi} = e^{\gamma d}$ with $\gamma \to (1 - \log 2)/2$, a singularity appears in the liquid free energy [153, 284]. This singularity has been identified with an instability of the liquid phase towards a modulated density profile, also called 'Kirkwood instability' [153, 155]. Note that the Kirkwood instability can also be detected by looking at the linear stability of the liquid against fluctuations of the density field $\rho(\mathbf{x})$ [242], which depends on the spectrum of the operator $S^{-1}(\mathbf{x} - \mathbf{y})$ defined in Eq. (2.21). However, as we will see in Chapter 3, the liquid phase becomes dynamically unstable towards a dynamically arrested glass phase much before then, when $\overline{\varphi} \propto d$, and the glass phase even ceases to exist when $\overline{\varphi} \sim d \log d$. Hence, what happens formally to the liquid free energy for $\overline{\varphi} \sim e^{\gamma d}$ is irrelevant for the rest of this book.

Liquid Theory Approximations and Numerical Results

In finite d, the virial series cannot be studied exactly, and even its radius of convergence is not known. However, by resumming some classes of diagrams, one can obtain closed integral equations for the radial distribution function $g(r)$. These

---

[9] Note that there are subtleties in the exchange of the limit $d \to \infty$ and the resummation of the series, which we do not discuss here in detail. For any finite d, in this region, crystallisation and the glass transition may happen, which should mathematically correspond to true (i.e., non-resummable) singularities in the virial series. These singularities, however, become weaker and weaker upon increasing d, as discussed in Chapter 1 for the ferromagnetic case, and can likely be neglected for $d \to \infty$.

provide approximations for $g(r)$ and for thermodynamic quantities. The most famous examples are the Percus-Yevick (PY) and hypernetted chain (HNC) approximations [175, chapter 4] that, in $d = 3$, give reasonable estimates of thermodynamic quantities with typical errors of $\approx 10\%-15\%$ in the dense liquid phase. In the case of hard spheres, the PY approximation has been solved in all odd [308] and even [2] dimensions, and in both cases, when $d \to \infty$, one recovers the scenario outlined in the preceding summary. For the HNC approximation, no exact solution is available, but an asymptotic analysis again suggests that it converges to the result given earlier for $d \to \infty$ [284]. Numerical simulations further show that the hard-sphere equation of state converges to Eq. (2.61) for all densities at which the liquid can be equilibrated – i.e., all the way to the glass transition [86]. Furthermore, the accuracy of the PY and HNC approximations in describing the radial distribution function obtained by numerical simulations improves quickly with dimension; see [2, 308] and references therein. All these results support the conclusions of the preceding summary.

2.3.2 Other Interaction Potentials

The analysis of Section 2.3.1 can be extended to many other interaction potentials. To illustrate this point, we start from the simplest example, a pure power-law soft-sphere potential of the form:

$$v(r) = \varepsilon \left(\frac{\ell}{r}\right)^{vd}, \qquad v > 1. \tag{2.70}$$

Here, ε is the interaction energy scale, ℓ sets the interaction range (an effective particle diameter) and v characterises the softness of the potential. The requirement that $v > 1$ ensures the temperedness of the potential (see Section 2.1.1). Under this condition, the potential energy also remains extensive; because $g(r) \to 1$ for $r \to \infty$, the integral in Eq. (2.28) is finite only if $\int_0^\infty dr \, r^{d-1} v(r)$ is convergent. A common choice is $v = 4$, which gives $v(r) = \varepsilon(\ell/r)^{12}$ in $d = 3$, corresponding to the repulsive part of a Lennard-Jones potential [175].

We have seen that for hard spheres, temperature plays a trivial role, and density is the only state parameter. The soft-sphere potential also has a special scale invariance property that simplifies its phase diagram: temperature and density are not independent state variables. The thermodynamic state of the system only depends on the combination $\Gamma = \rho/T^{1/v}$, which follows from the observation that the energy and length scales are not independent in Eq. (2.70). Changing the energy scale $\varepsilon \to \varepsilon\lambda$ is indeed equivalent to changing the length scale $\ell \to \ell\lambda^{1/(vd)}$.

To investigate the behaviour of the soft-sphere potential in large d, we start by computing the second virial coefficient according to Eq. (2.43), assuming that $\widehat{\beta} = \beta\varepsilon$ remains constant for $d \to \infty$. Changing variables, to $x = r/\ell$, one obtains

$$
B_2 = -\frac{\Omega_d}{2} \int_0^\infty dr \, r^{d-1} [e^{-\beta v(r)} - 1] = -\frac{\Omega_d \ell^d}{2} \int_0^\infty dx \, x^{d-1} [e^{-\widehat{\beta}/x^{vd}} - 1].
$$

(2.71)

The integrand is strongly peaked around $x = 1$. In fact, for $x < 1$, the Mayer function converges quickly to -1 while the factor x^{d-1} converges exponentially fast to zero. For $x > 1$, the integrand is approximated by $x^{d-1}[-\widehat{\beta}/x^{vd}]$, which also goes exponentially fast to zero. Changing the variable to h, such that $x = 1 + h/d$, and $r = \ell(1 + h/d)$, and taking the limit $d \to \infty$, we get

$$
B_2 = -\frac{V_d \ell^d}{2} \int_{-\infty}^\infty dh \, e^h [e^{-\widehat{\beta} e^{-vh}} - 1] = \frac{V_d \ell^d}{2} I(\widehat{\beta}, v) = B_2^{\text{HS}} \, I(\widehat{\beta}, v).
$$

(2.72)

The second virial coefficient is then given by the hard-sphere result, Eq. (2.50), multiplied by a finite integral $I(\widehat{\beta}, v)$ over the scaled variable h. Defining $\overline{\varphi} = \rho V_d \ell^d = 2^d \varphi$ as for hard spheres (but with ℓ now being an effective diameter), the hard-sphere asymptotic result Eq. (2.61) becomes

$$
-\beta f(\rho, T) = -\beta f^{\text{id}}(\rho, T) - \frac{\overline{\varphi}}{2} I(\widehat{\beta}, v),
$$

$$
p(\rho, T) = \frac{\beta P(\rho, T)}{\rho} = 1 + \frac{\overline{\varphi}}{2} I(\widehat{\beta}, v).
$$

(2.73)

Note that by changing variables in such a way that $\widehat{\beta}^{1/v} e^{-h} = e^{-z}$, one can show that $I(\widehat{\beta}, v) = \widehat{\beta}^{-1/v} I(1, v)$, which confirms that thermodynamic functions depend only on $\overline{\varphi} \widehat{\beta}^{-1/v} \propto \rho / T^{1/v} = \Gamma$.

This analysis shows that if the limit $d \to \infty$ is taken at constant $\widehat{\beta} = \beta \varepsilon$, the packing fraction $\varphi = \rho V_d (\ell/2)^d$ defined in terms of the interaction range, ℓ, then plays the same role as in hard spheres in controlling the behaviour of the virial expansion. The Mayer function indeed goes exponentially to -1 for $r < \ell$ and to 0 for $r > \ell$, and only differs from the hard-sphere Mayer function in a tiny region of the order of $1/d$ around $r = \ell$, where $r = \ell(1 + h/d)$. The leading exponential scalings in d of hard spheres therefore remain unchanged, and the virial coefficients are simply multiplied by finite factors that depend on the potential. For example, Eq. (2.52) remains valid, with the only difference that the Mayer functions do not impose $|\mathbf{r}_i - \mathbf{r}_j| < \ell$ strictly, but with a tolerance of order $1/d$. The conclusion is unchanged: adding an edge to a diagram imposes an additional constraint that lowers the value of $\det \hat{q}$ and suppresses exponentially the diagram. The main result of Section 2.3.1 for $d \to \infty$ thus remains correct. One can discard all the virial coefficients except B_2, as long as $\overline{\varphi} = e^{d\gamma}$ with $\gamma < (1 - \log 2)/2$, and the liquid free energy is given by the sum of the ideal gas term and the two-body contribution.

Similar results hold for any potential (or, equivalently, any Mayer function) that can be cast, for $d \to \infty$, in the form

$$v(r) = \bar{v}[d(r/\ell - 1)], \qquad\qquad f(r) = \bar{f}[d(r/\ell - 1)], \qquad (2.74)$$

with the function $\bar{v}(h)$ going to infinity for $h \to -\infty$, and to zero faster than e^{-h} for $h \to \infty$. Changing variables to $h = d(r/\ell - 1)$, the second virial coefficient becomes

$$B_2 = -\frac{V_d \ell^d}{2} \int_{-\infty}^{\infty} dh \, e^h \bar{f}(h) = B_2^{\mathrm{HS}} I(\bar{f}), \qquad I(\bar{f}) = -\int_{-\infty}^{\infty} dh \, e^h \bar{f}(h), \tag{2.75}$$

and all the other virial coefficients are subleading. The free energy is then

$$
\begin{aligned}
-\beta \mathrm{f}(\rho, T) &= -\beta \mathrm{f}^{\mathrm{id}}(\rho, T) - \frac{\overline{\varphi}}{2} I(\bar{f}), \\
-\beta \mathrm{f}^{\mathrm{id}}(\rho, T) &= -d \log(\Lambda/\ell) + 1 - \log(\rho \ell^d),
\end{aligned}
\tag{2.76}
$$

where the arguments of the logarithms in the ideal gas term have been made adimensional by using the reference length scale ℓ. Using Eq. (2.29) on the virial series truncated to second order, one obtains the radial distribution function of the liquid in $d \to \infty$,

$$g(\mathbf{r}) = e^{-\beta v(\mathbf{r})}. \tag{2.77}$$

The liquid energy (per particle) and reduced pressure are then given by

$$
\begin{aligned}
e(\rho, T) &= \frac{\partial(\beta \mathrm{f})}{\partial \beta} = \frac{\overline{\varphi}}{2} \int_{-\infty}^{\infty} dh \, e^{h - \beta \bar{v}(h)} \bar{v}(h), \\
p(\rho, T) &= \frac{\beta P(\rho)}{\rho} = 1 + \frac{\overline{\varphi}}{2} I(\bar{f}).
\end{aligned}
\tag{2.78}
$$

Examples of such sufficiently short-ranged potentials are [175, chapter 1] the following:

• The pure power-law soft-sphere potential discussed earlier, for which

$$v(r) = \varepsilon \left(\frac{\ell}{r}\right)^{vd}, \qquad \bar{v}(h) = \varepsilon e^{-vh}, \tag{2.79}$$

with $v > 1$.

• The Lennard-Jones potential, with

$$v(r) = \varepsilon \left[\left(\frac{\ell}{r}\right)^{vd} - \left(\frac{\ell}{r}\right)^{vd/2}\right], \qquad \bar{v}(h) = \varepsilon(e^{-vh} - e^{-vh/2}), \tag{2.80}$$

with the restriction $\nu > 2$ (the usual value in $d = 3$ is $\nu = 4$). This potential is usually a good model for simple atoms, especially inert atoms such as Helium, Neon and Argon.

- The Yukawa potential, with

$$v(r) = \begin{cases} -\frac{\varepsilon\ell}{r}e^{-\lambda d(r/\ell - 1)} & \text{if } r > \ell, \\ \infty & \text{if } r \le \ell, \end{cases} \qquad \bar{v}(h) = \begin{cases} -\varepsilon e^{-\lambda h} & \text{if } h > 0, \\ \infty & \text{if } h \le 0. \end{cases} \quad (2.81)$$

Note that the r in the denominator of $v(r)$ can be approximated with ℓ at the leading order in $d \to \infty$. Furthermore, $\lambda > 1$ is necessary for the convergence of the second virial coefficient. The Yukawa potential is a good model for the interaction of colloidal particles and of other screened charged systems.

- A square-well potential, with

$$v(r) = \begin{cases} 0 & \text{if } r \ge \ell(1 + h_0/d), \\ -\varepsilon & \text{if } \ell < r < \ell(1 + h_0/d), \\ \infty & \text{if } r \le \ell, \end{cases} \qquad \bar{v}(h) = \begin{cases} 0 & \text{if } h \ge h_0, \\ -\varepsilon & \text{if } 0 < h < h_0, \\ \infty & \text{if } h \le 0. \end{cases}$$

$$(2.82)$$

This potential finds many applications in the description of colloidal systems. The 'sticky sphere' limit corresponds to a vanishing range and infinite strength of the attractive potential, $h_0 \to 0$ and $\varepsilon \to \infty$, with $h_0 e^{\beta\varepsilon} = e^u$ in such a way that the integral $I(\bar{f})$ that enters in the second virial coefficient in Eq. (2.75),

$$I(\bar{f}) = 1 - \int_0^{h_0} dh\, e^h \left[e^{\beta\varepsilon} - 1 \right] = 1 - e^u, \qquad (2.83)$$

remains finite.

- The soft repulsive sphere potential, with

$$v(r) = \frac{\varepsilon d^\alpha}{\alpha} \left(\frac{r}{\ell} - 1 \right)^\alpha \theta(\ell - r), \qquad \bar{v}(h) = \frac{\varepsilon}{\alpha} h^\alpha \theta(-h). \qquad (2.84)$$

This potential describes soft spheres that interact repulsively only if they overlap ($r < \ell$), ℓ being their diameter. The choice $\alpha = 2$ corresponds to the harmonic soft-sphere potential, while $\alpha = 5/2$ corresponds to Hertzian soft spheres [195]. Both models are commonly used to describe materials such as pastes, emulsions, soft colloids, and soft macroscopic particles such as in granular materials [233].

- The Weeks–Chandler–Andersen (WCA) potential, given by

$$v(r) = \varepsilon \left[1 + \left(\frac{\ell}{r} \right)^{4d} - 2 \left(\frac{\ell}{r} \right)^{2d} \right] \theta(\ell - r),$$

$$\bar{v}(h) = \varepsilon \left(1 + e^{-4h} - 2e^{-2h} \right) \theta(-h). \qquad (2.85)$$

This is a Lennard–Jones potential modified to have its minimum in $r = \ell$, truncated to $r \leq \ell$ and shifted in such a way that $v(\ell) = 0$. Close to $h = 0$, it reduces to the harmonic sphere potential.

In the last two cases, taking the limit $\varepsilon \rightarrow \infty$ recovers the hard-sphere potential.

Note that an important requirement for any potential in $d \rightarrow \infty$ is that $I(\overline{f}) > 0$. Otherwise, the pressure is a decreasing function of density, which signals a thermodynamic instability. An example is the sticky sphere potential, for which, according to Eq. (2.83), one must have $u < 0$. The thermodynamic instability could be related to a gas–liquid phase separation. A detailed analysis for sticky spheres can be found in [259], where it is shown that in the infinite dimensional limit one recovers the standard Landau theory of the gas–liquid critical point. Another possibility is that the instability is due to a violation of the stability condition discussed in Section 2.1.1. Indeed, the thermodynamic limit does not exist (the free energy density diverges when $N \rightarrow \infty$ at fixed ρ) if the pair potential is always attractive. For a potential with an attractive and a repulsive part, the situation is more complex. Only if the repulsive part is sufficiently large is stability satisfied, but there are apparently harmless potentials for which the thermodynamic limit does not exist [310].

2.4 Wrap-Up

2.4.1 Summary

In this chapter, we have seen that

- In particle systems, the space-dependent density field $\rho(\mathbf{x})$ plays the role of magnetisation in magnetic systems. All the interesting thermodynamic observables can be written as a sum of integrals involving correlations of $\rho(\mathbf{x})$ (Section 2.1.2).
- The local density is thus the appropriate order parameter to construct a free energy functional $F[\rho(\mathbf{x})]$. The equilibrium phases of the system are minima of $F[\rho(\mathbf{x})]$ with the constraint $\int d\mathbf{x}\rho(\mathbf{x}) = N$ (Section 2.1.3). The fluid, gas and liquid phases are minima corresponding to uniform density, $\rho(\mathbf{x}) = \rho$. For attractive potentials, there is a unique fluid phase at high temperatures ($F[\rho]$ has a unique minimum), while at low temperatures, there are two phases (gas and liquid) separated by a first-order phase transition (in mean field, $F[\rho]$ has two local minima; see the general discussion of Chapter 1).
- The crystal phase is characterised by a periodically modulated $\rho(\mathbf{x})$ and is usually separated from the liquid by a first-order phase transition. Therefore, if an expression for $F[\rho(\mathbf{x})]$ is available, the crystal can be eliminated from the theoretical analysis by imposing uniformity of ρ.

- An explicit expression for $F[\rho(\mathbf{x})]$ can be obtained by means of the virial expansion (Section 2.2). Within this approach, $F[\rho(\mathbf{x})]$ is given by the ideal gas term plus a sum of diagrams that capture interactions between two, three and more particles; see Eq. (2.41).
- The virial expansion is both a high-temperature and a low-density expansion and can also be interpreted as a large-dimensional expansion (Section 2.3.1). For a hard-sphere liquid, in the limit $d \to \infty$, it can be truncated to the two-particle (second virial) term, resulting in Eq. (2.38), provided the scaled packing fraction $\overline{\varphi} = 2^d \varphi < \overline{\varphi}_0 \sim e^{d(1-\log 2)/2}$; see Eq. (2.67).
- The same result can be generalised to any potential that can be written for $d \to \infty$ in the form $v(r) = \bar{v}[d(r/\ell - 1)]$, where ℓ is an arbitrary length and the function $\bar{v}(h)$ goes to zero faster than e^{-h} for $h \to \infty$ and diverges for $h \to -\infty$ (Section 2.3.2).

2.4.2 Further Reading

We provide here a list of references that can be consulted to further explore the subjects discussed in this chapter, selected according to the criteria discussed in Section 1.6.2.

The thermodynamic theory of liquids is very well developed beyond the $d \to \infty$ limit. Historically, it was first derived by using the virial expansion as a low-density expansion and resumming classes of diagrams to obtain closed integro-differential equations for $g(r)$ in fixed dimension $d = 2, 3$. The limit $d \to \infty$ was considered much later. Famous examples are the hypernetted chain (HNC) and Percus-Yevick (PY) approximations, but many other schemes have been developed. While the reference textbook is Hansen and MacDonald, *Theory of Simple Liquids* [175], many other excellent textbooks are available. The reader is encouraged to gain familiarity with this literature, because similar methods can be applied to glasses in $d = 3$ using the replica method (Section 4.5.2). The solution of the HNC and PY approximations in large dimensions can be found in

- Parisi and Slanina, *Toy model for the mean-field theory of hard-sphere liquids* [284]
- Rohrmann, Robles, de Haro, et al., *Virial series for fluids of hard hyperspheres in odd dimensions* [308]
- Adda-Bedia, Katzav, and Vella, *Solution of the Percus-Yevick equation for hard hyperspheres in even dimensions* [2]

These approximation schemes reproduce correctly the $d \to \infty$ liquid properties. Molecular dynamics numerical simulations of liquids also provide a lot of information and allow one to precisely test the theory. An excellent textbook is Frenkel and

Smit, *Understanding molecular simulation: From algorithms to applications* [152], but also in this case many others are available.

A model for which the truncation of the virial series at its second term is exact in all dimensions was investigated in

• Kraichnan, *Stochastic models for many-body systems: I. Infinite systems in thermal equilibrium* [213]
• Mari and Kurchan, *Dynamical transition of glasses: From exact to approximate* [242]

This model contains a tunable parameter which allows one to interpolate continuously between the regular liquid and the truncated one, which has been proven very useful to compare systematically the mean field theory with the finite d problem.

It was shown in Section 2.3.2 that the soft-sphere potential has a unique control parameter, $\Gamma = \rho/T^{1/\nu}$. Hence, points in the plane (T, ρ) that correspond to the same Γ are isomorphic. While this property is not exactly true for other potentials, it has been shown in a variety of situations that it is approximately correct, with the correct choice of an effective parameter ν. The $d \to \infty$ solution has provided a justification for this fact, which is an interesting application of the large d limit in the liquid phase. These results are discussed in

• Dyre, *Simple liquids' quasiuniversality and the hard-sphere paradigm* [134]
• Maimbourg and Kurchan, *Approximate scale invariance in particle systems: A large-dimensional justification* [238]
• Costigliola, Schroder, and Dyre, *Studies of the Lennard-Jones fluid in 2, 3, and 4 dimensions highlight the need for a liquid-state $1/d$ expansion* [105]

Finally, a review of the statistical mechanics of long-range potentials, which are not covered in this book, can be found in Campa, Dauxois and Ruffo, *Statistical mechanics and dynamics of solvable models with long-range interactions* [74].

2.5 Appendix: Rotationally Invariant Integrals

We consider the integral of a rotationally invariant function $F(\{\mathbf{r}_a\})$, $a = 1, \ldots, n-1$, of $n-1$ vectors $\mathbf{r}_a$ having each d dimensions. For simplicity, we restrict here to the case $d > n - 1$, because we are mostly interested in the case where $d \to \infty$ and n is fixed.[10] By rotational invariance, $F(\{\mathbf{r}_a\}) = F(\hat{q})$, where $\hat{q}$ is the $(n-1) \times (n-1)$ matrix of scalar products $q_{ab} = \mathbf{r}_a \cdot \mathbf{r}_b$. We can write

---

[10] In the opposite case, $d \leq n - 1$, the matrix $\hat{U}\hat{U}^T$ has zero modes. The derivation remains correct, provided regular functions are replaced by distributions.

$$\int d\mathbf{r}_1 \ldots d\mathbf{r}_{n-1} F(\{\mathbf{r}_a\}) = \int d\hat{q} \, J(\hat{q}) F(\hat{q}),$$

$$J(\hat{q}) = \int d\mathbf{r}_1 \ldots d\mathbf{r}_{n-1} \prod_{a \le b}^{1, n-1} \delta(q_{ab} - \mathbf{r}_a \cdot \mathbf{r}_b),$$

(2.86)

where $J(\hat{q})$ is the Jacobian of the change of variable from the original vectors to the matrix $\hat{q}$. Here, $d\hat{q} = \prod_{a \le b}^{1, n-1} dq_{ab}$ because the matrix $\hat{q}$ is symmetric. To compute the Jacobian, we can write

$$J(\hat{q}) = \int d\mathbf{r}_1 \ldots d\mathbf{r}_{n-1} \prod_{a \le b}^{1, n-1} \delta(q_{ab} - \mathbf{r}_a \cdot \mathbf{r}_b) = \int d\hat{U} \delta\left[\hat{q} - \hat{U}\hat{U}^T\right], \quad (2.87)$$

where the matrix U is a $(n-1) \times d$ dimensional matrix with entries $U_{a\mu} = r_{a\mu}$ (i.e., each row of the matrix is a vector $\mathbf{r}_a$), and the measure $d\hat{U} = \prod_{a\mu} dU_{a\mu} = \prod_{a=1}^{n-1} d\mathbf{r}_a$.

Let us first suppose that $\hat{q}$ is diagonal, $\hat{q} = q_{aa}\delta_{ab}$. Then we have

$$J(\hat{q}) = \int d\hat{U} \delta\left[\hat{q} - \hat{U}\hat{U}^T\right]$$

$$= \int d\mathbf{r}_1 \ldots d\mathbf{r}_{n-1} \prod_{a=1}^{n-1} \delta\left(q_{aa} - |\mathbf{r}_a|^2\right) \prod_{a<b}^{1, n-1} \delta(\mathbf{r}_a \cdot \mathbf{r}_b).$$

(2.88)

The first delta function on the right-hand side of this equation fixes the length of the vectors $\{\mathbf{r}_a\}$, while the second imposes that they must be mutually orthogonal. In polar coordinates, we have

$$J(\hat{q}) = \int d\hat{\mathbf{r}}_1 \ldots d\hat{\mathbf{r}}_{n-1} \int_0^\infty dr_1 r_1^{d-1} \cdots \int_0^\infty dr_{n-1} r_{n-1}^{d-1}$$

$$\times \prod_{a=1}^{n-1} \delta\left(q_{aa} - r_a^2\right) \prod_{a<b}^{1, n-1} \delta(r_a r_b \hat{\mathbf{r}}_a \cdot \hat{\mathbf{r}}_b),$$

(2.89)

where $\hat{\mathbf{r}}_a = \mathbf{r}_a/r_a$ are unit vectors. Then

$$J(\hat{q}) = 2^{1-n} \int d\hat{\mathbf{r}}_1 \ldots d\hat{\mathbf{r}}_{n-1} \int_0^\infty dr_1 r_1^{d-1} \cdots \int_0^\infty dr_{n-1} r_{n-1}^{d-1}$$

$$\times \left(\prod_{a=1}^{n-1} \frac{1}{\sqrt{q_{aa}}} \delta\left(\sqrt{q_{aa}} - u_a\right)\right) \left(\prod_{a<b}^{1, n-1} \frac{1}{\sqrt{q_{aa} q_{bb}}}\right) \prod_{a<b}^{1, n-1} \delta(\hat{\mathbf{r}}_a \cdot \hat{\mathbf{r}}_b) \quad (2.90)$$

$$= C_{n,d} \prod_{a=1}^{n-1} q_{aa}^{(d-n)/2} = C_{n,d}(\det \hat{q})^{(d-n)/2}.$$

The constant

$$C_{n,d} = 2^{1-n} \int d\hat{\mathbf{r}}_1 \dots d\hat{\mathbf{r}}_{n-1} \prod_{a<b}^{1,n-1} \delta(\hat{\mathbf{r}}_a \cdot \hat{\mathbf{r}}_b) \tag{2.91}$$

can be computed recursively. If $n = 2$, there is just one unit vector and $C_{2,d} = \int d\hat{\mathbf{r}}_1 = \Omega_d$, where Ω_d is the d-dimensional solid angle. If $n = 3$, there are two unit vectors: the first can access the entire solid angle Ω_d, and the second can only access the $d - 1$ space orthogonal to the first vector, which gives Ω_{d-1}. Hence, $C_{3,d} = 2^{-1}\Omega_d\Omega_{d-1}$. Continuing this recursion, we get

$$C_{n,d} = 2^{1-n}\Omega_d\Omega_{d-1}\dots\Omega_{d-n+2}. \tag{2.92}$$

This completes the calculation when $\hat{q}$ is diagonal.

The generalisation to the non-diagonal case is straightforward. Because the matrix $\hat{q}$ is symmetric, it can be diagonalised by an orthogonal transformation:

$$\hat{q} = \hat{\Lambda}^{-1}\hat{q}_D\hat{\Lambda}, \qquad \det\hat{\Lambda} = 1, \qquad \det\hat{q}_D = \det\hat{q}, \qquad \hat{\Lambda}^T = \hat{\Lambda}^{-1}, \tag{2.93}$$

and, thus,

$$\begin{aligned} J(\hat{q}) &= \int d\hat{U}\delta[\hat{q} - \hat{U}\hat{U}^T] = \int d\hat{U}\delta[\hat{\Lambda}^{-1}\hat{q}_D\hat{\Lambda} - \hat{U}\hat{U}^T] \\ &= \int d\hat{U}\delta[\hat{q}_D - \hat{\Lambda}\hat{U}\hat{U}^T\hat{\Lambda}^{-1}]. \end{aligned} \tag{2.94}$$

Performing the unitary change of integration variables $\hat{V} = \hat{\Lambda}\hat{U}$, we get

$$\begin{aligned} J(\hat{q}) &= \int d\hat{U}\delta[\hat{q}_D - \hat{\Lambda}\hat{U}\hat{U}^T\hat{\Lambda}^T] = \int d\hat{V}\delta[\hat{q}_D - \hat{V}\hat{V}^T] \\ &= J(\hat{q}_D) = C_{n,d}(\det\hat{q})^{(d-n)/2}, \end{aligned} \tag{2.95}$$

recalling that $\det\hat{q} = \det\hat{q}_D$. We conclude that for any rotationally invariant function, we have

$$\int d\mathbf{r}_1 \cdots d\mathbf{r}_{n-1} F(\{\mathbf{r}_a\}) = C_{n,d} \int d\hat{q} \,(\det\hat{q})^{(d-n)/2} F(\hat{q}). \tag{2.96}$$

3

Atomic Liquids in Infinite Dimensions
Equilibrium Dynamics

The aim of this chapter is to review the dynamical properties of liquids. For simplicity, we will restrict ourselves to equilibrium dynamics – i.e., the study of a system that starts in equilibrium and maintains it at all subsequent times. The dynamics of low-density liquids is quite well understood through kinetic theory at the microscopic level and through hydrodynamics at large length and time scales, as discussed in [175]. Here, we are mostly interested in the properties of dense liquids in the region close to the glass transition, where kinetic theory does not apply and the hydrodynamic regime is pushed to extremely large scales [120]. Remember that, as discussed at the beginning of Chapter 2, in large dimensions, one can focus on the liquid phase without worrying about crystallisation.

The analytical study of the dynamics – e.g., of the correlation of two observables at different times – is much more complex than the study of thermodynamic properties – e.g., the correlation of these same observables at the same time in equilibrium. We have to include a new dimension, time, to the problem. For instance, the exact solution of the dynamics of large-dimensional liquids requires the use of advanced dynamical tools such as path integrals within the Martin-Siggia-Rose-De Dominicis-Janssen formalism [114, 187, 245]. These tools go beyond the scope of this book, and we refer to [239] for details of this particular derivation.

In Chapter 4, we will introduce a formalism that allows one to compute properties of the long time limit of dynamical observables, based on an extension of standard thermodynamics. This approach considerably simplifies the calculations, but understanding the motivations behind it requires some basic knowledge of dynamics. For this reason, in this chapter, we present some essential general notions about dynamics, and then we discuss, without a formal derivation, some important results obtained from the exact solution of the infinite-dimensional dynamics.

3.1 Properties of Equilibrium Dynamics

We consider here the same setting as in Chapter 2: N identical particles in d dimensions, described by a set of points $\underline{X} = \{\mathbf{x}_i\}$, which interact via a potential energy $V(\underline{X})$, with a total Hamiltonian $H(\underline{P}, \underline{X}) = \sum_i \frac{\mathbf{p}_i^2}{2m} + V(\underline{X})$. The Gibbs–Boltzmann equilibrium probability distribution at temperature T is proportional to $\exp[-\beta H(\underline{P}, \underline{X})]$, but the configurational equilibrium distribution, which is proportional to $\exp[-\beta V(\underline{X})]$, suffices to specify the probability distribution of observables $\mathcal{O}[\underline{X}]$ that depend solely on particle positions, integrating away the momenta.

We consider here dynamical equations such that the initial condition at time t_0 is extracted from the equilibrium probability distribution, and at all subsequent times, $t > t_0$, the probability distribution remains equal to the equilibrium one, $P_t(\underline{P}, \underline{X}) = P_{t_0}(\underline{P}, \underline{X}) \propto \exp[-\beta H(\underline{P}, \underline{X})]$. Notable simplifications arise in this case. For example, the average of observables that depend on a single time t – i.e., of the form $\mathcal{O}[\underline{X}(t)]$ – does not depend on t, and correlation functions of two observables at distinct times depend on the time difference only:

$$\langle \mathcal{O}_1[\underline{X}(t)] \mathcal{O}_2[\underline{X}(t')] \rangle = C_{\mathcal{O}_1, \mathcal{O}_2}(t - t'). \tag{3.1}$$

This property is called 'time-translational invariance'. By contrast, if the dynamics starts from an off-equilibrium configuration – e.g., from a configuration thermalised at a temperature $T_0 \neq T$ – the situation is much more complex, because one-time observables depend on t, and correlations depend separately on t, t'. Time-translation invariance is only recovered if the system reaches equilibrium at long times.

In the rest of this section, we provide examples of equilibrium dynamical equations, of relevant dynamical observables and of additional important properties of the equilibrium dynamics. Because these topics are already covered by several excellent books [152, 159, 168, 175, 343, 349], we limit our introduction to the basic notions needed for the subsequent sections.

3.1.1 Dynamical Equations

Many different dynamical equations preserve the equilibrium probability distribution. We present three examples in this section. Here, as in the rest of this chapter, a dot denotes a derivative with respect to time.

- **Hamiltonian dynamics** – The equations of motion are deterministic and derive from the Hamiltonian function, as

$$\dot{\mathbf{p}}_i = -\frac{\partial H(\underline{P}, \underline{X})}{\partial \mathbf{x}_i} = -\frac{\partial V(\underline{X})}{\partial \mathbf{x}_i}, \qquad \dot{\mathbf{x}}_i = \frac{\partial H(\underline{P}, \underline{X})}{\partial \mathbf{p}_i} = \frac{\mathbf{p}_i}{m}. \tag{3.2}$$

These are equivalent to the Newtonian dynamics

$$m\ddot{\mathbf{x}}_i(t) = -\frac{\partial V(\underline{X})}{\partial \mathbf{x}_i} = \mathbf{F}_i(\underline{X}) \tag{3.3}$$

of an isolated system exclusively composed of interacting particles. In this case, the Gibbs–Boltzmann probability distribution is left invariant by the time evolution, because the Hamiltonian is conserved by the dynamics and, according to Liouville's theorem, the time evolution also preserves volume elements in phase space.

- **Langevin dynamics** – The equations of motion are obtained by adding a friction term and a stochastic noise term to Eq. (3.3), resulting in

$$m\ddot{\mathbf{x}}_i(t) + \zeta\dot{\mathbf{x}}_i(t) = -\frac{\partial V(\underline{X})}{\partial \mathbf{x}_i} + \boldsymbol{\xi}_i(t), \tag{3.4}$$

where the stochastic variables $\xi_{i\mu}(t)$ are uncorrelated Gaussian white noises with zero mean and variance

$$\langle \xi_{i\mu}(t)\xi_{j\nu}(t')\rangle = 2T\zeta\delta_{ij}\delta_{\mu\nu}\delta(t - t'). \tag{3.5}$$

Physically, the Langevin equation describes a system of interacting Brownian particles, immersed in a solvent that acts as a thermal bath. The macroscopic interaction with the solvent is responsible for the frictional force $-\zeta\dot{\mathbf{x}}_i(t)$, while random collisions between the Brownian particles and the solvent particles provide the random force $\boldsymbol{\xi}_i(t)$. These two terms have a common physical origin: the interaction with the solvent. If the solvent is in equilibrium, the same friction coefficient ζ enters in both, and the prefactor T in Eq. (3.5) corresponds to the solvent temperature. Note that in the limit $\zeta \to 0$, the friction and noise terms disappear, and one recovers Newtonian dynamics. One can also consider the opposite limit in which inertia is negligible with respect to friction – i.e., $m \to 0$. In this case, one obtains the overdamped Langevin equation, which describes only the configurational degrees of freedom but not the momenta that are undefined in that limit. An explicit computation (see Section 3.6) shows that the Gibbs–Boltzmann probability distribution is left invariant by the Langevin evolution, provided that the value of β in the Gibbs–Boltzmann distribution is equal to $1/T$, where T is the temperature that appears in the noise variance, Eq. (3.5).

One can also consider a more general Langevin dynamics that depends on a kernel $K(t)$:

$$m\ddot{\mathbf{x}}_i(t) + \int_{t_0}^{t} dt' K(t - t')\dot{\mathbf{x}}_i(t') = -\frac{\partial V(\underline{X})}{\partial \mathbf{x}_i} + \boldsymbol{\xi}_i(t), \tag{3.6}$$

with

$$\langle \xi_{i\mu}(t)\xi_{j\nu}(t')\rangle = T\delta_{ij}\delta_{\mu\nu}K(t-t'). \tag{3.7}$$

In this case, $\xi_{i\mu}(t)$ are uncorrelated Gaussian coloured noises. Note that the kernel must be symmetric, $K(t) = K(-t)$, to be consistent with Eq. (3.7). The white noise case is recovered by choosing $K(t) = \zeta e^{-|t|/\tau}/\tau \to 2\zeta\delta(t)$ for $\tau \to 0$ and observing that $\int_{t_0}^{t} dt' K(t-t')\dot{x}(t') \to \zeta\dot{x}(t)$ in that same limit. For the coloured case also, if the same kernel appears both in the friction term and in the correlation of the coloured noise, the Gibbs–Boltzmann probability distribution is left invariant by the time evolution.[1]

• **Monte Carlo dynamics** – Here time is discrete, and the evolution is a Markov chain. The probability of going from one configuration at time t to another configuration at time $t + 1$, $P[\underline{X}(t+1)|\underline{X}(t)]$, does not depend on the previous history of the system. In the simplest Monte Carlo dynamics, one assumes that the Wegscheider-Einstein 'detailed balance' condition [349] is satisfied:

$$P[\underline{Y}|\underline{X}]e^{-\beta V(\underline{X})} = P[\underline{X}|\underline{Y}]e^{-\beta V(\underline{Y})}. \tag{3.8}$$

In this case, it is easy to check that the Gibbs–Boltzmann probability distribution is left invariant by the evolution, because of the relation

$$\int d\underline{X}\, P[\underline{Y}|\underline{X}]e^{-\beta V(\underline{X})} = \int d\underline{X}\, P[\underline{X}|\underline{Y}]e^{-\beta V(\underline{Y})} = e^{-\beta V(\underline{Y})}, \tag{3.9}$$

in which we used that $\int d\underline{X}\, P[\underline{X}|\underline{Y}] = 1$. Eq. (3.9) implies that if at a given time step the dynamics is in equilibrium, one additional step preserves that equilibrium.

3.1.2 Dynamical Observables

Several types of observables are of interest in liquid dynamics. We refer to [152, 168, 175] for an extended list. Here, we restrict the discussion to observables related to the equilibrium dynamics of an isotropic and homogeneous system – e.g., the liquid phase.

One-time observables are simply the dynamical version of the thermodynamic observables discussed in Section 2.1.2, and have the form

---

[1] This result has been known for a long time and is consistent with physical intuition. However, a simple formal mathematical proof is not easily found in the literature. One possibility is to represent $K(t)$ as a sum of exponentials and use the Ornstein-Uhlenbeck representation to map the coloured Langevin process into a white noise Langevin process in an extended space with additional degrees of freedom [159]. The equilibrium state in the extended space is a Gibbs–Boltzmann distribution, which can be marginalised over the additional degrees of freedom to recover the Gibbs–Boltzmann distribution for $\underline{X}$. Another possibility is to prove that the correlations and responses of Eq. (3.6), if the initial condition is in equilibrium, satisfy the fluctuation–dissipation relation (see Section 3.1.3) at all times [18], which (indirectly) implies that the equilibrium is preserved at all times.

$$\mathcal{O}[\underline{X}(t)] = \sum_i \mathcal{O}_1[\mathbf{x}_i(t)] + \sum_{i \neq j} \mathcal{O}_2[\mathbf{x}_i(t), \mathbf{x}_j(t)] + \cdots \tag{3.10}$$

These observables can be expressed in terms of a local density field

$$\rho[\mathbf{x}; \underline{X}(t)] = \sum_{i=1}^{N} \delta(\mathbf{x}_i(t) - \mathbf{x}), \qquad \rho(\mathbf{x}, t) = \langle \rho[\mathbf{x}; \underline{X}(t)] \rangle, \tag{3.11}$$

and its equal time spatial correlations, as discussed in Section 2.1.2 for the thermodynamic analysis. Note that here and in the following, brackets indicate an average over the ensemble of dynamical trajectories $\underline{X}(t)$ generated by the chosen model for the dynamics. Note also that for notational simplicity, we often use the shorthand $\mathcal{O}(t) = \mathcal{O}[\underline{X}(t)]$ in the following.

One can next consider observables of the form of Eq. (3.10), but with particle positions evaluated at different times. These observables can be expressed as functions of correlations of $\rho[\mathbf{x}; \underline{X}(t)]$ at different times and positions. We here focus on a particular class of two-time correlations that depend on both the configuration at time t and that at time t', of the form

$$\mathcal{O}(t, t') = \sum_{ij} \mathcal{O}_2[\mathbf{x}_i(t) - \mathbf{x}_j(t')]. \tag{3.12}$$

These correlations can be expressed in terms of the 'collective van Hove function' [175],

$$G(\mathbf{r}, t - t') = \frac{1}{\rho} \langle \rho[\mathbf{x} + \mathbf{r}; \underline{X}(t)] \rho[\mathbf{x}; \underline{X}(t')] \rangle$$

$$= \frac{1}{N} \sum_{ij} \langle \delta[\mathbf{r} - \mathbf{x}_i(t) + \mathbf{x}_j(t')] \rangle, \tag{3.13}$$

$$\langle \mathcal{O}(t, t') \rangle = N \int d\mathbf{r} \, \mathcal{O}_2(\mathbf{r}) G(\mathbf{r}, t - t'),$$

which represents equivalently the space–time correlation of the density field, or the (non-normalised) probability that a particle at time t is found at distance $\mathbf{r}$ from the position of another particle at time t'. To go from the first to the second line in Eq. (3.13), one makes use of translational invariance [175]. From the van Hove function, one can construct other observables such as the coherent scattering function or the dynamical structure factor [175].

In the limit of infinite dimensions, as we have seen in Chapter 1, correlations become extremely short ranged. As a consequence, all terms with $i \neq j$ in a correlation function such as Eq. (3.13) vanish. It then becomes useful to consider 'self' correlations, which depend on the time evolution of a single particle. One therefore introduces the distribution of single particle displacements, also called the 'self van Hove function', defined as

$$G_s(\mathbf{r}, t - t') = \frac{1}{N} \sum_{i=1}^{N} \langle \delta[\mathbf{r} - \mathbf{x}_i(t) + \mathbf{x}_i(t')] \rangle. \tag{3.14}$$

The self van Hove function contains only the diagonal terms with $i = j$ of Eq. (3.13). It represents the (normalised) probability that a single particle moves by $\mathbf{r}$ in time $t - t'$. From it, one can derive several interesting observables, such as the 'mean square displacement',

$$D(t) = \frac{1}{N} \sum_{i=1}^{N} \langle |\mathbf{x}_i(t) - \mathbf{x}_i(0)|^2 \rangle = \int d\mathbf{r} |\mathbf{r}|^2 G_s(\mathbf{r}, t), \tag{3.15}$$

and the 'self intermediate scattering function',

$$F_s(\mathbf{q}, t) = \frac{1}{N} \sum_{i=1}^{N} \langle e^{i\mathbf{q} \cdot [\mathbf{x}_i(t) - \mathbf{x}_i(0)]} \rangle = \int d\mathbf{r} \, e^{i\mathbf{q} \cdot \mathbf{r}} G_s(\mathbf{r}, t), \tag{3.16}$$

which are the second moment and the Fourier transform, respectively, of the self Van Hove function. In the rest of this chapter, we will focus particularly on $D(t)$.

Finally, we have seen in Chapter 1 that while correlations between different particles vanish in the limit $d \to \infty$, their volume integral, which defines a susceptibility, remains finite and can even diverge at a phase transition. This motivates the study of dynamical susceptibilities such as

$$\chi_4(t) = N \left(\langle \hat{D}^2(t) \rangle - \langle \hat{D}(t) \rangle^2 \right), \qquad \hat{D}(t) = \frac{1}{N} \sum_{i=1}^{N} |\mathbf{x}_i(t) - \mathbf{x}_i(0)|^2. \tag{3.17}$$

These functions can be written as volume integrals of correlation functions that depend on two spatial points and two times and are thus called 'four point correlation functions'. They are extremely useful for characterising the glass transition. We refer to [37] for a detailed review.

Finally, together with correlation functions, we consider the linear responses of observables to small perturbations. The perturbations are introduced by modifying the interaction potential

$$V_\epsilon[\underline{X}] = V[\underline{X}] - \epsilon(t)\mathcal{O}_2[\underline{X}], \tag{3.18}$$

and the variation of the average of an observable $\mathcal{O}_1(t)$ is considered in their presence. For small $\epsilon(t)$, the variation is linear and takes the most general form

$$\langle \mathcal{O}_1(t) \rangle_\epsilon = \langle \mathcal{O}_1(t) \rangle_{\epsilon=0} + \int_{-\infty}^{t} dt' R_{\mathcal{O}_1, \mathcal{O}_2}(t, t')\epsilon(t'), \tag{3.19}$$

which is equivalent to evaluating the functional derivative at vanishing perturbation:

$$R_{\mathcal{O}_1,\mathcal{O}_2}(t,t') = \left.\frac{\delta\langle\mathcal{O}_1(t)\rangle_\epsilon}{\delta\epsilon(t')}\right|_{\epsilon=0}. \tag{3.20}$$

By causality of the equations of motion, the response at time t cannot depend on times $t' > t$, which sets the upper bound of the integration in Eq. (3.19). Equilibrium response functions are also time-translationally invariant and, hence, only depend on the time difference $t - t'$.

Several examples of response functions that are relevant to liquid dynamics are given in [175, chapter 7]. The simplest example is obtained by choosing $\mathcal{O}_2[\underline{X}] = x_{i\mu}$, for an arbitrary particle i and spatial component μ, and $\mathcal{O}_1(t) = p_{i\mu}(t)/m = \dot{x}_{i\mu}(t)$. This choice physically corresponds to a force $\epsilon(t)$ being applied on a particle and its velocity being recorded. If at time t_0 a constant $\epsilon(t) = \epsilon\theta(t - t_0)$ is switched on, one then obtains

$$\langle\dot{x}_{i\mu}(t)\rangle_\epsilon = \epsilon\int_{t_0}^t dt'\, R_{\dot{x}_{i\mu},x_{i\mu}}(t - t'). \tag{3.21}$$

Assuming that the response function vanishes fast enough at long times, one then obtains

$$\mu_t = \lim_{\epsilon\to0}\lim_{t\to\infty}\frac{\langle\dot{x}_{i\mu}(t)\rangle_\epsilon}{\epsilon} = \int_0^\infty dt'\, R_{\dot{x}_{i\mu},x_{i\mu}}(t'). \tag{3.22}$$

Hence, at long times, the average velocity reaches a finite limit proportional to the applied force ϵ, via a 'transport coefficient' μ_t called translational 'mobility'. Transport coefficients are generally expressed as integrals of response functions, as in Eq. (3.22). Other important examples of transport coefficients are the electrical conductivity and the shear viscosity [175, chapters 7 and 8].

3.1.3 Reversibility and Fluctuation–Dissipation Relation

A very interesting general property of the equilibrium dynamical equations introduced in Section 3.1.1 is that the correlations are invariant under time reflections. First we note that, due to time-translation invariance, correlations satisfy the relation

$$C_{\mathcal{O}_1,\mathcal{O}_2}(t) = \langle\mathcal{O}_1(t)\mathcal{O}_2(0)\rangle = \langle\mathcal{O}_1(0)\mathcal{O}_2(-t)\rangle = C_{\mathcal{O}_2,\mathcal{O}_1}(-t). \tag{3.23}$$

If $\mathcal{O}_1[\underline{X}]$ and $\mathcal{O}_2[\underline{X}]$ are invariant under time reversal, reversibility implies the additional relation

$$C_{\mathcal{O}_1,\mathcal{O}_2}(t) = C_{\mathcal{O}_1,\mathcal{O}_2}(-t) \quad\Leftrightarrow\quad C_{\mathcal{O}_1,\mathcal{O}_2}(t) = C_{\mathcal{O}_2,\mathcal{O}_1}(t), \tag{3.24}$$

where the equivalence is due to Eq. (3.23). This commutation symmetry of the correlations is a direct consequence of the microscopic reversibility of the dynamical equations in the case of Hamiltonian dynamics and of the detailed balance condition (3.8) in the case of Monte Carlo dynamics. It may, however, look surprising for Langevin dynamics, because the microscopic equations are not time reversible. They indeed describe the motion of particles in presence of friction and noise. This microscopic arrow of time nonetheless bears no consequence on the equilibrium behaviour. A proof of this fact is given in Section 3.6.

Another general property of equilibrium dynamics is the fluctuation–dissipation relation. This relation is a consequence of the zeroth law of thermodynamics, which states that two bodies in thermal contact acquire the same temperature in equilibrium. In fact, the fluctuation–dissipation relation can be derived from the theory of harmonic thermometers [110, 111]. A basic principle of thermodynamics is that, when a thermometer is weakly coupled to a much larger system, it reaches the temperature of the larger system. If the thermometer is a harmonic oscillator, one obtains the temperature of the system by measuring the average energy of the harmonic oscillator, which is equal to its temperature by the equipartition relation.

Consider then a harmonic oscillator $x(t)$ of frequency ω, coupled to an observable $\mathcal{O}(t) = \mathcal{O}[\underline{X}(t)]$ of a much larger system, through a total Hamiltonian

$$H_{\text{tot}}(p, x; \underline{P}, \underline{X}) = \frac{p^2}{2} + \frac{\omega^2 x^2}{2} - \epsilon x \mathcal{O}(\underline{X}) + H[\underline{P}, \underline{X}], \qquad (3.25)$$

with a coupling constant ϵ. The equation of motion of the harmonic oscillator is then

$$\ddot{x}(t) - \omega^2 x(t) + \epsilon \mathcal{O}(t) = 0. \qquad (3.26)$$

The oscillator perturbs the system slightly by the coupling term $-\epsilon x(t)\mathcal{O}(\underline{X})$. Assuming that the coupling is small and that for simplicity $\langle \mathcal{O}(t) \rangle = 0$ in absence of coupling, linear response theory (Section 3.1.2) gives

$$\mathcal{O}_\epsilon(t) = \mathcal{O}_\epsilon^f(t) + \epsilon \int_{-\infty}^t dt' \, R_{\mathcal{O},\mathcal{O}}(t - t')x(t'). \qquad (3.27)$$

The first term is the fluctuating part of $\mathcal{O}(t)$, which has zero average, and the second term describes the shift of the average of $\mathcal{O}(t)$ due to the coupling with the oscillator. The dynamical equation for the oscillator then becomes

$$\ddot{x}(t) - \omega^2 x(t) + \epsilon \mathcal{O}_\epsilon^f(t) + \epsilon^2 \int_{-\infty}^t dt' \, R_{\mathcal{O},\mathcal{O}}(t - t')x(t') = 0. \qquad (3.28)$$

This linear equation is a particular case of Eq. (3.6) for a single variable $x(t)$ with a harmonic potential and a friction kernel[2] $\epsilon^2 R_{O,O}(t) = -\theta(t)\dot{K}(t)$. The noise $\epsilon O_\epsilon^f(t)$ has autocorrelation $\epsilon^2 \langle O_\epsilon^f(t)O_\epsilon^f(0)\rangle = \epsilon^2 C_{O,O}(t) = TK'(t)$. Imposing that the kernels $K(t)$ and $K'(t)$ be the same, as in Eq. (3.6), guarantees that, at long times, the oscillator reaches equilibrium at temperature T independently of its frequency. This condition leads to the relation

$$R_{O,O}(t) = -\beta\theta(t)\dot{C}_{O,O}(t), \tag{3.29}$$

which is the fluctuation–dissipation relation. Alternatively, Eq. (3.28) can be solved explicitly to compute the average energy of the oscillator. See [110] for details. The energy of the oscillator is thus equal to T, independently of its frequency ω, only if Eq. (3.29) holds. This argument shows that the fluctuation–dissipation relation is a physical requirement for a system to be in equilibrium. If Eq. (3.29) does not hold, then a thermometer coupled to the system measures a different temperature depending on its characteristic frequency [110].

The fluctuation–dissipation relation, Eq. (3.29), can also be mathematically derived by calculating directly the response function, considering the variation of the dynamical equations under the perturbation; see, e.g., [175, 282, 343, 349]. This derivation can also be generalised to two different observables, leading to a fluctuation–dissipation relation of the form

$$R_{O_1,O_2}(t) = -\beta\theta(t)\dot{C}_{O_1,O_2}(t). \tag{3.30}$$

The fluctuation–dissipation relation has many important applications. In particular, it can be used to express transport coefficients in terms of correlation functions. For example, using the fluctuation–dissipation relation and the reversibility property[3] expressed by Eq. (3.24), one can rewrite Eq. (3.22) as [175, chapter 7]

$$\mu_t = \beta \int_0^\infty dt \, \langle \dot{x}_{i\mu}(t)\dot{x}_{i\mu}(0)\rangle = \beta D, \tag{3.31}$$

where

$$D = \lim_{t\to\infty} \frac{D(t)}{2dt} \tag{3.32}$$

is the 'diffusion constant'. The mobility is then expressed as the time integral of the autocorrelation of the particle velocity, which is an example of a 'Green–Kubo relation' [175, 343, 349]. The relation between mobility and the diffusion constant is known as 'Einstein relation'.

---

[2] The derivative is there because in Eq. (3.28) the response function is coupled to $x(t)$ instead of $\dot{x}(t)$. An integration by parts of $\int_{-\infty}^t dt' \, \dot{K}(t-t')x(t')$ allows one to obtain Eq. (3.6).
[3] Note that because $\dot{x}_{i\mu}$ is odd under time reversal, there is an additional minus sign in Eq. (3.24).

3.2 Langevin Dynamics of Liquids in Infinite Dimensions

We now consider a system of particles evolving under the Langevin dynamics defined by Eqs. (3.4) and (3.5). The aim of this section is to show that the Langevin equation of N interacting particles, taking the thermodynamic limit $N \to \infty$ followed by the limit $d \to \infty$, can be mapped into the dynamics of a single effective particle, coupled to a coloured thermal bath that represents the interaction of the effective particle with all the other particles. The kernel of the effective thermal bath can then be determined self-consistently. More precisely, any dynamical correlation function of the original many-body Langevin process can be computed as a dynamical correlation of the effective process. This mapping is a general feature of mean field systems. An example is the spin glass case discussed in [111, 254].

It is important to stress that while this mapping will be discussed for a Langevin dynamics with friction and noise, one can also consider the limit $\zeta \to 0$ at which the noise disappears and the Langevin dynamics becomes Newtonian. This limit, however, should be taken after the limits $N \to \infty$ and $d \to \infty$. In the limit $\zeta \to 0$, the coloured noise in the effective Langevin process does not disappear. Newtonian dynamics is then also described by a self-consistent effective Langevin dynamics, for which noise is generated by interactions.

The idea of deriving effective equations for the dynamics of liquids in $d \to \infty$ was first introduced by Kirkpatrick and Wolynes [206]. The exact solution by mapping onto an effective process was derived in full detail in [239], but this derivation requires advanced path integral dynamical tools [79, 111, 114, 187, 200, 245] that will not be discussed in this book. Here, for simplicity, we will not give a full derivation but a series of physically reasonable arguments, which were first proposed by Szamel [339]. The advantage of the simplified derivation is that it gives a clear understanding of the physical properties that underlie the full derivation, and it also leads to the final result in the simplest possible way. The disadvantage, however, is that some steps are not fully justified, and the correctness of the derivation can only be verified a posteriori by comparing the result with that obtained via the path integral formalism [3].

For additional simplicity, in the rest of this section, we consider the overdamped case, $m = 0$, but the inertial term can be reinserted at any time in the discussion. We also recall that, as in all this chapter, dynamics starts in equilibrium at the initial time, chosen here to be $t_0 = 0$.

3.2.1 Single-Particle Effective Process

The Langevin equation for particle i has the form

$$\zeta \dot{\mathbf{x}}_i(t) = \mathbf{F}_i(t) + \boldsymbol{\xi}_i(t),$$

$$\mathbf{F}_i(t) = -\frac{\partial V[\underline{X}(t)]}{\partial \mathbf{x}_i} = -\sum_{j(\neq i)} \frac{\partial v(|\mathbf{x}_i(t) - \mathbf{x}_j(t)|)}{\partial \mathbf{x}_i} = \sum_{j(\neq i)} \mathbf{F}_{j \to i}(t), \qquad (3.33)$$

where $F_{j\to i}(t)$ is the force that particle j exerts on i at time t. In finite d, because forces are short ranged, particle i only interacts with a finite number of neighbours. However, when $d \to \infty$, the number of neighbours also diverges,[4] and $F_i(t)$ is the sum of an infinite number of terms. This situation is very similar to what happens when d is finite, but particle i is much bigger than the other fluid particles and thus undergoes a Brownian motion. We are going to make use of this analogy to derive, in $d \to \infty$, an effective equation for particle i that has the form of a generalised Brownian motion.

First, we consider the case in which particle i is immobile ($\dot{\mathbf{x}}_i = 0$). Then, by isotropy of the liquid phase, the force $F_i(t)$ exerted on i by the other particles has zero average. Furthermore, because in $d \to \infty$ the number of neighbours also diverges, the forces $F_{j\to i}(t)$ can be considered as uncorrelated, exactly as in the Ising model, see Chapter 1. This is illustrated in Figure 3.1. Because $F_i(t) = \sum_{j(\neq i)} F_{j\to i}(t)$ is then the sum of many weakly correlated terms, its fluctuations are Gaussian by the central limit theorem. The autocorrelation of one of its components

$$M(t - t') = \langle F_{i\mu}(t) F_{i\mu}(t') \rangle = \sum_{j,k(\neq i)} \langle F_{j\to i,\mu}(t) F_{k\to i,\mu}(t') \rangle, \qquad (3.34)$$

does not depend on i nor μ, because all particles are identical and the fluid is isotropic. The contributions of different particles being uncorrelated, one can further simplify this expression by removing the terms with $j \neq k$ to obtain

$$M(t - t') = \sum_{j(\neq i)} \langle F_{j\to i,\mu}(t) F_{j\to i,\mu}(t') \rangle = \frac{1}{N} \sum_{i\neq j} \langle F_{j\to i,\mu}(t) F_{j\to i,\mu}(t') \rangle, \qquad (3.35)$$

where in the second equality we used the fact that particles are identical.

Next, we consider a moving particle. As in the case of standard Brownian motion, to the fluctuating part discussed earlier, we should now add an average force $F_i^{av}(t)$ that is proportional to the particle velocity, because the rest of the fluid acts as a frictional medium. In equilibrium, the frictional force must be related to the noise, as discussed in Section 3.1.1, and, hence, $F_i^{av}(t) = -\beta \int_0^t dt' M(t - t') \dot{\mathbf{x}}_i(t')$. We thus obtain that, in the limits $N \to \infty$ and $d \to \infty$, the dynamics of any given particle i is governed by a single-particle Langevin process with coloured noise

---

[4] This intuitive statement can be supported by two observations. First, the 'kissing number', which is the maximal number of non-overlapping spheres that can be put in contact with a central sphere, diverges with d [103]. Second, for a finite range potential, the average number of neighbours of any particle in the liquid phase can be defined as the integral for $r \leq \ell$ of the radial distribution function given in Eq. (2.77),

$$z = \rho \Omega_d \int_0^\ell dr g(r) = \rho \Omega_d \int_0^\ell dr e^{-\beta v(r)} = \overline{\varphi} \int_{-\infty}^0 dh\, e^{h - \beta \overline{v}(h)}.$$

Because the dynamical glass transition happens for $\overline{\varphi} \propto d$ (see Section 3.3.3) and the integral over h is finite, the number of neighbours is proportional to d.

$\Xi_i(t) = \xi_i(t) + F_i^{\text{fl}}(t)$, where the fluctuating force is Gaussian, independent from $\xi_i(t)$, and has a memory kernel $M(t - t')$:

$$\zeta \dot{\mathbf{x}}_i(t) = -\beta \int_0^t dt' M(t - t') \dot{\mathbf{x}}_i(t') + \Xi_i(t),$$

$$\langle \Xi_{i\mu}(t) \Xi_{i\nu}(t') \rangle = \delta_{\mu\nu} [2T\zeta\delta(t - t') + M(t - t')].$$

(3.36)

Eq. (3.36) is certainly not exact in finite d, except in the case where particle i is much bigger than the other particles. In that case, particle i indeed undergoes Brownian motion. The Brownian motion description is appropriate for a much bigger particle because it interacts with many small particles at the same time, and the interactions can then be described using the central limit theorem, which leads to Eq. (3.36). But in infinite d, the number of neighbours is very large even for a particle of the same size as its neighbours, which justifies Eq. (3.36). Note that Eq. (3.36) depends only on $\dot{\mathbf{x}}_i(t)$, and, therefore, the initial condition for $\mathbf{x}_i(t)$ is completely irrelevant.

3.2.2 Two-Particle Effective Process and Equation for the Memory Kernel

Equation (3.35) expresses the memory kernel as an autocorrelation function of the interparticle force. Its computation requires writing an effective process for the dynamics of two particles, as illustrated in Figure 3.1. The original Langevin equation for particles i and j reads

$$\zeta \dot{\mathbf{x}}_i(t) = \mathbf{F}_{j \to i}(t) + \mathbf{F}_i^{(j)}(t) + \xi_i(t),$$

$$\zeta \dot{\mathbf{x}}_j(t) = -\mathbf{F}_{j \to i}(t) + \mathbf{F}_j^{(i)}(t) + \xi_j(t),$$

(3.37)

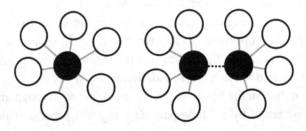

Figure 3.1 Illustration of the one-particle (left) and two-particles (right) effective stochastic processes. In the one-particle case, the central particle interacts with a large number of neighbours, diverging with d, which are assumed to be uncorrelated for $d \to \infty$. The resulting fluctuating and friction forces provide an effective thermal bath. In the two-particle case, the pair of particle exchange a force, and each particle also interacts with a large number of neighbours, which are assumed to be independent and thus provide two independent effective thermal baths, each identical to a one-particle process.

where $F_i^{(j)}(t)$ denotes the total force on particle i, without the contribution coming from particle j. In the limit $d \to \infty$, the number of terms in $F_i^{(j)}(t)$ diverges with d, and removing one contribution is a small correction of order $1/d$. We can thus apply to $F_i^{(j)}(t)$ the same treatment as that applied to the total force in Section 3.2.1, to obtain

$$\zeta \dot{\mathbf{x}}_i(t) = -\beta \int_0^t dt' M(t - t')\dot{\mathbf{x}}_i(t') + \Xi_i(t) + \mathbf{F}_{j\to i}(t),$$

$$\zeta \dot{\mathbf{x}}_j(t) = -\beta \int_0^t dt' M(t - t')\dot{\mathbf{x}}_j(t') + \Xi_j(t) - \mathbf{F}_{j\to i}(t),$$

(3.38)

where the noises $\Xi_i(t)$ and $\Xi_j(t)$ have the same statistics as in Eq. (3.36).

In Eq. (3.38), the terms $\mathbf{F}_{j\to i}(t)$ are small corrections to the noise of order $1/d$, as discussed earlier, because the force between particle i and j is very weakly correlated with the positions $\mathbf{x}_i(t)$ and $\mathbf{x}_j(t)$ of these particles. Yet the relative displacement $\mathbf{r}(t) = \mathbf{x}_i(t) - \mathbf{x}_j(t)$ satisfies the equation obtained by taking the difference of the first and second lines in Eq. (3.38) and dividing by 2 for later convenience,

$$\frac{\zeta}{2}\dot{\mathbf{r}}(t) = -\frac{\beta}{2} \int_0^t dt' M(t - t')\dot{\mathbf{r}}(t') + \Xi(t) + \mathbf{F}(\mathbf{r}(t)),$$

$$\langle \Xi_\mu(t)\Xi_\nu(t')\rangle = \delta_{\mu\nu}\left[T\zeta\delta(t - t') + \frac{1}{2}M(t - t')\right],$$

(3.39)

where

$$\mathbf{F}(\mathbf{r}) = -\frac{\partial v(|\mathbf{r}|)}{\partial \mathbf{r}}.$$

(3.40)

Now, the term $\mathbf{F}(\mathbf{r}(t))$ is strongly correlated with $\mathbf{r}(t)$ because, for central potentials, the force is parallel to the distance. Then this term cannot be neglected[5] in Eq. (3.39).

The initial condition for $\mathbf{r}(t)$ in Eq. (3.39), which we denote by $\mathbf{r}_0 = \mathbf{r}(0)$, should be taken at random from the equilibrium distribution of interparticle distances, which is proportional to $g(\mathbf{r}_0) = e^{-\beta v(|\mathbf{r}_0|)}$ in $d \to \infty$, as discussed in Chapter 2. Indeed, one can check that this distribution is left invariant by Eq. (3.39). However, because $g(\mathbf{r}_0) \to 1$ for $|\mathbf{r}_0| \to \infty$, this distribution is not normalisable. To determine the correct normalisation, we consider the value of $M(t - t')$ at equal times, $t = t'$. Using the thermodynamic results of Section 2.1.2, Eq. (3.35) gives

---

[5] When $d \to \infty$, the projection of $\mathbf{F}(\mathbf{r}(t))$ on a random direction is much smaller than its projection on $\mathbf{r}(t)$, which explains why it can be neglected in Eq. (3.38) while it must be kept in Eq. (3.39). This statement will be made precise in Section 3.3.3.

$$M(0) = \frac{1}{Nd} \sum_{i \neq j} \langle |\mathbf{F}_{j \to i}(0)|^2 \rangle = \frac{\rho^2}{Nd} \int d\mathbf{x} d\mathbf{y} g(\mathbf{x} - \mathbf{y}) |\mathbf{F}(\mathbf{x} - \mathbf{y})|^2$$

$$= \frac{\rho}{d} \int d\mathbf{r}_0 g(\mathbf{r}_0) |\mathbf{F}(\mathbf{r}_0)|^2 . \tag{3.41}$$

At different times, by continuity, one therefore obtains

$$M(t) = \frac{1}{Nd} \sum_{i \neq j} \langle \mathbf{F}_{j \to i}(t) \cdot \mathbf{F}_{j \to i}(0) \rangle = \frac{\rho}{d} \int d\mathbf{r}_0 g(\mathbf{r}_0) \langle \mathbf{F}(\mathbf{r}(t)) \rangle \cdot \mathbf{F}(\mathbf{r}_0), \tag{3.42}$$

which reduces to Eq. (3.41) for $t = 0$. The dynamical average $\langle \mathbf{F}(\mathbf{r}(t)) \rangle$ in Eq. (3.42) is originally over the many-body dynamics but can be replaced by an average over the effective process in Eq. (3.39) that encodes the dynamical evolution of interparticle distances. Therefore, one should consider every possible initial condition $\mathbf{r}(0) = \mathbf{r}_0$, evolve it according to Eq. (3.39) to compute $\langle \mathbf{F}(\mathbf{r}(t)) \rangle$ averaged over the noise $\Xi(t)$ and, finally, integrate over $\mathbf{r}_0$ to obtain the memory kernel. Note that because $\mathbf{F}(\mathbf{r}_0)$ decays quickly to zero for $|\mathbf{r}_0| \to \infty$, the integral over $\mathbf{r}_0$ in Eq. (3.42) is convergent.

Eqs. (3.39) and (3.42) then constitute a closed system of equations for $M(t)$. One could solve it numerically by starting with a guess for $M(t)$, solving the process in Eq. (3.39), using the result to compute a new guess for $M(t)$ using Eq. (3.42) and repeating until convergence.

3.2.3 Mean Square Displacement and the Diffusion Constant

Once the self-consistent equation for $M(t)$ has been solved, one can use the effective particle dynamics given by Eq. (3.36) to obtain dynamical observables. As an example, we derive here the equation for the mean square displacement $D(t)$ introduced in Eq. (3.15). To this end, let us rewrite Eq. (3.36) as

$$\int_0^\infty dt' \Gamma(t - t') \dot{\mathbf{x}}(t') = \Xi(t),$$

$$\langle \Xi_\mu(t) \Xi_\nu(t') \rangle = \delta_{\mu\nu} T[\Gamma(t - t') + \Gamma(t' - t)], \tag{3.43}$$

where we dropped the suffix i because all particles follow the same identical equation, and we introduced

$$\Gamma(t - t') = \theta(t - t')[2\zeta\delta(t - t') + \beta M(t - t')]. \tag{3.44}$$

We then have

$$\dot{\mathbf{x}}(t) = \int_0^\infty dt' \Gamma^{-1}(t - t') \Xi(t'),$$

$$\mathbf{x}(t) - \mathbf{x}(0) = \int_0^t dt' \dot{\mathbf{x}}(t') = \int_0^t dt' \int_0^\infty dt'' \Gamma^{-1}(t' - t'') \Xi(t''), \tag{3.45}$$

where $\Gamma^{-1}(t)$ is the operator inverse of $\Gamma(t)$. Hence,

$$\dot{D}(t) = 2 \langle [\mathbf{x}(t) - \mathbf{x}(0)] \cdot \dot{\mathbf{x}}(t) \rangle \tag{3.46}$$

$$= 2 \int_0^t dt' \int_0^\infty dt'' \int_0^\infty dt''' \Gamma^{-1}(t' - t'') \Gamma^{-1}(t - t''') \langle \Xi(t'') \cdot \Xi(t''') \rangle.$$

Using Eq. (3.43) for the noise correlation, we have

$$\dot{D}(t) = 2dT \int_0^t dt' \int_0^\infty dt'' \int_0^\infty dt''' \Gamma^{-1}(t' - t'') \Gamma^{-1}(t - t''')$$

$$\times [\Gamma(t'' - t''') + \Gamma(t''' - t'')]$$

$$= 2dT \int_0^t dt' \left[\Gamma^{-1}(t - t') + \Gamma^{-1}(t' - t) \right] \tag{3.47}$$

$$= 2dT \int_0^\infty dt' \Gamma^{-1}(t - t'),$$

where in the last step we used that $\Gamma^{-1}(t - t')$ vanishes for $t' > t$, because it is the inverse of an operator that has the same property.[6]

Equation (3.47) can be integrated over time using $D(0) = 0$ to obtain an explicit expression of $D(t)$ in terms of $M(t)$, but this requires inverting $\Gamma(t)$. It is more convenient to derive an equation for $D(t)$ by writing

$$\int_0^\infty dt' \Gamma(t - t') \dot{D}(t') = 2dT \int_0^\infty dt' \Gamma(t - t') \int_0^\infty dt'' \Gamma^{-1}(t' - t'') = 2dT, \tag{3.48}$$

which can be written more explicitly as

$$\zeta \dot{D}(t) = -\beta \int_0^t dt' M(t - t') \dot{D}(t') + 2dT. \tag{3.49}$$

Equation (3.49) is the desired result that expresses $D(t)$ as a function of $M(t)$. Introducing Laplace transforms, $\widehat{D}(s) = \int_0^\infty dt\, e^{-st} D(t)$, it can also be written as

$$\widehat{D}(s) = \frac{2dT}{s^2[\zeta + \beta \widehat{M}(s)]}. \tag{3.50}$$

It is reasonable to assume on very general grounds that in the liquid, $D(t)$ is diffusive – i.e., linear at large times, $D(t) \sim 2dDt$ – and $M(t)$ decays to zero

---

[6] The operators $\Gamma(t - t')$ that vanish for $t' > t$ form a closed algebra with identity, and, therefore, the inverse $\Gamma^{-1}(t' - t)$ also vanishes for $t' > t$. These operators are the continuum analog of upper triangular matrices, which also have the same properties.

sufficiently fast when $t \to \infty$. We then have $\lim_{t \to \infty} \int_0^t dt' \, M(t - t')\dot{D}(t') = 2dD$ $\int_0^\infty dt \, M(t)$, and we obtain from Eq. (3.49) an expression for the diffusion coefficient:

$$\zeta D = T - \beta D \int_0^\infty dt \, M(t) \quad \Rightarrow \quad D = \frac{T}{\zeta + \beta \int_0^\infty dt \, M(t)}. \quad (3.51)$$

The same expression is equivalently obtained from Eq. (3.50) by observing that for $s \to 0$, $\hat{D}(s) \sim 2dD/s^2$, while $\hat{M}(s)$ has a finite limit corresponding to the integral of $M(t)$. Because the memory kernel $M(t)$ originates from interactions, the absence of interactions at low density gives $M(t) = 0$, as it is clear from Eq. (3.42). One then recovers the Einstein expression of the diffusion coefficient in Eq. (3.31), $D = T/\zeta = T\mu_t$ with mobility $\mu_t = 1/\zeta$, for independent particle dynamics. Upon increasing density, $M(t)$ increases and the diffusion coefficient decreases.

3.3 Dynamical Glass Transition

The self-consistent expression for the memory kernel $M(t)$ provided by Eqs. (3.39) and (3.42) predicts that the system dynamically arrests at high density or low temperature. Dynamical arrest corresponds to the memory kernel not decaying to zero in the limit $t \to \infty$. We can thus write

$$M(t) = M_f(t) + M_\infty, \qquad \lim_{t \to \infty} M_f(t) = 0, \quad (3.52)$$

and assume that $M_f(t)$ decays to zero sufficiently fast to be integrable. If $M_\infty > 0$, then the integral of $M(t)$ diverges and according to Eq. (3.51) the diffusion constant vanishes.[7] The system is then dynamically arrested. As a result, the mean square displacement also reaches a plateau[8] at long times – i.e., $D(t) \to D$ for $t \to \infty$. Using the decomposition in Eq. (3.52) into Eq. (3.49), and recalling that $D(0) = 0$, we obtain

$$\zeta \dot{D}(t) = -\beta \int_0^t dt' M_f(t - t')\dot{D}(t') - \beta M_\infty D(t) + 2dT. \quad (3.53)$$

When $t \to \infty$, both $\dot{D}(t)$ and $M_f(t)$ quickly decay to zero. For large t, in the integral $\int_0^t dt' M_f(t - t')\dot{D}(t')$, the variable t' can be either finite, in which case

---

[7] Note that the diffusion constant can also vanish if $M(t) \sim t^{-\alpha}$ for $t \to \infty$, with $0 < \alpha < 1$. In this case, $\hat{M}(s) \sim s^{\alpha-1}$ for $s \to 0$. From Eq. (3.50), the mean square displacement behaves as $\hat{D}(s) \sim s^{-1-\alpha}$; hence, $D(t) \sim t^\alpha$ – i.e., it is sub-diffusive. While this situation can happen at some dynamical critical points [168], we do not consider it in this book.

[8] We avoid adding a suffix ∞ to the plateau value, D, to simplify the notation in Chapter 4. We also stress that D represents the diffusion coefficient, while D represents the plateau of the mean square displacement, which are two different quantities.

$M_f(t - t') \to 0$, or of order t, in which case $\dot{D}(t') \to 0$; the integral therefore goes to zero for $t \to \infty$. Using these results, we obtain from Eq. (3.53) the relation

$$-\beta M_\infty D + 2dT = 0 \qquad \Rightarrow \qquad D = \frac{2d}{\beta^2 M_\infty}, \qquad (3.54)$$

between the plateau of the mean square displacement and the plateau of the memory kernel. The same result can be obtained from Eq. (3.50) by observing that for $s \to 0$, $\hat{M}(s) \sim M_\infty/s$ and $\hat{D}(s) \sim D/s$. In the following, we derive a closed equation for M_∞ and, from it, an equation for the dynamical transition point.

3.3.1 Calculation of the Plateau Value of Correlation Functions

Using the decomposition in Eq. (3.52), we can write Eq. (3.39) as

$$\frac{\zeta}{2}\dot{\mathbf{r}}(t) = -\frac{\beta}{2}\int_0^t dt' M_f(t - t')\dot{\mathbf{r}}(t') + \Xi_f(t)$$

$$\qquad - \frac{\partial v}{\partial \mathbf{r}}(|\mathbf{r}(t)|) - \frac{\beta}{2}M_\infty[\mathbf{r}(t) - \mathbf{r}(0)] + \Xi_\infty, \qquad (3.55)$$

$$\left\langle \Xi_{f,\mu}(t)\Xi_{f,\nu}(t') \right\rangle = \delta_{\mu\nu}\left[T\zeta\delta(t - t') + \frac{1}{2}M_f(t - t')\right],$$

$$\left\langle \Xi_{\infty,\mu}\Xi_{\infty,\nu} \right\rangle = \delta_{\mu\nu}\frac{1}{2}M_\infty.$$

Note that we decomposed the noise into two independent components: a constant one, Ξ_∞, whose correlation is related to M_∞, and a fast one, $\Xi_f(t)$, whose correlation is related to $M_f(t)$ and decays quickly over time. The two variables Ξ_∞ and $\Xi_f(t)$ are assumed to be independent and Gaussian, with zero average; their sum $\Xi(t) = \Xi_\infty + \Xi_f(t)$ thus has zero average and the correct variance, as requested by Eq. (3.39).

Eq. (3.55) describes the dynamics of an effective particle, $\mathbf{r}(t)$, moving in the potential

$$w(\mathbf{r}) = v(|\mathbf{r}|) + \frac{\beta}{4}M_\infty(\mathbf{r} - \mathbf{r}_0)^2 - \Xi_\infty \cdot \mathbf{r} \qquad (3.56)$$

and subjected to a retarded friction and coloured noise with fast-decaying kernels that also satisfy the fluctuation–dissipation relation. Note that, thanks to the quadratic term proportional to M_∞, this potential is confining around $\mathbf{r}_0$. Therefore, at long times, the effective particle position equilibrates in the potential $w(\mathbf{r})$, and its distribution is given by the equilibrium result,

$$P(\mathbf{r}|\mathbf{s}) = \frac{e^{-\beta w(\mathbf{r})}}{\int d\mathbf{r}' e^{-\beta w(\mathbf{r}')}} = \frac{e^{-\beta v(|\mathbf{r}|) - \frac{\beta^2}{4}M_\infty|\mathbf{r}|^2 + \beta \mathbf{s}\cdot\mathbf{r}}}{\int d\mathbf{r}' e^{-\beta v(|\mathbf{r}'|) - \frac{\beta^2}{4}M_\infty|\mathbf{r}'|^2 + \beta \mathbf{s}\cdot\mathbf{r}'}}, \qquad (3.57)$$

with $\mathbf{s} = \Xi_\infty + \beta M_\infty \mathbf{r}_0/2$. The average force, $\langle F(\mathbf{r}(t)) \rangle$, then converges to the average force over the equilibrium distribution and the noise

$$\langle F(\mathbf{r}) \rangle = \int d\Xi_\infty P(\Xi_\infty) \int d\mathbf{r} P(\mathbf{r}|\Xi_\infty + \beta M_\infty \mathbf{r}_0/2) F(\mathbf{r}), \qquad (3.58)$$

where $P(\Xi_\infty) = e^{-\frac{|\Xi_\infty|^2}{M_\infty}} / (\pi M_\infty)^{d/2}$ is the distribution of the Gaussian noise Ξ_∞. Plugging Eq. (3.58) in Eq. (3.42) gives a self-consistent equation for M_∞,

$$M_\infty = \frac{\rho}{d} \int d\mathbf{r}_0 g(\mathbf{r}_0) F(\mathbf{r}_0) \cdot \langle F(\mathbf{r}) \rangle. \qquad (3.59)$$

Note that $M_\infty = 0$ – i.e., the solution corresponding to the liquid phase – is always a solution of Eq. (3.59). In fact, for $M_\infty = 0$, one has $\Xi_\infty = \mathbf{0}$ and then $\mathbf{s} = \mathbf{0}$. Eq. (3.58) then becomes $\langle F(\mathbf{r}) \rangle = \int d\mathbf{r} F(\mathbf{r}) e^{-\beta v(|\mathbf{r}|)} / \int d\mathbf{r} e^{-\beta v(|\mathbf{r}|)}$. In absence of the confining term provided by M_∞, the integral in the denominator, $\int d\mathbf{r} e^{-\beta v(|\mathbf{r}|)}$, is divergent because $e^{-\beta v(|\mathbf{r}|)} \to 1$ at large $|\mathbf{r}|$. The integral in the numerator is instead finite, because $F(\mathbf{r})$ is short ranged. It follows that $\langle F(\mathbf{r}) \rangle = \mathbf{0}$, which can be inserted in Eq. (3.59) and self-consistently shows that $M_\infty = 0$ is a solution.

Finding non-zero solutions of Eq. (3.59) is not easy because numerically evaluating its right-hand side requires calculating several integrals over d-dimensional vectors that are not rotationally invariant. It can, however, be simplified by assuming that $g(\mathbf{r}_0) = \exp[-\beta v(|\mathbf{r}_0|)]$, which is exact for $d \to \infty$, as discussed in Chapter 2. After some manipulations[9] (details can be found in [339]), one finds that Eq. (3.59) can be equivalently expressed in terms of D, as

$$\frac{d}{\rho V_d} = F_1(\mathsf{D}; \beta),$$

$$F_1(\mathsf{D}; \beta) = -d\,\mathsf{D} \int_0^\infty dr\, r^{d-1} \log[q(\mathsf{D}, \beta; r)] \frac{\partial q(\mathsf{D}, \beta; r)}{\partial \mathsf{D}}, \qquad (3.60)$$

$$q(\mathsf{D}, \beta; r) = \int d\mathbf{u} \frac{e^{-\frac{d|\mathbf{u}|^2}{2\mathsf{D}}}}{(2\pi \mathsf{D}/d)^{d/2}} e^{-\beta v(|\mathbf{r}+\mathbf{u}|)}.$$

Eq. (3.60) is much easier than Eq. (3.59) to evaluate numerically, because the function $q(\mathsf{D}, \beta; r)$ is the convolution of a Gaussian with a rotationally invariant function, and, therefore, it is itself rotationally invariant. It can thus be transformed into a one-dimensional integral either by moving to Fourier space or by using bipolar coordinates, which gives [292]

---

[9] Note that the equivalence of Eq. (3.60) and Eq. (3.59) only holds when $M_\infty > 0$, which means that the liquid solution $M_\infty = 0$ is lost in going from one equation to the next. However, this limitation is not severe because we already know that the solution $M_\infty = 0$ is always present.

$$q(D,\beta;r) = \int_0^\infty du \, e^{-\beta v(u)} \left(\frac{u}{r}\right)^{\frac{d-1}{2}} \frac{e^{-\frac{d(r-u)^2}{2D}}}{\sqrt{2\pi \, D/d}} \left[e^{-\frac{d\,ru}{D}} \sqrt{2\pi \frac{d\,ru}{D}} I_{\frac{d-2}{2}} \left(\frac{d\,ru}{D}\right) \right],$$

$$(3.61)$$

where $I_n(x)$ is the modified Bessel function of the first kind. The derivative of $q(D,\beta;r)$ with respect to D is then easily computed, and $F_1(D;\beta)$ can be obtained via another one-dimensional integral. The solution of Eq. (3.60) for D provides the plateau value of mean square displacement and that of the memory kernel via Eq. (3.54).

3.3.2 Equation for the Dynamical Glass Transition

A dynamical glass transition can emerge from Eq. (3.60) if the density is high enough or if the temperature is low enough. Because the function $F_1(D;\beta)$ does not depend on density, which only appears in the left-hand side of the first Eq. (3.60), the case in which temperature is held constant and density is increased is easier to discuss graphically. Thus, we here specialise to this case, but similar results can be obtained at constant density upon lowering the temperature.

One can show that for simple interaction potentials, the function $F_1(D;\beta)$ is positive, vanishes both for $D \to 0$ and $D \to \infty$ and therefore has an absolute maximum in between.[10] Therefore, besides the liquid solution with $M_\infty = 0$ and $D = \infty$ that always satisfies Eq. (3.59), no additional solution can be found if $\rho < \rho_d(\beta)$, where

$$\frac{d}{\rho_d(\beta) V_d} = \max_D F_1(D;\beta),$$

$$(3.62)$$

as illustrated in Figure 3.2. This condition allows one to determine the dynamical transition density $\rho_d(\beta)$ for each temperature. For $\rho > \rho_d(\beta)$, Eq. (3.60) admits at least two solutions, which correspond to the intersections of the function $F_1(D;\beta)$ with a horizontal line at level $d/(\rho V_d)$. For reasons that will be better understood from the discussion of Chapter 4, the solution that correctly describes the arrested glass phase is that with the smaller D.

Eq. (3.62) can be solved in any spatial dimension d to obtain the dynamical transition, but in low d, it overestimates the glass transition because the rich structure of the liquid is neglected[11] by the approximation $g(r) = \exp[-\beta v(r)]$. In general,

---

[10] For some particular potentials, there can be multiple local maxima, leading to multiple glass phases and even to glass–glass transitions [240, 325]. We will not further discuss this possibility in this book.

[11] Better approximation schemes that give more reasonable estimates in $d = 3$ have been discussed in [241]. Eqs. (3.60) and (3.62) also provide a quantitative, although approximate, description of a particular model, the Mari-Krzakala-Kurchan model [242, 243], where distances between particles are randomly shifted to induce a mean field like interaction in finite d. See [90] for a detailed discussion.

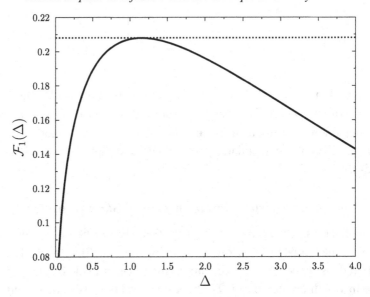

Figure 3.2 The function $\mathcal{F}_1(\Delta) = \lim_{d\to\infty} F_1(D = \ell^2\Delta/d)$ corresponding to the hard-sphere potential (for which the dependence on β is absent). The horizontal dashed line corresponds to $1/\widehat{\varphi}_d$ with $\widehat{\varphi}_d = 4.8067\ldots$ For $\widehat{\varphi} < \widehat{\varphi}_d$, Eq. (3.66) has no solution at finite Δ. At $\widehat{\varphi} = \widehat{\varphi}_d$, $1/\widehat{\varphi}$ coincides with the maximum of $\mathcal{F}_1(\Delta)$ and a solution appears, indicating dynamical arrest. For $\widehat{\varphi} > \widehat{\varphi}_d$, there are two solutions of Eq. (3.66). The plateau of the mean square displacement is the smaller of the two.

one expects that Eq. (3.62) becomes exact only in the limit $d \to \infty$ [239]. We then consider the limit $d \to \infty$ for the class of potentials introduced in Section 2.3.2, such that Eq. (2.74) holds – i.e., $v(r) = \bar{v}[d(r/\ell - 1)]$ with a typical interaction scale ℓ. The numerical solution [241] indicates that the dynamical transition density and plateau, expressed in terms of adimensional quantities, have the following scaling when $d \to \infty$:

$$\varphi = \rho V_d \left(\frac{\ell}{2}\right)^d = \frac{d\widehat{\varphi}}{2^d}, \qquad D = \frac{\ell^2\Delta}{d}, \tag{3.63}$$

where $\widehat{\varphi}$ and Δ are finite quantities. This scaling can also be checked by proving that Eq. (3.62) has a finite limit when $d \to \infty$ if and only if Eq. (3.63) is obeyed. Plugging Eq. (2.74) and Eq. (3.63) in Eq. (3.61) and using the asymptotic properties of Bessel functions (details can be found in [292]), one obtains that $q(D, \beta; r)$ tends, for $h = d(r/\ell - 1)$, to

$$q(\Delta, \beta; h) = \int_{-\infty}^{\infty} dz \frac{e^{-\frac{(z-h)^2}{2\Delta}}}{\sqrt{2\pi\Delta}} e^{-\beta\bar{v}(z+\Delta/2)}. \tag{3.64}$$

From this, one can show that $F_1(D;\beta)$ tends to

$$\mathcal{F}_1(\Delta;\beta) = -\Delta \int_{-\infty}^{\infty} dh\, e^h \log[q(\Delta,\beta;h)] \frac{\partial q(\Delta,\beta;h)}{\partial \Delta}, \qquad (3.65)$$

and Eq. (3.62) becomes

$$\frac{1}{\widehat{\varphi}_d(\beta)} = \max_{\Delta} \mathcal{F}_1(\Delta;\beta). \qquad (3.66)$$

For the hard-sphere potential, for which β is irrelevant, this equation gives a dynamical transition at $\widehat{\varphi}_d = 4.8067\ldots$, above which the dynamics is arrested (see Figure 3.2). A more detailed discussion and results for other potentials are given in Chapter 4.

Note that the dynamical transition happens when the scaled packing fraction $\overline{\varphi} = 2^d \varphi = d\widehat{\varphi}$ is proportional to d, which according to the results of Chapter 2, belongs to the region in which the liquid thermodynamics is well defined and given by the ideal gas result plus the first virial correction.

3.3.3 Large-Dimensional Scaling of the Dynamical Equations

Having determined the correct scaling of the density and of the mean square displacement in the limit $d \to \infty$, in Eq. (3.63), we can write the full dynamical equations, Eq. (3.39), in this same limit. We begin by representing the distance vector between two particles in 'polar coordinates' as $\mathbf{r}(t) = r(t)\hat{\mathbf{r}}(t)$, where $r(t) = |\mathbf{r}(t)|$ is the modulus and $\hat{\mathbf{r}}(t) = \mathbf{r}(t)/r(t)$ is an angular unit vector. One can then show that in the limit $d \to \infty$, under the appropriate scaling, the angular vector remains constant on the relevant time scales and that the dynamics can then be reduced to a one-dimensional equation for the modulus, or more precisely, the variable $h(t) = d[r(t)/\ell - 1]$.

The stochastic Eq. (3.39) can be written in the form

$$\frac{\zeta}{2}\dot{\mathbf{r}}(t) = G(\mathbf{r}(t)) + \Xi(t),$$

$$G(\mathbf{r}(t)) = F(\mathbf{r}(t)) - \frac{\beta}{2}\int_0^t dt'\, M(t-t')\dot{\mathbf{r}}(t'). \qquad (3.67)$$

To write equations for $r(t)$ and $\hat{\mathbf{r}}(t)$, we use the well-known Itô's formula (or lemma), which provides the correct way of performing a change of coordinates in a stochastic process with white noise [159, section 4.3.3], to change to polar coordinates. Eq. (3.67) then becomes

$$\frac{\zeta}{2}\dot{r}(t) = \frac{(d-1)T}{r(t)} + \hat{\mathbf{r}}(t) \cdot [G(\mathbf{r}(t)) + \Xi(t)], \tag{3.68}$$

$$\frac{\zeta}{2}\frac{d\hat{\mathbf{r}}(t)}{dt} = -\frac{(d-1)T}{r(t)^2}\hat{\mathbf{r}}(t) + \frac{1}{r(t)}[G(\mathbf{r}(t)) + \Xi(t)] \tag{3.69}$$

$$-\frac{1}{r(t)^2}\hat{\mathbf{r}}(t)\mathbf{r}(t) \cdot [G(\mathbf{r}(t)) + \Xi(t)].$$

Note that because the norm of $\hat{\mathbf{r}}(t)$ is constant, one has $\hat{\mathbf{r}}(t) \cdot \dot{\mathbf{r}}(t) = \dot{r}(t)$. Furthermore,

$$\hat{\mathbf{r}} \cdot F(\mathbf{r}) = -\hat{\mathbf{r}} \cdot \frac{\partial v(|\mathbf{r}|)}{\partial \mathbf{r}} = -v'(r) = -\frac{d}{\ell}\bar{v}'(h), \tag{3.70}$$

Using these relations, from Eq. (3.68) we can derive the equation for $h(t) = d[r(t)/\ell - 1]$ at leading order in $1/d$,

$$\frac{\zeta\ell}{2d}\dot{h}(t) = \frac{d}{\ell}\{T - \bar{v}'[h(t)]\} - \frac{\beta\ell}{2d}\int_0^t dt' M(t - t')\dot{h}(t') + \frac{d}{\ell}\Xi(t), \tag{3.71}$$

having defined $\Xi(t) = (\ell/d)\hat{\mathbf{r}} \cdot \Xi(t)$, which is the projection of the noise over a given coordinate (rescaled by d for a reason that will become immediately clear). Its correlation is deduced from Eq. (3.39):

$$\langle\Xi(t)\Xi(t')\rangle = 2T\frac{\ell^2}{2d^2}\zeta\delta(t - t') + \frac{\ell^2}{2d^2}M(t). \tag{3.72}$$

Eqs. (3.71) and (3.72) show that in order to obtain a finite limit when $d \to \infty$, one needs to define rescaled variables,

$$\widehat{\zeta} = \frac{\ell^2}{2d^2}\zeta, \qquad \mathcal{M}(t) = \frac{\ell^2}{2d^2}M(t). \tag{3.73}$$

The scaling of the friction coefficient ζ is needed for the derivative $\dot{h}(t)$ and force $\bar{v}'(h)$ terms to be of the same order in Eq. (3.71); the scaling of the memory function $M(t)$ is needed to keep the two noise terms of the same order in Eq. (3.72). Eq. (3.71) then becomes

$$\widehat{\zeta}\dot{h}(t) = T - \bar{v}'[h(t)] - \beta\int_0^t dt' \mathcal{M}(t - t')\dot{h}(t') + \Xi(t), \tag{3.74}$$

$$\langle\Xi(t)\Xi(t')\rangle = 2T\widehat{\zeta}\delta(t - t') + \mathcal{M}(t - t'),$$

where all quantities remain finite in the limit $d \to \infty$. Having defined the scaling of ζ and $M(t)$ for large d, one can also check that Eq. (3.69) gives $\frac{d\hat{\mathbf{r}}(t)}{dt} \propto \frac{1}{d}$, which implies that $\hat{\mathbf{r}}(t)$ is constant in time (for finite t).[12]

Finally, we can write in this limit the self-consistent equation that determines the memory kernel, Eq. (3.42). Under the assumption that $\hat{\mathbf{r}}(t)$ is constant, the angular direction of $\boldsymbol{F}[\mathbf{r}(t)]$, which is parallel to $\hat{\mathbf{r}}(t)$, is also constant. Using Eq. (3.70), we get

$$
\begin{aligned}
\mathcal{M}(t) &= \frac{\ell^2 \rho}{2d^3} \int d\mathbf{r}_0 e^{-\beta v(|\mathbf{r}_0|)} \langle \boldsymbol{F}(\mathbf{r}(t)) \rangle \cdot \boldsymbol{F}(\mathbf{r}_0) \\
&= \frac{\widehat{\varphi}}{2} \int_{-\infty}^{\infty} dh_0 e^{h_0 - \beta \bar{v}(h_0)} \langle \bar{v}'[h(t)] \rangle \bar{v}'(h_0).
\end{aligned}
\tag{3.75}
$$

The two equations, Eqs. (3.74) and (3.75), provide a self-consistent equation for $\mathcal{M}(t)$ that holds in the limit $d \to \infty$, with the appropriate scaling of all relevant quantities. Finally, Eq. (3.49), which relates the memory kernel with the mean square displacement, can be immediately rescaled by defining, consistently with Eq. (3.63),

$$
D(t) = \frac{\ell^2}{d} \Delta(t) \qquad \Rightarrow \qquad \widehat{\zeta}\dot{\Delta}(t) = -\beta \int_0^t dt' \mathcal{M}(t - t')\dot{\Delta}(t') + T, \tag{3.76}
$$

which allows one to compute $\Delta(t)$ from $\mathcal{M}(t)$. Note that Eqs. (3.74), (3.75) and (3.76) have also been derived via an exact solution of the dynamics in [239].

3.4 Critical Properties of the Dynamical Glass Transition

We conclude this chapter with a brief discussion of the critical behaviour of dynamical correlators upon approaching the glass transition from the liquid phase.

3.4.1 Stress Correlations, Viscosity and Stokes–Einstein Relation

The shear viscosity of the liquid can be deduced from the autocorrelation function of the pressure tensor[13] via a Green–Kubo relation; see [83, 186] for details. In the overdamped $m \to 0$ limit, the pressure tensor (at zero momentum) is defined by

---

[12] The proof goes, schematically, as follows. Assuming $\frac{d\hat{\mathbf{r}}(t)}{dt} \propto \frac{1}{d}$, then

$$
\dot{\mathbf{r}}(t) = r(t)\frac{d\hat{\mathbf{r}}(t)}{dt} + \hat{\mathbf{r}}(t)\dot{r}(t) \propto \frac{1}{d}.
$$

As a consequence of this scaling and of the scaling in Eq. (3.73), $\boldsymbol{G}(t) \propto d$ and $\Xi(t) \propto d$. The right-hand side of Eq. (3.69) is then proportional to d, while the left-hand side, with $\widehat{\zeta} \propto d^2$, is proportional to $d^2 \frac{d\hat{\mathbf{r}}(t)}{dt}$. This proves self-consistently the original assumption.

[13] The stress tensor, commonly used in mechanics, is the negative of the pressure tensor. In the overdamped limit, its kinetic contribution can be neglected.

$$\Pi_{\mu\nu}(t) = -\frac{1}{2} \sum_{i\neq j} \frac{r_{ij\mu}(t)r_{ij\nu}(t)}{r_{ij}(t)} v'(r_{ij}(t)), \tag{3.77}$$

where $\mathbf{r}_{ij} = \mathbf{x}_i - \mathbf{x}_j$ is the interparticle distance. Denoting by $\mu \neq \nu$ two distinct (and otherwise arbitrary, by isotropy of the liquid) spatial components of this tensor, the shear viscosity is then given by

$$\eta_s = \frac{\beta}{V} \int_0^\infty dt \left\langle \Pi_{\mu\nu}(t)\Pi_{\mu\nu}(0) \right\rangle. \tag{3.78}$$

In the limit $d \to \infty$, following similar steps as in Section 3.2.1, one obtains

$$\begin{aligned}
\left\langle \Pi_{\mu\nu}(t)\Pi_{\mu\nu}(0) \right\rangle &= \frac{1}{4} \sum_{i\neq j, k\neq l} \left\langle \frac{r_{ij\mu}(t)r_{ij\nu}(t)}{r_{ij}(t)} v'(r_{ij}(t)) \frac{r_{kl\mu}(0)r_{kl\nu}(0)}{r_{kl}(0)} v'(r_{kl}(0)) \right\rangle \\
&\sim \frac{1}{2d^2} \sum_{i\neq j} \left\langle \frac{|\mathbf{r}_{ij}(t) \cdot \mathbf{r}_{ij}(0)|^2}{r_{ij}(t)\,r_{ij}(0)} v'(r_{ij}(t))v'(r_{ij}(0)) \right\rangle \tag{3.79} \\
&= \frac{\rho^2 V}{2d^2} \int d\mathbf{r}_0\, g(\mathbf{r}_0) \left\langle \frac{|\mathbf{r}(t) \cdot \mathbf{r}_0|^2}{r(t)\,r_0} v'(r(t))v'(r_0) \right\rangle,
\end{aligned}$$

where, in the second line, it has been assumed that only pairs $\langle ij \rangle = \langle kl \rangle$ or $\langle ij \rangle = \langle lk \rangle$ are correlated, and an average has been taken over all directions $\mu\nu$ (including $\mu = \nu$, which gives subdominant $1/d$ corrections), while in the third line, the average over the many-body Langevin dynamics has been replaced by an average over the effective process defined in Section 3.2.2.

Finally, in Section 3.3.3, it was shown that in the limit $d \to \infty$, the unit vector $\hat{\mathbf{r}}(t)$ is constant on the relevant time scales, and dynamics happens only in the direction parallel to it. Also, at leading order, $r(t) \sim \ell$, with corrections of order $1/d$. As a consequence, $|\mathbf{r}(t) \cdot \mathbf{r}_0|^2/(r(t)\,r_0) \sim \ell^2$, and

$$\begin{aligned}
\frac{1}{V} \left\langle \Pi_{\mu\nu}(t)\Pi_{\mu\nu}(0) \right\rangle &\sim \frac{\rho^2 \ell^2}{2d^2} \int d\mathbf{r}_0\, g(\mathbf{r}_0) \left\langle v'(r(t))v'(r_0) \right\rangle \\
&\sim \rho \frac{\ell^2 M(t)}{2d} = \rho\, d\, \mathcal{M}(t), \tag{3.80}
\end{aligned}$$

using the expression for $M(t)$ in Eq. (3.42), under the same constant $\hat{\mathbf{r}}(t)$ hypothesis. In the limit $d \to \infty$, the stress correlation thus coincides with the memory kernel, and

$$\eta_s = \frac{\beta \rho \ell^2}{2d} \int_0^\infty dt\, M(t) = \beta \rho\, d \int_0^\infty dt\, \mathcal{M}(t). \tag{3.81}$$

Eq. (3.51) then becomes a relation between the diffusion coefficient and the shear viscosity,

$$D = \frac{T}{\zeta + \beta \int_0^\infty dt\, M(t)} = \frac{T}{\zeta + \frac{2d\,\eta_s}{\rho\ell^2}}. \tag{3.82}$$

At low densities, where $\eta_s \to 0$ and $D \to T/\zeta$, one recovers the ideal gas (overdamped Langevin) kinetics. Upon approaching the glass transition, $M(t)$ develops an infinite plateau. The viscosity thus diverges, $\eta_s \to \infty$, the diffusion coefficient vanishes, $D \to 0$, and one obtains an effective Stokes–Einstein relation [120] of the form

$$D\eta_s = \frac{T\ell^2\rho}{2d} = \frac{T}{\zeta_{\text{eff}}}, \qquad \zeta_{\text{eff}} = \frac{2d}{\rho\ell^2} = \frac{\ell^{d-2}}{\widehat{\varphi}} \frac{2\pi^{d/2}}{\Gamma(d/2+1)}. \tag{3.83}$$

This scaling of the effective Stokes drag ζ_{eff} is very close to that obtained via a hydrodynamic treatment [83] and can be also derived via a proper treatment of the dynamics of a single Brownian particle in a solvent, in the limit $d \to \infty$ [84].

3.4.2 Dynamical Critical Exponents and Dynamical Susceptibility

The structure of the dynamical equations in $d \to \infty$ is qualitatively similar to those of the mode-coupling theory (MCT) of glasses, which has long been used to describe dynamical arrest. We refer to the books [168, 175] for a detailed discussion of this theory.

Like Eq. (3.49), MCT is a set of equations that relate dynamical correlators to memory kernels, derived in fixed d via a series of approximations of the exact dynamics. In MCT, however, the memory kernels are expressed as polynomial functions of the correlators. This differs from the exact $d \to \infty$ solution, in which the memory kernel is self-consistently determined via an average over the effective process, as discussed in Section 3.2.2. Quantitative differences thus arise between the two theories. For example, the dynamical glass transition of hard spheres predicted by MCT has an incorrect scaling in the limit $d \to \infty$, as $\varphi_d^{\text{MCT}} \propto d^2/2^d$ [183, 317]. The scaling of the plateau of the mean square displacement and the Stokes–Einstein relations obtained from MCT also differ from the exact solution in $d \to \infty$. The qualitative structure of the two theories is nonetheless similar. In particular, the exact $d \to \infty$ theory displays precisely the same critical scaling of the dynamical correlators as MCT in the vicinity of the dynamical transition, which is illustrated in Figure 3.3. A direct derivation of these critical properties from the dynamical equations is possible within MCT. The same derivation is technically much more involved in the case of the $d \to \infty$ equations because the equation that determines the memory kernel is implicit and involves an average over the effective process. Regardless of the nature of the memory kernel, however, the existence of a

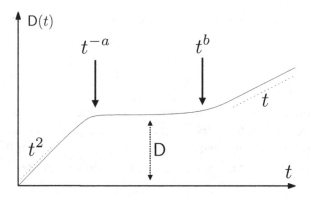

Figure 3.3 Illustration of the critical behaviour of the mean square displacement in $d \to \infty$, in the liquid phase very close to the dynamical glass transition. The figure is in log–log scale. At short times, for Newtonian dynamics, one has $D(t) \sim t^2$, while at very long times one has diffusive behaviour, $D(t) \sim Dt$. At intermediate times, a plateau emerges at $D(t) \sim D$. The approach to the plateau is a power law, $D(t) - D \propto t^{-a}$, and the departure from the plateau is also a power law, $D(t) - D \propto t^b$. Upon approaching the dynamical glass transition, the diffusion constant D vanishes, and the plateau extends to infinite times.

long-time plateau simplifies the dynamical equations in this regime, leading to schematic MCT equations near the plateau [288]. In the following, we only briefly review the critical scaling results, without derivation.

As an illustration, we consider the hard-sphere case, for which the control parameter is the packing fraction φ. For other systems, the dynamical glass transition is generically also controlled by temperature or other parameters. The same scaling results then hold by replacing φ with the appropriate control parameter. As discussed in Section 3.3.2, for $\varphi \geq \varphi_d$, dynamical correlations have an infinite plateau at long times, $\lim_{t \to \infty} D(t) = D$ and $\lim_{t \to \infty} M(t) = M_\infty$. The critical regime is observed when $\varphi \to \varphi_d$. For $\varphi \to \varphi_d^-$, a very long plateau emerges at time $t \sim t_p$ with $D(t) \sim D$, followed by a diffusive regime with a very small diffusion constant D. The plateau value D tends continuously to the true, infinite plateau solution of Eq. (3.60), when $\varphi \to \varphi_d^-$. The mean square displacement $D(t)$ scales as power law upon approaching and leaving the plateau:

$$D(t) - D \propto \begin{cases} -t^{-a} & \text{for } t_m \ll t \ll t_p, \\ t^b & \text{for } t \gg t_p, \end{cases} \tag{3.84}$$

where t_m is a microscopic time scale, given by $t_m \sim \ell^2 \zeta / T$ for Langevin dynamics and $t_m \sim \ell \sqrt{m/T}$ for Newtonian dynamics. The two dynamical critical exponents a and b that appear in Eq. (3.84) are related by [168]

$$\lambda = \frac{\Gamma(1-a)^2}{\Gamma(1-2a)} = \frac{\Gamma(1+b)^2}{\Gamma(1+2b)}, \tag{3.85}$$

where $\Gamma(x)$ is the Gamma function. The parameter λ is not universal. It depends on the system under investigation and is different within MCT and the $d \to \infty$ solution. A calculation of λ from thermodynamics is also possible [72, 288]. For hard spheres in $d \to \infty$, one finds $\lambda = 0.70698\ldots$, which implies that $a = 0.32402\ldots$ and $b = 0.62915\ldots$ [221]. The dynamical equations additionally predict a power-law divergence of the shear viscosity η_s and vanishing of the diffusion constant D with the same critical exponent γ (as implied by the Stokes–Einstein relation of Section 3.4.1):

$$\eta_s \sim D^{-1} \sim |\varphi - \varphi_d|^{-\gamma}, \qquad \varphi \to \varphi_d^-, \qquad \gamma = \frac{1}{2a} + \frac{1}{2b}. \tag{3.86}$$

For hard spheres, this relation gives $\gamma = 2.33786\ldots$ [221]. On the other side of the dynamical transition, when $\varphi \to \varphi_d^+$, one has $D(t) - D \propto t^{-a}$, with the same exponent a given by Eq. (3.85), but the exponent b is not defined because the plateau extends to infinite time.

At a standard second-order phase transition, the correlation function of the order parameter displays a divergent correlation length, and the susceptibility defined by its volume integral diverges, as discussed in Chapter 1. By analogy, in the case of the dynamical transition, one can study the dynamical susceptibility defined in Eq. (3.17), which is the volume integral of a four-point dynamical correlation [37, 56, 146, 204]. It encodes the fluctuations of dynamical correlators, here represented by the mean square displacement. In the dynamically arrested phase, the long-time limit of this susceptibility goes to a constant – i.e., $\lim_{t \to \infty} \chi_4(t) = \chi$ – that diverges upon approaching the dynamical transition with $\varphi \to \varphi_d^+$ as $\chi \sim |\varphi - \varphi_d|^{-1/2}$. Approaching the dynamical point from the liquid side instead, $\chi_4(t)$ has a peak at $t \sim 1/D$, with $\chi = \chi_4(1/D)$ similarly diverging [37, 56, 146, 204]. In addition, the dynamical correlation length associated with χ_4 diverges as $\xi_4 \sim |\varphi - \varphi_d|^{-1/4}$ near the transition [56, 145, 149].

3.5 Wrap-Up

3.5.1 Summary

In this chapter, we have seen that

- There exist several models of equilibrium dynamics. Examples are Hamiltonian, Langevin and Monte Carlo dynamics. For these dynamics, if the initial condition is described by the equilibrium Gibbs–Boltzmann distribution, the same distribution describes the system at all times. The dynamics then satisfies reversibility and fluctuation–dissipation relations, and the transport coefficients can be written in terms of equilibrium correlations via Green–Kubo relations (Section 3.1).
- In the limit $d \to \infty$, each particle performs an independent Langevin dynamics, in the presence of a thermal bath (encoded by effective noise and friction terms)

that describes the average (mean field) interaction with all the other particles. The memory kernel $M(t)$ that describes the thermal bath is self-consistently determined as an autocorrelation of the force, in the effective Langevin dynamics (Sections 3.2.1 and 3.2.2).

- Once $M(t)$ is determined, all the other dynamical observables, such as the mean square displacement $\mathsf{D}(t)$ and the diffusion constant D, can be expressed in terms of $M(t)$ (Section 3.2.3).
- The effective $d \to \infty$ dynamical equations predict that at a sufficiently high density or low temperature, the dynamics of the liquid phase becomes arrested. The memory kernel does not decay to zero in the arrested phase, and the diffusion constant vanishes. Explicit equations describe the dynamical transition point and the plateau value of the memory kernel (Section 3.3.1).
- From these equations, one can identify the proper scaling of all the dynamical quantities for $d \to \infty$. The most important scaling forms are those of the mean square displacement, $\mathsf{D}(t) = \ell^2 \Delta(t)/d$, and of the packing fraction, $\varphi = d\widehat{\varphi}/2^d$ (Section 3.3.2).
- Using this scaling, the equations can be simplified in the $d \to \infty$ limit and reduced to a one-dimensional effective Langevin equation. The final result given by Eqs. (3.74), (3.75) and (3.76) coincides with what has been obtained via an exact solution of the dynamics in [239] (Section 3.3.3).
- One can also show that the viscosity diverges at the dynamical transition point, that an effective Stokes–Einstein relation holds in the vicinity of dynamical arrest and that the dynamical correlations display power-law scalings close to the plateau, as in the mode-coupling theory of glasses. A dynamical four-point susceptibility diverges upon approaching the dynamical transition (Section 3.4).

3.5.2 Further Reading

We provide here a list of references that can be consulted to further explore the subjects discussed in this chapter, selected according to the criteria discussed in Section 1.6.2.

Introductory reviews on the phenomenology of liquid dynamics close to the glass transition are

- Debenedetti, *Metastable liquids: Concepts and principles* [120]
- Ediger, *Spatially heterogeneous dynamics in supercooled liquids* [135]
- Donth, *The glass transition: Relaxation dynamics in liquids and disordered materials* [130]
- Henkel, Pleimling and Sanctuary (eds), *Ageing and the glass transition* [176]
- Hunter and Weeks, *The physics of the colloidal glass transition* [181]
- Binder and Kob, *Glassy materials and disordered solids: An introduction to their statistical mechanics* [53]

Mode-coupling theory (MCT) is by far the most accurate microscopic theory to describe liquid dynamics in $d = 3$. Good introductory reviews to MCT are

- Reichman and Charbonneau, *Mode-coupling theory* [301]
- Janssen, *Mode-coupling theory of the glass transition: A primer* [188]
- Götze, *Complex dynamics of glass-forming liquids: A mode-coupling theory* [168]
- Götze, *Recent tests of the mode-coupling theory for glassy dynamics* [169]

In particular, the third reference [168] provides all the technical details of the theory, while the fourth [169] reviews the numerical and experimental tests of MCT. A systematic numerical investigation of the supercooled liquid dynamics in dimensions $d = 3, \dots, 12$, which focuses in particular on the comparison with MCT and on the Stokes–Einstein relation, can be found in

- Charbonneau, Ikeda, Parisi et al., *Glass transition and random close packing above three dimensions* [86]
- Charbonneau, Ikeda, Parisi et al., *Dimensional study of the caging order parameter at the glass transition* [85]
- Charbonneau, Charbonneau, Jin et al., *Dimensional dependence of the Stokes–Einstein relation and its violation* [83]

The complete derivation of the $d \to \infty$ dynamical solution via path integrals is based on the Martin–Siggia–Rose–De Dominicis–Janssen formalism. Pedagogical introductions to this formalism can be found in

- Cugliandolo, *Dynamics of glassy systems* [111]
- Castellani and Cavagna, *Spin-glass theory for pedestrians* [79]
- Kamenev, *Field theory of non-equilibrium systems* [200]

Both MCT and the $d \to \infty$ dynamical equations derived in this chapter fall into a general class of 'dynamical mean field equations'. These equations describe the dynamics of a variety of mean field systems and are discussed in references [79, 111]. A general discussion of the connection between MCT and dynamical mean field equations and the application of dynamical mean field equations to several regimes of the out-of-equilibrium dynamics of liquids can be found in

- Cugliandolo and Kurchan, *Analytical solution of the off-equilibrium dynamics of a long-range spin-glass model* [109]
- Bouchaud, Cugliandolo, Kurchan, et al., *Out of equilibrium dynamics in spin-glasses and other glassy systems* [65]
- Berthier, Barrat and Kurchan, *A two-time-scale, two-temperature scenario for nonlinear rheology* [36]
- Berthier and Kurchan, *Non-equilibrium glass transitions in driven and active matter* [41]

A general discussion of the analogy between dynamical mean field equations and replica equations, which is used in the calculation of dynamical critical exponents presented in Section 3.4.2, can be found in

- Kurchan, *Supersymmetry, replica and dynamic treatments of disordered systems: A parallel presentation* [218]
- Parisi and Rizzo, *Critical dynamics in glassy systems* [288]

All these ingredients are behind the solution of the dynamics of particle systems in $d \to \infty$. Detailed derivations can be found in

- Maimbourg, Kurchan and Zamponi, *Solution of the dynamics of liquids in the large-dimensional limit* [239]
- Agoritsas, Maimbourg and Zamponi, *Out-of-equilibrium dynamical equations of infinite-dimensional particle systems* [3]

In particular, the second reference [3] contains an extension of both the approach presented in this chapter, and the path integral derivation, to the out-of-equilibrium dynamics.

It is well established that the mean field sharp dynamical transition becomes (at best) a dynamical crossover in $d = 2, 3$. The disappearance of the dynamical transition is due to fluctuations of different nature, which are not taken into account by the mean field theory. Theories of the fluctuations around the dynamical transitions have been proposed in

- Schweizer and Saltzman, *Entropic barriers, activated hopping, and the glass transition in colloidal suspensions* [320]
- Bhattacharyya, Bagchi and Wolynes, *Facilitation, complexity growth, mode coupling, and activated dynamics in supercooled liquids* [47]
- Franz, Parisi, Ricci-Tersenghi et al., *Field theory of fluctuations in glasses* [145]
- Rizzo and Voigtmann, *Qualitative features at the glass crossover* [304]
- Janssen and Reichman, *Microscopic dynamics of supercooled liquids from first principles* [189]
- Charbonneau, Jin, Parisi et al., *Hopping and the Stokes–Einstein relation breakdown in simple glass formers* [90]

At temperatures well below the dynamical transition, dynamical relaxation is believed to be due to activated events, and many glass forming materials show some sort of Arrhenius (or modified Arrhenius) behaviour of their relaxation time. The reviews mentioned at the beginning of this section provide a good introduction to this dynamical regime; additional references are given in Section 4.5.2.

3.6 Appendix: Reversibility for Langevin Dynamics

We briefly discuss here how to prove reversibility at equilibrium for a Langevin equation. In order to simplify the notation, we consider a one-dimensional example (a single particle at position x evolving in a potential $V(x)$); hence, the overdamped Langevin equation is

$$\dot{x} = -\frac{dV(x)}{dx} + \xi(t), \qquad \langle \xi(t)\xi(t') \rangle = 2T\delta(t - t'). \tag{3.87}$$

The probability that the particle be at x at time t, $P(x,t)$, then satisfies the Fokker-Planck equation [159]

$$\dot{P}(x,t) = \frac{d}{dx}\left(\frac{dV(x)}{dx}\frac{dP(x,t)}{dx}\right) + T\frac{d^2P(x,t)}{dx^2}. \tag{3.88}$$

If we choose an initial condition in equilibrium with probability

$$P(x,0) = P_{\text{eq}}(x) \propto e^{-\beta V(x)}, \tag{3.89}$$

we can check that $\dot{P}(x,0) = 0$, as expected because the equilibrium distribution is a stationary solution of Eq. (3.88). We could also introduce the transition probability $T(x,y;t)$, defined by

$$P(x,t) = \int dy\, T(x,y;t)P(y,0), \tag{3.90}$$

which is the probability of finding the particle at x at time t, provided that it is at y at time 0. The function $T(x,y;t)$ is not symmetric as a consequence of the microscopic irreversibility.

To prove reversibility at equilibrium, it is convenient to introduce the function $\rho(x) = \sqrt{P_{\text{eq}}(x)}$ and the function $Q(x,t)$ defined by

$$\rho(x)Q(x,t) = P(x,t). \tag{3.91}$$

From Eq. (3.88), it follows that

$$\dot{Q}(x,t) = -\mathcal{H}Q(x,t),$$

$$\mathcal{H} = -T\frac{d^2}{dx^2} + W(x), \qquad W(x) = \frac{1}{4T}\left(\frac{dV(x)}{dx}\right)^2 - \frac{1}{2}\frac{d^2V(x)}{dx^2}. \tag{3.92}$$

The operator $\mathcal{H}$ is self-adjoint; hence, we have

$$Q(x,t) = \int dy\, G(x,y;t)Q(y,0), \tag{3.93}$$

with a symmetric $G(x, y; t) = G(y, x; t)$. Comparison of Eq. (3.90) and Eq. (3.93) leads to

$$G(x, y; t) = T(x, y; t)\rho(x)/\rho(y) = e^{-\frac{1}{2}\beta V(x)} T(x, y; t) e^{\frac{1}{2}\beta V(y)}. \tag{3.94}$$

We finally arrive to the needed expression of the equilibrium correlations:

$$\langle \mathcal{O}_1(x(t)) \mathcal{O}_2(x(0)) \rangle = \int dx dy \mathcal{O}_1(x) T(x, y; t) \mathcal{O}_2(y) P_{eq}(y)$$

$$= \int dx dy \rho(x) \mathcal{O}_1(x) G(x, y; t) \mathcal{O}_2(y) \rho(y), \tag{3.95}$$

which is clearly symmetric under the exchange of $\mathcal{O}_1$ and $\mathcal{O}_2$.

Notice that Eq. (3.94) implies that $e^{-\frac{1}{2}\beta V(x)} T(x, y; t) e^{\frac{1}{2}\beta V(y)}$ is a symmetric function, which is a sort of balance equation. The detailed balance Eq. (3.8) can also be written as

$$e^{-\frac{1}{2}\beta V(X_{t+1})} P(X_{t+1}|X_t) e^{\frac{1}{2}\beta V(X_t)} = e^{-\frac{1}{2}\beta V(X_t)} P(X_t|X_{t+1}) e^{\frac{1}{2}\beta V(X_{t+1})}. \tag{3.96}$$

4

Thermodynamics of Glass States

In Chapter 3, we have seen that simple liquids in the limit $d \to \infty$ generically undergo a dynamical phase transition to an arrested phase upon lowering temperature or increasing density. In this phase, the long-time limit of the mean square displacement is finite. In other words, the system is trapped into a restricted portion of phase space, selected by the initial condition, and is unable to diffuse away from it.

This restricted portion of phase space defines a 'glass state'. We can then invoke an ergodic hypothesis, restricted to a single glass state, in order to convert the dynamical long time average of observables within the glass into an appropriate thermodynamical average. In this way, the long time behaviour can be described by a thermodynamic framework called the 'state following' formalism. This construction requires the introduction of replicas of the original system, which brings about a new symmetry into the problem – namely the permutation symmetry between replicas, or 'replica symmetry'. In this chapter, we introduce the state following construction, we obtain the main formulae that describe the restricted glass thermodynamics, and we derive the phase diagram for standard models of glasses within the simplest scheme, in which replica symmetry is unbroken.

4.1 Arrested Dynamics and Restricted Thermodynamics

In Chapter 3, the equations that describe the dynamics of simple liquids in high dimensions have been discussed. In the liquid phase, close to the dynamical transition, the equilibrium dynamics happens on two well-separated time scales (Section 3.4.2): at short times, particles vibrate around an amorphous glassy structure, while at long times, they diffuse away from the initial structure. In the infinite-dimensional limit, upon approaching the dynamical transition, the diffusion constant D vanishes as a power law, given in Eq. (3.86). Beyond the dynamical transition, diffusion is arrested ($D = 0$), and the mean square displacement reaches a finite plateau in the long-time limit.

In a nutshell, the dynamical arrest corresponds to the formation of a metastable state [204, 205, 206, 207, 208]. We have discussed metastability for the ferromagnet in Chapter 1, but the physics is different here. In ferromagnets, there are only one equilibrium state and one metastable state, while in glasses, there is a quite large number of states (exponentially large in the number of particles, as we shall see later). The aim of this chapter is to generalise the discussion of metastability presented in Chapter 1 in order to recover and extend the results of Chapter 3, without solving the dynamics of the system but remaining in the framework of the Gibbs–Boltzmann statistical mechanics, in which no explicit reference to the dynamics is made.

Studying dynamical arrest without solving the dynamics may look like a contradiction, but it is not. In fact, we have already seen in Section 1.5 that, for ferromagnets in the $d \to \infty$ limit, the metastable state with positive magnetic field, $B > 0$, and negative magnetisation, $m < 0$, can be obtained by (1) studying the system in equilibrium with the standard Gibbs–Boltzmann distribution at $B < 0$, (2) computing the observables we are interested in, (3) and then analytically continuing their values to $B > 0$. In other words, metastable states can always be reached by starting from a stable equilibrium state and changing adiabatically (i.e., analytically continuing on) a control parameter. This scheme thus provides a well-defined statistical mechanics procedure to compute observables in a metastable state.[1]

We have thus shown that the construction of statistical mechanics for metastable states is based on adding an appropriate term in the Hamiltonian that stabilises the metastable state and then gradually removing it. Of course, different metastable states could be reached by changing the form of the Hamiltonian. To study dynamical arrest, we want to study the metastable states in which the system explores a small region of phase space around an initial equilibrium liquid configuration [144, 205, 206, 252, 260]. To construct these states, we can add a new term in the Hamiltonian that increases the energy of configurations that are too far away from the initial configuration. If this new term is sufficiently large, in the new equilibrium the system is confined around the initial configuration. We can then perform bona fide equilibrium statistical mechanics computations. When we remove the extra term, the system remains trapped in the metastable state defined by this procedure.

Note that one can choose as a starting point an equilibrium configuration at a given temperature and repeat the aforementioned procedure at another temperature.

---

[1] One should keep in mind, however, that the analytical continuation is only well defined in $d \to \infty$, while in finite dimensions, an essential singularity appears at $B = 0$ [282]. This singularity is weak enough that the analytical continuation remains quite unambiguous if B remains sufficiently close to zero.

This procedure is called 'state following' [144, 214]. It allows one to compute physical observables of a glass phase prepared by cooling or compressing an initial equilibrium configuration. These physical ideas are quite simple, and we shall implement them in the rest of the chapter, hoping that the technical difficulties do not obscure their intrinsic simplicity.

As in Chapters 2 and 3, we consider here a system of N identical particles, enclosed in a volume V, with number density $\rho = N/V$, with positions $\underline{X} = \{x_i\}$, interacting via a potential $V(\underline{X})$ with a typical interaction scale ℓ and strength ε. We consider the thermodynamic limit $N \to \infty$ and $V \to \infty$ at constant ρ, in which the boundary conditions become irrelevant, and there are only two adimensional control parameters, the scaled temperature T/ε and the packing fraction $\varphi = \rho V_d (\ell/2)^d$ defined in terms of the interaction range ℓ (see Section 2.3.2). In the following, we will be interested in comparing two configurations $\underline{X}$ and $\underline{Y}$ at different density. It will be convenient to take N and V (and thus ρ) to be the same for the two configurations so that particle positions, x_i and y_i, can be directly compared.[2] A change in packing fraction, with constant N and V, can be conveniently achieved by changing the interaction scale ℓ, as we will do in the following, while for temperature, it is simpler to vary T at fixed ε.

4.1.1 Arrested Dynamics in the Glass Phase

Consider the dynamics starting from an equilibrium configuration $\underline{Y} \equiv \{y_i\}$, extracted from the Gibbs–Boltzmann distribution, at packing fraction φ_g (with interaction scale ℓ_g) and temperature T_g that fall into the dynamically arrested region – i.e., $T_g < T_d(\varphi_g)$. We call $\underline{X}(t) \equiv \{x_i(t)\}$ the configuration of the system at time t, such that $\underline{X}(0) = \underline{Y}$. The mean square displacement

$$D(\underline{X}, \underline{Y}) = \frac{1}{N} \sum_{i=1}^{N} |x_i - y_i|^2, \qquad (4.1)$$

is a natural measure of similarity between two configurations $\underline{X}, \underline{Y}$.

We first consider the case in which the dynamics is run at the same state point (φ_g, T_g) at which the initial condition is prepared, as illustrated in Figure 4.1. In a dynamically arrested phase, the diffusion coefficient vanishes, and

$$\lim_{t \to \infty} \langle D(\underline{X}(t), \underline{Y}) \rangle = D_r, \qquad (4.2)$$

---

[2] In fact, if the volume is changed, the particle coordinates should be rescaled accordingly, and if the particle number is changed, no one-to-one correspondence between the x_i and the y_i is possible.

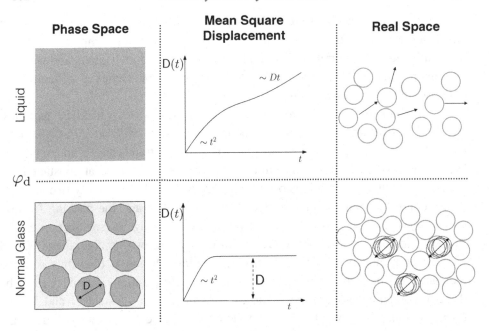

Figure 4.1 At low densities or high temperatures (top row), the phase space is composed by only one ergodic component, which coincides with the liquid state. For Newtonian dynamics, the mean square displacement behaves ballistically at short times and diffusively at long times (here, in log-log scale). Beyond the dynamical transition (bottom row), the phase space clusters in a set of metastable glass states, as a consequence of ergodicity breaking. The long time limit of the mean square displacement is then finite, which signals that particles are caged by their neighbours.

where D_r is a finite constant,[3] and the brackets represent either a dynamical noise average, for Brownian dynamics, or an average over a large enough time window around t, for Newtonian dynamics. By contrast to the liquid phase, in which particles diffuse away from their initial positions and $D(\underline{X}(t), \underline{Y})$ grows with time, in the dynamically arrested region, the system remains close, at all times, to its initial starting point. Furthermore, because the initial condition $\underline{Y}$ is sampled at equilibrium, the dynamics is, on average, time-translationally invariant

$$\langle D(\underline{X}(t+\tau), \underline{X}(t))\rangle = D(\tau), \qquad \forall t > 0 ; \qquad \lim_{\tau \to \infty} D(\tau) = D = D_r . \quad (4.3)$$

The behaviour of the system at long times is therefore described by a well-defined quantity, D_r, that represents both the mean square displacement between $\underline{X}(t)$ and $\underline{Y} = \underline{X}(0)$ at large t and the mean square displacement between two configurations

---

[3] In equilibrium, as it will be shown below, $D_r = D$, so one recovers the notation of Chapter 3.

$\underline{X}(t)$, $\underline{X}(t + \tau)$, separated by a long time window $\tau \to \infty$. A priori, the value of D_r depends on the initial configuration $\underline{Y}$. However, $\mathsf{D}_r(\underline{Y})$ is an intensive quantity, and its fluctuations obey the central limit theorem[4]

$$\mathsf{D}_r(\underline{Y}) = \overline{\mathsf{D}_r} + O\left(N^{-1/2}\right); \tag{4.4}$$

hence, the dependence on $\underline{Y}$ disappears in the thermodynamic limit. A quantity with this property is said to be 'self-averaging' [254]. Note that here and in the rest of this chapter, an overbar denotes an average over the initial configuration $\underline{Y}$, sampled from a Gibbs–Boltzmann distribution at (φ_g, T_g).

We now generalise the previous discussion to the case in which the system starts from an initial configuration $\underline{Y}$, prepared in equilibrium at (φ_g, T_g) as before, but evolves at a different state point $(\varphi, T) \neq (\varphi_g, T_g)$. This procedure corresponds to preparing a glass state at some temperature and packing fraction and then instantaneously bringing that glass to another temperature and packing fraction, where it is then kept.[5] If the new state point (φ, T) is such that the system remains in the glass phase, the diffusion constant still vanishes, and Eq. (4.2) still holds. Under the assumption that there is no phase transition that intervenes during the process of glass cooling or compression (this condition is not always satisfied, as will be discussed in Chapter 6), several additional properties are verified. First, one has

$$\lim_{t \to \infty} \langle \mathsf{D}(\underline{X}(t + \tau), \underline{X}(t)) \rangle = \mathsf{D}(\tau); \qquad \lim_{\tau \to \infty} \mathsf{D}(\tau) = \mathsf{D} \neq \mathsf{D}_r. \tag{4.5}$$

In other words, in the long time limit, the dynamics at the new temperature and packing fraction becomes stationary and is described by two quantities D_r and D, which correspond respectively to the average mean square displacement between $\underline{Y} = \underline{X}(0)$ and $\underline{X}(t)$ for large t, and between two configurations $\underline{X}$ at two very different and both very large times. This situation is illustrated in Figure 4.2. Second, the fluctuations of both D and D_r over the initial condition $\underline{Y}$ follow a central limit theorem analogous to Eq. (4.4), and, hence, D and D_r are self-averaging. Third, the long time properties are independent of how the system has been brought from (φ_g, T_g) to (φ, T). We described earlier an instantaneous change of the control parameters, but any other process (e.g., a slow cooling from T_g to T) would give the same results for times much larger than the preparation time.

---

[4] Unless there are long-range correlations in the system.
[5] An instantaneous quench can be realised, in practice, by using Langevin dynamics with noise at temperature $T \neq T_g$, or by using Newtonian dynamics with the initial velocities being rescaled by an appropriate factor to bring the system from T_g to T. An instantaneous compression is realised by preparing the system with a potential with length scale ℓ_g, and running the dynamics with a different potential with $\ell \neq \ell_g$.

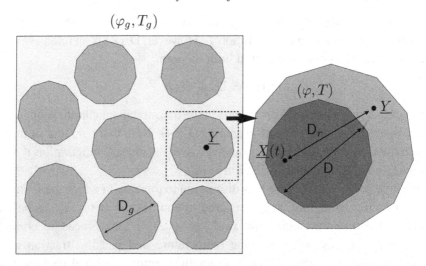

Figure 4.2 A schematic picture of the state following procedure. A glass state is selected by the configuration $\underline{Y}$ that is extracted in equilibrium at (φ_g, T_g). The configuration $\underline{Y}$ can thus be regarded as a pinning field for selecting a particular metastable state, as discussed in Chapter 1. (Left) If $\underline{Y}$ is dynamically evolved at the same state point (φ_g, T_g), then the long time limit of the mean square displacement is D_g. (Right) If the dynamics is instead evolved starting from $\underline{Y}$ but at different (φ, T), the state selected by $\underline{Y}$ at (φ_g, T_g) (light grey in the right panel) evolves as $\underline{X}(t)$ into a slightly different state (dark grey). The long time limit of the mean square displacement between $\underline{Y}$ and $\underline{X}(t)$ is D_r while the typical distance between two configurations separated by a long time window is D.

4.1.2 Restricted Thermodynamics of the Glass State

The idea of using a thermodynamic formalism to describe dynamical arrest in liquids dates back to the work of Kirkpatrick, Thirumalai and Wolynes [204, 205, 206, 329] and is at the root of the random first-order transition approach to the glass transition [40, 80, 208, 357]. The 'state following construction', which we introduce in this section, is based on this idea and was originally developed by Franz and Parisi in the context of spin glasses in [28, 144], where the equivalence with the long time limit of the dynamics was also discussed [27]. It is also called the 'Franz–Parisi construction', and it has been adapted to particles in $d \to \infty$ in [59, 299].

The method is based on the following idea. In equilibrium statistical physics, the long time limit of dynamical observables can be computed (in the thermodynamic limit) using a probabilistic approach that relies on the use of appropriate statistical ensembles [158, 310]. In equilibrium, entropy is maximised under the constraint that the total energy of the system takes a prescribed value. This leads to the Gibbs–Boltzmann distribution of the canonical ensemble [158]. By analogy, the state following construction aims to describe metastable glass states by a probability distribution that maximises entropy under a minimal set of constraints.

In Section 4.1.1, it was shown that in the dynamically arrested phase, the dynamics starting from an equilibrium configuration $\underline{Y}$ remains confined to a portion of phase space around the initial configuration (Figure 4.1). While the configuration $\underline{Y}$ selects one of the possible glass states (Figure 4.1), at large times, the configuration $\underline{X}(t)$ ergodically samples the restricted region of phase space defined by $\langle D(\underline{X}, \underline{Y}) \rangle = D_r$. This constraint defines a region of phase space that we call a 'glass state'. We further assume that the stationary long time average of dynamical observables coincides with a restricted statistical average over this glass state. The simplest assumption to construct such a restricted statistical average is that, besides the energy, the only other constraint to be enforced concerns the mean square displacement. We thus seek a probability distribution over $\underline{X}$ that maximises the entropy under the two constraints of fixed energy and fixed mean square displacement from $\underline{Y}$. Because the additional constraint does not affect the kinetic energy, the distribution of the momenta remains Maxwellian; hence, we omit it in the following. Focusing on the configurational part, we obtain a generalisation of the canonical distribution

$$P(\underline{X}|\varphi, \beta; \underline{Y}, \lambda) \propto e^{-\beta V[\underline{X}, \ell] - \lambda D(\underline{X}, \underline{Y})}, \tag{4.6}$$

in which the dependence of $V[\underline{X}, \ell]$ on ℓ is indicated explicitly to highlight the dependence of the probability on the packing fraction. Here the inverse temperature β is a Lagrange multiplier coupled to the energy, while λ is coupled to the mean square displacement. Just like β is determined by requiring a given average energy, λ is determined by requiring that the average of $D(\underline{X}, \underline{Y})$ coincides with D_r. Equivalently, one can use a different ensemble, in which the mean square displacement constraint is strictly enforced:

$$P(\underline{X}|\varphi, \beta; \underline{Y}, D_r) = \frac{1}{Z[\varphi, \beta; \underline{Y}, D_r]} e^{-\beta V[\underline{X}, \ell]} \delta(D_r - D(\underline{X}, \underline{Y})),$$

$$Z[\varphi, \beta; \underline{Y}, D_r] = \int d\underline{X} e^{-\beta V[\underline{X}, \ell]} \delta(D_r - D(\underline{X}, \underline{Y})). \tag{4.7}$$

where the factor $Z[\varphi, \beta; \underline{Y}, D_r]$ is a 'restricted configurational integral' that generalises Eq. (2.2) to the restricted equilibrium construction. The two ensembles defined in Eqs. (4.6) and (4.7) are equivalent in the thermodynamic limit for the same reason that the canonical and microcanonical ensembles are equivalent [158].

The free energy of the glass state selected by $\underline{Y}$ and brought to (φ, T) is then given by

$$f(\varphi, T; \underline{Y}, D_r) = -\frac{T}{N} \log Z[\varphi, \beta; \underline{Y}, D_r]. \tag{4.8}$$

Here again, this free energy fluctuates with the configuration $\underline{Y}$ but is self-averaging in the thermodynamic limit, in which it does not fluctuate unless the system undergoes a phase transition [254]. The average of the free energy over $\underline{Y}$ is thus representative of the typical value of $f(\varphi, T; \underline{Y}, D_r)$ and is given by

$$
\begin{aligned}
f_g(\varphi, T; \varphi_g, T_g, D_r) &= \overline{f(\varphi, T; \underline{Y}, D_r)} \\
&= -\frac{T}{N} \int \frac{d\underline{Y}}{Z[\varphi_g, \beta_g]} e^{-\beta_g V[\underline{Y}, \ell_g]} \log Z[\varphi, \beta; \underline{Y}, D_r],
\end{aligned}
\tag{4.9}
$$

where (φ_g, T_g) are the parameters that control the distribution of the initial condition $\underline{Y}$, and $Z[\varphi_g, \beta_g] = \int d\underline{Y} e^{-\beta_g V[\underline{Y}, \ell_g]}$ is the standard configurational integral at (φ_g, T_g). This expression defines the 'Franz–Parisi potential', or 'glass free energy'. It represents the free energy of a typical glass state prepared at (φ_g, T_g) and followed to the state point (φ, T), for a given choice of D_r.

Note that the glass free energy depends explicitly on D_r, which was introduced as a dynamical quantity in Eq. (4.2). In principle, one should take D_r from the solution of the dynamics and use it in the calculation of the glass free energy. This approach is, however, unsatisfying. One would like to have a fully thermodynamical computation of the glass free energy without having to solve the dynamics first. To resolve this problem, one can use the general principles of statistical mechanics [223]. The constrained equilibrium free energy, expressed as a function of D_r, can be thought of as a large deviation function for the fluctuations of D_r in an unconstrained ensemble. This free energy should then be minimised over D_r. Because the global minimum is found at $D_r \to \infty$ and corresponds to the liquid state, to describe the glass, one should instead choose the local minimum at finite D_r (see Figure 4.3 for an example). In other words, the proper value of D_r in Eq. (4.2) is a local minimum of the Franz–Parisi potential at finite D_r, keeping all other parameters fixed.

In fact, as we show explicitly in the rest of this chapter, at high T and low φ, the Franz–Parisi potential has a unique minimum when $D_r \to \infty$. This corresponds to the liquid phase in which the dynamics is not arrested. When T is low enough and φ is high enough (provided the preparation values φ_g and T_g fall in the dynamically arrested region), a new local minimum of the Franz–Parisi potential appears at finite D_r. We will show that the value of D_r at this local minimum precisely corresponds to the value of D_r selected by the dynamics. One thus concludes that the free energy of a typical glass state prepared at (φ_g, T_g) and followed to (φ, T) is given by

$$
f_g(\varphi, T; \varphi_g, T_g) = \min_{D_r : D_r < \infty} f_g(\varphi, T; \varphi_g, T_g, D_r),
\tag{4.10}
$$

which provides a fully thermodynamical expression for the glass free energy, without any reference to the dynamics.

The structure of Eq. (4.9) is also interesting. One has first to compute the free energy of the system $\underline{X}$, which depends on the configuration $\underline{Y}$, and then average the free energy over $\underline{Y}$. One can think to the system $\underline{X}$ as feeling an external disordered potential, due to the configuration $\underline{Y}$, that is fixed in time (or 'quenched'). The configuration $\underline{X}$ evolves in presence of this external potential and samples its equilibrium distribution. One then averages the free energy over the disorder, represented by $\underline{Y}$. In other words, the configuration $\underline{Y}$ acts as a quenched disorder for the system $\underline{X}$. This formulation therefore reduces to a standard problem in the statistical physics of disordered systems [254], and the techniques developed in this field can then be used in the context of structural glass problems without quenched disorder, as we now describe.

4.1.3 The Replica Method

It is usually very difficult to average the logarithm of the disorder-dependent free energy [254]. Disorder indeed explicitly breaks translational invariance, which is a crucial symmetry for most theoretical physics methods, from mean field methods to perturbative expansions. In the case of Eq. (4.9), more specifically, one cannot apply the virial expansion to the $\underline{Y}$-dependent free energy $f(\varphi, T; \underline{Y}, D_r)$, because of the presence of the disordered space-dependent external potential.

The replica method is designed to solve this problem. It makes use of the simple identity $\log x = \lim_{s\to0} \partial_s x^s$, using the shorthand notation $\partial_s = d/ds$. If x is a random variable, then

$$\overline{\log x} = \lim_{s\to0} \partial_s \overline{x^s}. \tag{4.11}$$

Applying this identity to Eq. (4.9), we get

$$f_g(\varphi, T; \varphi_g, T_g, D_r) = -\lim_{s\to0} \frac{T}{N} \partial_s \int \frac{d\underline{Y}}{Z[\varphi_g, \beta_g]} e^{-\beta_g V[\underline{Y}, \ell_g]} Z[\varphi, \beta; \underline{Y}, D_r]^s. \tag{4.12}$$

Computing this expression for a general value of s is as difficult as computing the logarithm, but a big simplification occurs if s is an integer. We can then write

$$Z[\varphi, \beta; \underline{Y}, D_r]^s = \prod_{a=2}^{s+1} \int d\underline{X}^a e^{-\beta V[\underline{X}^a, \ell]} \delta\left(D_r - D(\underline{X}^a, \underline{Y})\right), \tag{4.13}$$

and, defining $\underline{X}^1 = \underline{Y}$, we obtain

$$f_g(\varphi, T; \varphi_g, T_g, D_r) = -\frac{T}{N} \lim_{s\to0} \frac{1}{Z[\varphi_g, \beta_g]} \partial_s Z_{s+1}[\varphi_a, \beta_a, D_r], \tag{4.14}$$

$$Z_{s+1}[\varphi_a, \beta_a, D_r] = \int \left(\prod_{a=1}^{s+1} d\underline{X}^a e^{-\beta_a V[\underline{X}^a, \ell_a]}\right) \left(\prod_{a=2}^{s+1} \delta\left(D_r - D(\underline{X}^a, \underline{X}^1)\right)\right),$$

where $\beta_1 = \beta_g$, $\varphi_1 = \varphi_g$, $\beta_{a\geq 2} = \beta$ and $\varphi_{a\geq 2} = \varphi$. We thus have to compute the configurational integral of a system of $s + 1$ replicas of the original system, such that all replicas $a \geq 2$ are coupled to replica 1 by the mean square displacement constraint. This constraint imposes that particle i in each replica must be close to particle i in all other replicas, $\mathbf{x}_i^a \sim \mathbf{x}_i^b$ for all a, b. The $s + 1$ particles $\mathbf{x}_i^a$ for each given i thus form a 'molecule', bound by the mean square displacement constraint, and $Z_{s+1}[\varphi_a, \beta_a, \mathbf{D}_r]$ corresponds to the configurational integral of a 'molecular liquid'.

The fundamental simplification introduced by the replica method for each integer value of s is that the molecular liquid is translationally and rotationally invariant. A global translation of all replicas, $\mathbf{x}_i^a \rightarrow \mathbf{x}_i^a + \mathbf{X}$, leaves the partition function invariant,[6] as does a global rotation. This symmetry allows us to use the virial expansion and the methods introduced in Chapter 2 to compute $Z_{s+1}[\varphi_a, \beta_a, \mathbf{D}_r]$ for each integer s. It is important to stress that, in principle, even the molecular liquid could have glassy or crystalline phases in which translational invariance is spontaneously broken, but we are not interested in such phases. Averaging over all equilibrium configurations $\underline{Y}$ (as illustrated in Figure 4.2) corresponds to sampling all possible glass states, which restores the translational and rotational invariance of the system. Therefore, the ensemble of glass states is described (once replicas are introduced) by a molecular liquid that remains in its translationally and rotationally invariant – i.e., liquid – phase.

Assuming that one is able to find an expression for $Z_{s+1}[\varphi_a, \beta_a, \mathbf{D}_r]$ that can be analytically continued from integer s to real s, the Franz–Parisi potential can be extracted as follows. First, we note that $Z_1[\varphi_a, \beta_a, \mathbf{D}_r] = Z[\varphi_g, \beta_g]$. Then, we define a replicated free energy[7]

$$-\mathsf{f}_{s+1}(\varphi_a, T_a, \mathbf{D}_r) = \frac{1}{N} \log Z_{s+1}[\varphi_a, \beta_a, \mathbf{D}_r]$$
$$= -\beta_g \mathsf{f}(\varphi_g, T_g) - s\beta \mathsf{f}_g(\varphi, T; \varphi_g, T_g, \mathbf{D}_r) + O(s^2),$$

(4.15)

which also implies

$$\beta \mathsf{f}_g(\varphi, T; \varphi_g, T_g, \mathbf{D}_r) = \lim_{s \to 0} \partial_s \mathsf{f}_{s+1}(\varphi_a, T_a, \mathbf{D}_r).$$

(4.16)

From the expansion of the replicated free energy in powers of s, we can thus derive the glass free energy. In Section 4.2, we discuss how to obtain an exact expression for the replicated free energy in the infinite dimensional limit.

---

[6] Provided periodic boundary conditions are used for a finite volume V.

[7] Note that usually $-\beta \mathsf{f}$ is the logarithm of the partition function. Here, we omitted the factor β because the replicas are at different temperatures β_a. As a consequence, $\mathsf{f}_{s+1}(\varphi_a, T_a, \mathbf{D}_r)$ is adimensional. Another possible convention, which we do not follow here, would be to multiply the partition function by any of the replica temperatures $T_a = 1/\beta_a$.

4.2 Restricted Thermodynamics in Infinite Dimensions

In Section 4.1, we have seen that in order to describe the thermodynamics of a glass one needs to introduce identical replicas of the original atomic system. In the replicated system, the copies of a same particle in different replicas are close to one another and, thus, form a molecule. In this section, we derive an exact expression for the glass free energy in infinite dimensions. First, we generalise the discussion of Section 2.3.1 to a molecular liquid. Then, we use the translational and rotational invariance of the molecular liquid to express its free energy in terms of the matrix of scalar products of particle displacements in a molecule. Finally, we consider the scaling in the limit $d \to \infty$ and we obtain the expression for the Franz–Parisi potential.

In this section, we provide an exact derivation of the replicated free energy in $d \to \infty$, adapted from the original papers [88, 220, 221] but using the important simplifications obtained in [59]. Its content is more technical than the rest of the book. In a first reading, or if the reader is not interested in the technical details, it is possible to jump directly to Section 4.2.4, where the final result is given in compact form. A simpler derivation is also possible by means of a Gaussian ansatz, which, however, introduces an additional assumption [88]. More details on this simpler derivation are given in Section 4.6.3.

4.2.1 Virial Expansion for a Molecular System

We begin by considering a generic molecular liquid, in which a basic constituent is not a simple atom (labelled by its position $\mathbf{x}_i$) but a molecule, composed of n atoms, and labelled $\underline{\mathbf{x}}_i = \{\mathbf{x}_i^1, \ldots, \mathbf{x}_i^n\}$, where $\mathbf{x}_i^a$ is the position of atom $a = 1, \ldots, n$ in molecule $i = 1, \ldots, N$. The total configuration of the system is then $\underline{X} = \{\underline{\mathbf{x}}_i\}_{i=1,\ldots,N}$. The interaction potential has the form

$$V(\underline{X}) = \sum_i w(\underline{\mathbf{x}}_i) + \sum_{i<j} \underline{v}(\underline{\mathbf{x}}_i - \underline{\mathbf{x}}_j), \qquad (4.17)$$

where $w(\underline{\mathbf{x}})$ is the interaction between atoms inside a same molecule, which is responsible for its binding, while $\underline{v}(\underline{\mathbf{x}} - \underline{\mathbf{y}})$ is the interaction between molecules $\underline{\mathbf{x}}$ and $\underline{\mathbf{y}}$. Although we ultimately want to consider one special replica at temperature T_g and density φ_g, and s identical replicas at different temperature T and density φ, as in Section 4.1, for notational simplicity, we first consider the case in which all replicas are at the same (φ, T).

Identifying $V(\underline{X})$ with Eq. (4.17), all the definitions of Section 2.1.1 remain identical, except that the total number of degrees of freedom is now ndN instead of dN. Therefore, one should replace $\Lambda^{dN} \to \Lambda^{ndN}$ in the denominator of Q_N in Eq. (2.3).

This modification only affects the ideal gas term of the free energy, in which the kinetic term $dT \log \Lambda$ is multiplied by n. The definitions of Section 2.1.2 are also straightforwardly extended to the molecular liquid by replacing $\mathbf{x} \to \underline{\mathbf{x}}$ everywhere. The local density thus becomes a local 'molecular density' $\rho(\underline{\mathbf{x}})$, normalised by $\int d\underline{\mathbf{x}} \rho(\underline{\mathbf{x}}) = N$, and is proportional to the probability of finding a molecule with atom 1 in position $\mathbf{x}^1$, atom 2 in position $\mathbf{x}^2$ and so on. Similarly, the definitions and results of Section 2.1.3 can be straightforwardly extended to the molecular liquid by replacing $\mathbf{x} \to \underline{\mathbf{x}}$ and replacing the definition of $z(\mathbf{x})$ in Eq. (2.16) by

$$z(\underline{\mathbf{x}}) = \frac{e^{\beta\mu - \beta w(\underline{\mathbf{x}}) - \beta\phi(\underline{\mathbf{x}})}}{\Lambda^{ndN}}. \tag{4.18}$$

Introducing the molecular Mayer function

$$\underline{f}(\underline{\mathbf{x}} - \underline{\mathbf{y}}) = e^{-\beta\underline{v}(\underline{\mathbf{x}} - \underline{\mathbf{y}})} - 1, \tag{4.19}$$

the virial expansion can be obtained as in Section 2.2 (once again replacing $\mathbf{x} \to \underline{\mathbf{x}}$). At second order in $\rho(\underline{\mathbf{x}})$, the free energy of the molecular liquid is

$$-\beta F[\rho] = -ndN \log \Lambda + \int d\underline{\mathbf{x}} \rho(\underline{\mathbf{x}})[1 - \log \rho(\underline{\mathbf{x}})]$$
$$+ \frac{1}{2} \int d\underline{\mathbf{x}} d\underline{\mathbf{y}} \rho(\underline{\mathbf{x}}) \rho(\underline{\mathbf{y}}) \underline{f}(\underline{\mathbf{x}} - \underline{\mathbf{y}}). \tag{4.20}$$

The third and higher-order virial terms have diagrammatic forms identical to those of Section 2.2.3, but with integrations over molecular coordinates $\underline{\mathbf{x}}$, with a molecular density $\rho(\underline{\mathbf{x}})$ on each vertex and a molecular Mayer function $\underline{f}(\underline{\mathbf{x}} - \underline{\mathbf{y}})$ on each edge.

4.2.2 The Liquid Phase: Rotational and Translational Invariance

We are interested in the liquid phase of the molecular liquid, which is rotationally and translationally invariant. We first discuss some important preliminary consequences of these symmetries.

Translational Invariance

Translational invariance implies that the probability of finding a molecule in $\underline{\mathbf{x}} = \{\mathbf{x}^1, \ldots, \mathbf{x}^n\}$ is the same as that of finding it in a translated configuration[8] $\underline{\mathbf{x}} - \mathbf{X} = \{\mathbf{x}^1 - \mathbf{X}, \ldots, \mathbf{x}^n - \mathbf{X}\}$. Therefore, the density satisfies $\rho(\underline{\mathbf{x}}) = \rho(\underline{\mathbf{x}} - \mathbf{X})$. A convenient way to take advantage of this property is to introduce a change of variables that takes atom 1 as reference – i.e., $\mathbf{X} = \mathbf{x}^1$ – and then defines

---

[8] Note that this expression defines, in our notation, the operation of summing a single vector and a molecular configuration.

$$u^a = x^a - x^1, \qquad \underline{u} = \{0, u^2, \ldots, u^n\},$$

$$d\underline{x} = dX\, d\underline{u}, \qquad d\underline{u} = \prod_{a=2}^{n} du^a. \qquad (4.21)$$

Because $\rho(\underline{x})$ is translationally invariant, it does not depend on X; hence, $\rho(\underline{x}) = \rho(\underline{x} - X) = \rho(\underline{u})$. We define $\pi(\underline{u}) = \rho(\underline{u})/\rho$, where $\rho = N/V$ is the molecular number density, normalised by

$$1 = \frac{1}{N} \int d\underline{x}\, \rho(\underline{x}) = \frac{\rho}{N} \int dX\, d\underline{u}\, \pi(\underline{u}) = \int d\underline{u}\, \pi(\underline{u}), \qquad (4.22)$$

where in the last equality we used that $\rho \int dX = \rho V = N$. Hence, $\pi(\underline{u})$ can be interpreted as the probability distribution of displacements $\underline{u}$ in a molecule, relative to the first atom. With this change of variable, the second virial term becomes, for instance,

$$\bullet\!\!-\!\!\bullet = \frac{1}{2} \int d\underline{x}\, d\underline{y}\, \rho(\underline{x}) \rho(\underline{y}) \underline{f}(\underline{x} - \underline{y}) \qquad (4.23)$$

$$= \frac{\rho^2}{2} \int dX\, dY\, d\underline{u}\, d\underline{v}\, \pi(\underline{u}) \pi(\underline{v}) \underline{f}(X - Y + \underline{u} - \underline{v}) = N\frac{\rho}{2} \int dR\, f_{\text{eff}}(R),$$

where

$$f_{\text{eff}}(R) = \int d\underline{u}\, d\underline{v}\, \pi(\underline{u}) \pi(\underline{v}) \underline{f}(R + \underline{u} - \underline{v}) = \left\langle \underline{f}(R + \underline{u} - \underline{v}) \right\rangle_{\underline{u}, \underline{v}}. \qquad (4.24)$$

In this section, $\langle \bullet \rangle_{\underline{u}}$ denotes an average over the internal fluctuations of a molecule. The function $f_{\text{eff}}(R)$ can be interpreted as an effective interaction between the atoms in replica 1, once internal molecular degrees of freedom associated to the other replicas have been averaged out.

The second virial coefficient B_2 is then given by Eq. (2.43), with the Mayer function $f_{\text{eff}}(R)$. Similarly, the third virial coefficient can be written as

$$\triangle = \frac{1}{6} \int d\underline{x}\, d\underline{y}\, d\underline{z}\, \rho(\underline{x}) \rho(\underline{y}) \rho(\underline{z}) \underline{f}(\underline{x} - \underline{y}) \underline{f}(\underline{x} - \underline{z}) \underline{f}(\underline{y} - \underline{z})$$

$$= N\frac{\rho^2}{6} \int dR_1\, dR_2\, f_{\text{eff}}^{(3)}(R_1, R_2), \qquad (4.25)$$

where we defined $R_1 = X - Z$ and $R_2 = Y - Z$ with $X - Y = R_1 - R_2$, and

$$f_{\text{eff}}^{(3)}(R_1, R_2) = \left\langle \underline{f}(R_1 - R_2 + \underline{u} - \underline{v}) \underline{f}(R_1 + \underline{u} - \underline{w}) \underline{f}(R_2 + \underline{v} - \underline{w}) \right\rangle_{\underline{u}, \underline{v}, \underline{w}}. \qquad (4.26)$$

This function represents an effective three-particle interaction between reference coordinates. In general, it cannot be factorised as a product of two-particle interactions.

Rotational Invariance

Rotational invariance implies that the probability distribution function of displacements $\pi(\underline{\mathbf{u}})$ is invariant under a global rotation of all vectors $\mathbf{u}^a$; it then becomes a function of $q_{ab} = \mathbf{u}_a \cdot \mathbf{u}_b$ only, i.e. $\pi(\underline{\mathbf{u}}) = \pi(\hat{q})$. Rotational invariance also implies that $\langle \mathbf{u}^a \rangle_{\underline{\mathbf{u}}} = 0$. One can introduce a change of variables from the displacements $\underline{\mathbf{u}}$ to the scalar products matrix $\hat{q}$. Recall that here $a, b = 2, \ldots, n$ and the matrix $\hat{q}$ is thus a $(n-1) \times (n-1)$ matrix. The corresponding Jacobian is

$$J(\hat{q}) = \int d\underline{\mathbf{u}} \prod_{a \leq b}^{2,n} \delta(q_{ab} - \mathbf{u}^a \cdot \mathbf{u}^b) = C_{n,d} \left(\det \hat{q}\right)^{(d-n)/2}, \tag{4.27}$$

as proven in Section 2.5. The averages over $d\underline{\mathbf{u}}\pi(\underline{\mathbf{u}})$ can then be replaced by averages over $d\hat{q} J(\hat{q})\pi(\hat{q})$. Note that both distributions are, in fact, correctly normalised to unity.

Finally, note that rotational invariance implies that $f_{\text{eff}}(R)$ depends only on the modulus $R = |\mathbf{R}|$ and that $f_{\text{eff}}^{(3)}(\mathbf{R}_1, \mathbf{R}_2)$ depends only on $|\mathbf{R}_1|$, $|\mathbf{R}_2|$, and $|\mathbf{R}_1 - \mathbf{R}_2|$. The third virial coefficient B_3 is thus given by Eq. (2.44) with the replacement $f(|\mathbf{r}_1|)f(|\mathbf{r}_2|)f(|\mathbf{r}_1 - \mathbf{r}_2|) \rightarrow f_{\text{eff}}^{(3)}(|\mathbf{R}_1|, |\mathbf{R}_2|, |\mathbf{R}_1 - \mathbf{R}_2|)$. This procedure can be further generalised to higher-order virial coefficients.

4.2.3 Scaling in the Large-Dimensional Limit

We now focus on the limit $d \rightarrow \infty$, using a specific scaling that is relevant for our problem, and is motivated by the discussions of Chapters 2 and 3.

Scaling of the Molecular Liquid in $d \rightarrow \infty$

We take the limit $d \rightarrow \infty$ of the molecular liquid under two assumptions:

1. The inter-molecular interaction has the form

$$\underline{v}(\underline{\mathbf{x}} - \underline{\mathbf{y}}) = \sum_{a=1}^{n} v(|\mathbf{x}^a - \mathbf{y}^a|), \qquad v(r) = \bar{v}[d(r/\ell - 1)]. \tag{4.28}$$

This amounts to assuming that only atoms of the same type a interact directly, which is the relevant situation for computing the Franz–Parisi potential (Section 4.1.3) and that the interaction potential satisfies the correct scaling to obtain a non-trivial liquid phase in $d \rightarrow \infty$ (Section 2.3.2). In the following, for simplicity, we consider that all the replicas have the same potential $\bar{v}(h)$, the same interaction scale ℓ, and the same inverse temperature β. We will later generalise the result to the case of different interaction scales and temperatures, which is required for computing the Franz–Parisi potential.

2. We define a $(n-1) \times (n-1)$ adimensional matrix $\hat{\alpha}$ from[9]

$$\langle \mathbf{u}^a \rangle = 0, \qquad \frac{\alpha_{ab}\ell^2}{d} = \langle \mathbf{u}^a \cdot \mathbf{u}^b \rangle = \int d\underline{\mathbf{u}}\,\pi(\underline{\mathbf{u}})\mathbf{u}^a \cdot \mathbf{u}^b, \qquad (4.29)$$

for $a, b = 2, \ldots, n$. Because of rotational invariance, one has

$$\langle u^a_\mu u^b_\mu \rangle = \frac{1}{d}\langle \mathbf{u}^a \cdot \mathbf{u}^b \rangle = \frac{\alpha_{ab}\ell^2}{d^2}, \qquad (4.30)$$

for each component $\mu = 1, \ldots, d$. We assume that $\hat{\alpha}$ remains finite when $d \to \infty$, which corresponds to molecules being strongly localised. In fact, Eq. (4.30) with $a = b$ implies that the projection of $\mathbf{u}^a$ along an arbitrary unit vector $\mathbf{e}$ is a random variable with zero mean and variance $\alpha_{aa}\ell^2/d$. Its typical value thus scales as $\mathbf{u}^a \cdot \mathbf{e} \propto \sqrt{\alpha_{aa}}\ell/d$. We can also extend the matrix $\hat{\alpha}$ to a $n \times n$ matrix $\hat{\gamma}$ such that $\gamma_{ab} = \alpha_{ab}$ for $a, b = 2, \ldots, n$ and $\gamma_{a1} = \gamma_{1b} = 0$, which is consistent with $\mathbf{u}^1 = 0$. Then we define the $n \times n$ mean square displacement matrix $\hat{\Delta}$ by the relation

$$\frac{\Delta_{ab}\ell^2}{d} = \langle |\mathbf{u}^a - \mathbf{u}^b|^2 \rangle \qquad \Leftrightarrow \qquad \Delta_{ab} = \gamma_{aa} + \gamma_{bb} - 2\gamma_{ab}. \qquad (4.31)$$

The assumption that $\hat{\alpha}$ remains finite when $d \to \infty$ is equivalent to the assumption that $\hat{\Delta}$ remains finite – i.e., that the mean square displacement between atoms in the molecule is proportional to $1/d$. This assumption is consistent with the discussion of Section 3.3.2, which shows that the long time limit of the mean square displacement in the glass phase (here encoded by the displacement between different replicas) is indeed proportional to $1/d$. See, in particular, Eq. (3.63).

Saddle Point Method

In the limit $d \to \infty$, the averages over the internal displacements of a molecule can be calculated via the saddle point method. To illustrate this point, consider the normalisation of $\pi(\underline{\mathbf{u}})$ together with Eq. (4.29):

$$1 = \int d\underline{\mathbf{u}}\,\pi(\underline{\mathbf{u}}) = \int d\hat{q}\, J(\hat{q})\pi(\hat{q}) = C_{n,d}\int d\hat{q}\, e^{\frac{d-n}{2}\log\det\hat{q}+\log\pi(\hat{q})},$$

$$\frac{\alpha_{ab}\ell^2}{d} = \langle q_{ab} \rangle = \int d\hat{q}\, J(\hat{q})\pi(\hat{q})q_{ab} = C_{n,d}\int d\hat{q}\, e^{\frac{d-n}{2}\log\det\hat{q}+\log\pi(\hat{q})}q_{ab}. \qquad (4.32)$$

---

[9] In this section, we drop the suffix $\underline{\mathbf{u}}$ in the averages $\langle \bullet \rangle_{\underline{\mathbf{u}}}$ to simplify the notation.

When $d \to \infty$, the term in the exponent is proportional to d at leading order.[10] According to the saddle point method, both integrals in Eq. (4.32) are then dominated by the value of $\hat{q}$ that maximises this term, i.e.,

$$\hat{q}^* = \underset{\hat{q}}{\mathrm{argmax}} \left[\frac{d}{2} \log \det \hat{q} + \log \pi(\hat{q}) \right]. \tag{4.33}$$

The first of Eqs. (4.32) also imposes that, at leading exponential order in d,

$$1 = C_{n,d} e^{\frac{d-n}{2} \log \det \hat{q}^* + \log \pi(\hat{q}^*)} \qquad \Rightarrow$$

$$\Rightarrow \qquad 0 = \frac{d(n-1)}{2} \log \left(\frac{2\pi e}{d} \right) + \frac{d}{2} \log \det \hat{q}^* + \log \pi(\hat{q}^*), \tag{4.34}$$

where the second line is obtained by taking the logarithm and keeping only terms that are proportional to d, using the asymptotic relation $\log C_{n,d} \sim (n-1) \log \Omega_d \sim \frac{d(n-1)}{2} \log(2\pi e/d)$. Then, the second of Eqs. (4.32) becomes

$$\frac{\alpha_{ab}\ell^2}{d} = C_{n,d} e^{\frac{d-n}{2} \log \det \hat{q}^* + \log \pi(\hat{q}^*)} q_{ab}^* = q_{ab}^*. \tag{4.35}$$

A similar result holds for any function $F(\hat{q})$ that does not depend exponentially on d. We thus obtain

$$\langle F(\hat{q}) \rangle \underset{d \to \infty}{\to} F(\hat{q}^*) = F(\hat{\alpha}\ell^2/d). \tag{4.36}$$

Truncation of the Virial Expansion

We now show that when $d \to \infty$, the virial expansion of the molecular liquid can be truncated at second order, as for atomic liquids. The center-of-mass Mayer function, defined in Eq. (4.24), can be written explicitly as

$$f_{\mathrm{eff}}(R) = \int \mathrm{d}\underline{u}\, \mathrm{d}\underline{v}\, \pi(\mathbf{u}) \pi(\mathbf{v}) \left[e^{-\beta \sum_{a=1}^{n} \bar{v}[d(|\mathbf{R} + \mathbf{u}^a - \mathbf{v}^a|/\ell - 1)]} - 1 \right]. \tag{4.37}$$

When $|h| \to \infty$, one has $h = d(R/\ell - 1) \approx d(|\mathbf{R} + \mathbf{u}^a - \mathbf{v}^a|/\ell - 1)$, $\forall a$, and therefore, $\bar{v}[d(|\mathbf{R} + \mathbf{u}^a - \mathbf{v}^a|/\ell - 1)] \approx \bar{v}(h)$. This equivalence holds because the component of each $\mathbf{u}^a - \mathbf{v}^a$ along the direction of $\mathbf{R}$ is of order $1/d$. From Eq. (4.37), it then follows that for $|h| \to \infty$,

$$f_{\mathrm{eff}}(R) \approx e^{-\beta n \bar{v}(h)} - 1 \to \begin{cases} -1 & h \to -\infty, \\ 0 & h \to \infty. \end{cases} \tag{4.38}$$

---

[10] This statement relies on the assumption that $\log \pi(\hat{q}) \propto d$, because the two terms must have the same scaling to obtain a non-trivial equation for $\hat{q}$. The validity of this assumption is strongly suggested by Eq. (4.34), which shows that it holds at least for $\hat{q} = \hat{q}^*$ (with $\log d$ corrections). It can also be checked for $\hat{q} \neq \hat{q}^*$, see [221] for details.

The function $f_{\text{eff}}(R)$ thus increases from -1 to 0 when h increases from $-\infty$ to ∞ – i.e., over an interval of order $1/d$ around $R = \ell$. The same scaling is obtained for the potentials considered in Section 2.3.2. As a consequence of these hypotheses, the Mayer function scales in the same way as the potential:

$$f_{\text{eff}}(R) = \overline{f}_{\text{eff}}[d(R/\ell - 1)] = \overline{f}_{\text{eff}}(h) . \tag{4.39}$$

An explicit expression of $\overline{f}_{\text{eff}}(h)$ is given next. We can then follow the procedure described in Section 2.3.2 to show that the second virial coefficient has the same form as in Eq. (2.75),

$$B_2 = -\frac{V_d \ell^d}{2} \int_{-\infty}^{\infty} dh \, e^h \overline{f}_{\text{eff}}(h) = B_2^{\text{HS}} I(\pi, \bar{v}) , \tag{4.40}$$

with $I(\pi, \bar{v})$ being a finite integral that depends on the details of the interaction potential $\bar{v}(h)$ and the molecular distribution $\pi(\mathbf{u})$. We thus conclude that B_2 has the same leading exponential scaling with d as for hard spheres and all the potentials of Section 2.3.2.

A similar analysis can be applied to all the other diagrams that contribute to the higher-order virial coefficients. The expression in Eq. (4.26), together with the hypotheses made earlier that lead to Eq. (4.39), imply that $f_{\text{eff}}^{(3)}(|\mathbf{R}_1|, |\mathbf{R}_2|, |\mathbf{R}_1 - \mathbf{R}_2|)$ coincides with a product of three hard-sphere Mayer functions if at least one of the three arguments differs from ℓ by a quantity of order ℓ/d. This implies that the third virial coefficient of the molecular liquid has the same leading exponential scaling in d as the third virial coefficient of hard spheres, and the same is true for higher-order virial coefficients. We conclude that the analysis of Section 2.3.1 holds for the molecular liquid; the virial series can be truncated at the second order, and Eq. (4.20) becomes exact in the large-dimensional limit. We now compute the two terms of Eq. (4.20).

The Ideal Gas Free Energy

Recalling that $\rho(\mathbf{x}) = \rho\pi(\mathbf{u})$ and $d\mathbf{x} = d\mathbf{X}d\mathbf{u}$, we can write the ideal gas contribution to the free energy per particle as

$$-\beta f^{\text{id}} + nd \log \Lambda = \frac{1}{N} \int d\mathbf{x}\rho(\mathbf{x})[1 - \log \rho(\mathbf{x})] = 1 - \log \rho - \langle \log \pi(\mathbf{u}) \rangle . \tag{4.41}$$

The average of $\log \pi(\mathbf{u})$ can be computed by changing variable to $\hat{q}$ and applying Eq. (4.36), because we assumed that $\log \pi(\mathbf{u})$ is not exponential in d. At leading order in d, we get from Eq. (4.34) that

$$\langle \log \pi(\hat{q}) \rangle = \log \pi(\hat{\alpha}\ell^2/d) = -\frac{d(n-1)}{2} \log\left(\frac{2\pi e}{d}\right) - \frac{d}{2} \log \det(\hat{\alpha}\ell^2/d),$$

$$(4.42)$$

and therefore, still at leading order in d,

$$-\beta f^{\mathrm{id}} = -nd \log \Lambda - \log \rho + \frac{d(n-1)}{2} \log\left(\frac{2\pi e}{d}\right) + \frac{d}{2} \log \det(\hat{\alpha}\ell^2/d)$$

$$= -nd \log(\Lambda/\ell) - \log(\rho \ell^d) + \frac{d(n-1)}{2} \log\left(\frac{2\pi e}{d^2}\right) + \frac{d}{2} \log \det \hat{\alpha}.$$

$$(4.43)$$

Note that in the second line, all factors have been made adimensional by using the reference length scale ℓ.

The Excess Free Energy

We now derive an expression for the Mayer function – and, therefore, for the second virial coefficient that determines the excess free energy – as a function of $\hat{q}$. Recalling that $\mathbf{u}^1 = \mathbf{v}^1 = \mathbf{0}$, Eq. (4.39) can be written as

$$f_{\mathrm{eff}}(R) = e^{-\beta \bar{v}[d(R/\ell-1)]} \left\langle e^{-\beta \sum_{a=2}^{n} \bar{v}[d(|\mathbf{R}+\mathbf{u}^a-\mathbf{v}^a|/\ell-1)]} \right\rangle - 1. \qquad (4.44)$$

In order to evaluate this expression, we need to compute the distribution of random variables $x_a = |\mathbf{R} + \mathbf{u}^a - \mathbf{v}^a|$, which depend on random variables[11] $\mathbf{w}^a = \mathbf{u}^a - \mathbf{v}^a$. In order to characterise the statistics of x_a, it is convenient to introduce yet another random variable,

$$y_a = x_a^2 - R^2 = |\mathbf{w}^a|^2 + 2\mathbf{R} \cdot \mathbf{w}^a = \sum_{\mu=1}^{d} [(w_\mu^a)^2 + 2R_\mu w_\mu^a]. \qquad (4.45)$$

Each y_a is the sum of d random variables. By the central limit theorem, when $d \to \infty$, y_a then become Gaussian random variables specified by their mean and covariance, which can be computed as follows. For the mean, we note that $\langle \mathbf{w}_a \rangle = 0$ and $\langle \mathbf{u}_a \cdot \mathbf{v}_a \rangle = \langle \mathbf{u}_a \rangle \cdot \langle \mathbf{v}_a \rangle = 0$. We then have

$$\langle y_a \rangle = \langle |\mathbf{w}_a|^2 \rangle = \langle |\mathbf{u}_a|^2 \rangle + \langle |\mathbf{v}_a|^2 \rangle = 2\alpha_{aa}\ell^2/d. \qquad (4.46)$$

The term $|\mathbf{w}^a|^2$ is a sum of d positive terms, each of order $1/d^2$. Its mean is thus of order $1/d$, and its standard deviation is of order $1/d^{3/2}$. Its fluctuations can then be

---

[11] The distribution of $\mathbf{w}^a$ can be determined by recalling that $\underline{u}$ and $\underline{v}$ are independently and identically distributed according to $\pi(\underline{u})$, but we do not need here its explicit form.

neglected, $|\mathbf{w}^a|^2 \approx \langle|\mathbf{w}^a|^2\rangle$. To compute the covariance, we thus consider at leading order that $y_a - \langle y_a\rangle \approx 2\mathbf{R}\cdot\mathbf{w}^a$, and we obtain

$$\langle y_a y_b\rangle - \langle y_a\rangle\langle y_b\rangle = 4\sum_{\mu\nu} R_\mu R_\nu \langle w_\mu^a w_\nu^b\rangle = 8R^2\alpha_{ab}\ell^2/d^2, \qquad (4.47)$$

where we used that $\langle w_\mu^a w_\nu^b\rangle = \langle u_\mu^a u_\nu^b\rangle + \langle v_\mu^a v_\nu^b\rangle = \delta_{\mu\nu}\,2\alpha_{ab}\ell^2/d^2$ by rotational invariance and Eq. (4.30).

Equivalently, we can introduce random Gaussian variables $\underline{z} = \{z_a\}_{a=2,...,n}$ with probability measure

$$\mathcal{D}\underline{z} = d\underline{z}\,\frac{\exp\left\{-\frac{1}{2}\sum_{a,b=2}^n\left[(2\hat{\alpha})^{-1}\right]_{ab}z_a z_b\right\}}{\sqrt{(2\pi)^{n-1}\det(2\hat{\alpha})}}, \qquad d\underline{z} = \prod_{a=2}^n dz_a, \qquad (4.48)$$

such that $\langle z_a\rangle = 0$ and $\langle z_a z_b\rangle = 2\alpha_{ab}$, and express y_a as

$$y_a = 2\alpha_{aa}\ell^2/d + 2R\ell z_a/d, \qquad (4.49)$$

which ensures that y_a are Gaussian variables with the correct mean and covariance. Eq. (4.49) shows that $y_a \sim 1/d$. Recalling from Eq. (4.45) that $x_a = \sqrt{R^2 + y_a}$, and defining $h = d(R/\ell - 1)$, we obtain at leading order in $1/d$:

$$d(x_a/\ell - 1) = d(\sqrt{R^2 + y_a}/\ell - 1) \sim d\left[\frac{R}{\ell}\left(1 + \frac{y_a}{2R^2}\right) - 1\right]$$

$$\sim h + \frac{d\,y_a}{2\ell^2} = h + \alpha_{aa} + z_a. \qquad (4.50)$$

Eq. (4.44) can then be written as

$$\overline{f}_{\text{eff}}(h) = e^{-\beta\bar{v}(h)}\left\langle e^{-\beta\sum_{a=2}^n \bar{v}(h+\alpha_{aa}+z_a)}\right\rangle_{\underline{z}} - 1$$

$$= e^{-\beta\bar{v}(h)}\int\mathcal{D}\underline{z}\left[e^{-\beta\sum_{a=2}^n \bar{v}(h+\alpha_{aa}+z_a)}\right] - 1. \qquad (4.51)$$

Using the general identity in Eq. (4.88), the Gaussian convolution can be rewritten in a differential form, and Eq. (4.51) becomes

$$\overline{f}_{\text{eff}}(h) = e^{-\beta\bar{v}(h)}e^{\sum_{a,b=2}^n \alpha_{ab}\frac{\partial^2}{\partial h_a \partial h_b}}\left[e^{-\beta\sum_{a=2}^n \bar{v}(h_a+\alpha_{aa})}\right]\Big|_{h_a=h} - 1. \qquad (4.52)$$

Plugging this result in Eq. (4.40) allows us to write the second virial coefficient and the excess free energy per particle as

$$-\beta f^{\text{ex}} = -\rho B_2 \qquad (4.53)$$

$$= \frac{\overline{\varphi}}{2}\int_{-\infty}^{\infty} dh\,e^h\left\{e^{-\beta\bar{v}(h)}e^{\sum_{a,b=2}^n \alpha_{ab}\frac{\partial^2}{\partial h_a \partial h_b}}\left[e^{-\beta\sum_{a=2}^n \bar{v}(h_a+\alpha_{aa})}\right]\Big|_{h_a=h} - 1\right\},$$

where $\overline{\varphi} = \rho V_d \ell^d = 2^d\varphi$ is defined as in Chapter 2.

Total Free Energy

In summary, in $d \to \infty$, the total free energy per particle, $f = f^{id} + f^{ex}$, of a molecular liquid can be written as

$$-\beta f = -nd \log(\Lambda/\ell) - \log(\rho\ell^d) + \frac{d(n-1)}{2} \log\left(\frac{2\pi e}{d^2}\right) + \frac{d}{2} \log \det \hat{\alpha}$$

$$+ \frac{\overline{\varphi}}{2} \int_{-\infty}^{\infty} dh \, e^h \left\{ e^{-\beta \bar{v}(h)} e^{\sum_{a,b=2}^{n} \alpha_{ab} \frac{\partial^2}{\partial h_a \partial h_b}} \left[e^{-\beta \sum_{a=2}^{n} \bar{v}(h_a + \alpha_{aa})} \right] \Big|_{h_a = h} - 1 \right\}. \quad (4.54)$$

This expression clearly depends on density, temperature and interaction potential. It also depends on the molecular distribution $\pi(\mathbf{u})$, but only through the matrix $\hat{\alpha}$ defined in Eq. (4.29). This matrix encodes the average scalar products of the internal displacements of a molecule. Minimising the free energy with respect to $\pi(\mathbf{u})$ thus amounts to minimising it with respect to $\hat{\alpha}$.

Eq. (4.54) has been derived by using replica 1 as reference, which is clearly arbitrary; an equivalent result would be obtained by choosing any other replica as reference. To make this symmetry more explicit, we can express Eq. (4.54) as a function of the scaled mean square displacement matrix $\hat{\Delta}$ defined in Eq. (4.31). In Section 4.6.1, it is shown that

$$\det \hat{\alpha} = 2 \det(-\hat{\Delta}/2)(-\underline{1}^T \hat{\Delta}^{-1} \underline{1}), \quad (4.55)$$

where $\underline{1} = \{1, \ldots, 1\}$ is the vector of all ones. This allows us to express the ideal gas term as a function of $\hat{\Delta}$. For the excess free energy term – introducing, as before, a $n \times n$ matrix $\hat{\gamma}$ such that $\gamma_{ab} = \alpha_{ab}$ for $a, b = 2, \ldots, n$ and $\gamma_{a1} = \gamma_{1b} = 0$ – we can use the identity in Eq. (4.90) together with Eq. (4.31) to obtain

$$-\beta f = -nd \log(\Lambda/\ell) - \log(\rho\ell^d) + \frac{d(n-1)}{2} \log\left(\frac{2\pi e}{d^2}\right)$$

$$+ \frac{d}{2} \log[2 \det(-\hat{\Delta}/2)(-\underline{1}^T \hat{\Delta}^{-1} \underline{1})] \quad (4.56)$$

$$+ \frac{\overline{\varphi}}{2} \int_{-\infty}^{\infty} dh \, e^h \left\{ e^{-\frac{1}{2} \sum_{a,b=1}^{n} \Delta_{ab} \frac{\partial^2}{\partial h_a \partial h_b}} \left[e^{-\beta \sum_{a=1}^{n} \bar{v}(h_a)} - 1 \right] \right\}_{h_a = h}.$$

Both Eqs. (4.54) and (4.56) show that the molecular liquid free energy does not depend, in the limit $d \to \infty$, on the full shape of the molecular distribution $\pi(\mathbf{u})$ but only on its second moments. A Gaussian ansatz for $\pi(\mathbf{u})$ would thus provide an equivalent result, as discussed in Section 4.6.3. Minimising the free energy over $\pi(\mathbf{u})$ then also becomes equivalent, in the limit $d \to \infty$, to minimising Eq. (4.54) over $\hat{\alpha}$ or, equivalently, Eq. (4.56) over $\hat{\Delta}$.

4.2.4 Replicated Free Energy in Infinite Dimensions

The result obtained in Section 4.2.3 can be used to compute exactly the replicated free energy introduced in Eq. (4.15) and the Franz–Parisi potential, with some minor adaptations. We need to consider $n = s + 1$ replicas with the following properties:

- The first replica has density $\varphi_1 = \varphi_g$ and temperature $T_1 = T_g$, while the others have $\varphi_a = \varphi$ and $T_a = T$. The limit $d \to \infty$, as discussed in Chapter 3, should be taken with a constant scaled packing fraction $\widehat{\varphi} = 2^d \varphi / d$. In the general case, each replica has a different interaction range ℓ_a, with all $\widehat{\varphi}_a$ finite. We then need to scale the ranges as $\ell_a = \ell_g (1 + \eta_a / d)$, in such a way that

$$\widehat{\varphi}_a = \widehat{\varphi}_g (1 + \eta_a / d)^d \to \widehat{\varphi}_g e^{\eta_a}. \tag{4.57}$$

Here $\ell_g = \ell_1$, which corresponds to $\widehat{\varphi}_g = \widehat{\varphi}_1$, is used as reference unit of length. In order to compute the glass free energy, we are eventually interested in setting $\eta_1 = 0$ and $\eta_a = \eta$ for $a \geq 2$.

- The interaction potential of replica a is, following the analysis of Section 2.3.2,

$$v(r_a) = \bar{v} \left[d \left(\frac{r_a}{\ell_a} - 1 \right) \right]$$

$$= \bar{v} \left[d \left(\frac{r_a}{\ell_g (1 + \eta_a / d)} - 1 \right) \right] \xrightarrow[d \to \infty]{} \bar{v}(h_a - \eta_a), \tag{4.58}$$

where $h_a = d(r_a / \ell_g - 1)$ is the scaled interparticle distance for replica a, using ℓ_g as reference unit of length.

- Following the analysis of Section 4.2.3, the mean square displacement between replicas a and b should be scaled as $\Delta_{ab} = (d/\ell_g^2) \langle |\mathbf{u}^a - \mathbf{u}^b|^2 \rangle = (d/\ell_g^2) D_{ab}$, where $D_{ab} = \langle D(\underline{X}^a, \underline{X}^b) \rangle$. The constraint in Eq. (4.14) imposes that $D_{1a} = D_{a1} = D_r$, $\forall a \neq 1$, which implies $\Delta_{1a} = \Delta_{a1} = (d/\ell_g^2) D_r \equiv \Delta_r$. The first row and column of Δ_{ab} are therefore fixed by the constraint. The other matrix elements Δ_{ab}, with $a \neq 1$ and $b \neq 1$, which encode the distance between the replicas different from 1, are left free and, according to the analysis of Section 4.2.3, the free energy should be minimised over them.

To summarise, we can start from Eq. (4.56), with $\ell = \ell_g$ and $n = s+1$, with replica a having temperature β_a and an interaction potential $\bar{v}(h_a - \eta_a)$, which corresponds to $\widehat{\varphi}_a = \widehat{\varphi}_g e^{\eta_a}$, and impose the constraint that $\Delta_{1a} = \Delta_{a1} = \Delta_r$. In Section 4.2.3, it was assumed for simplicity that all replicas have the same $\beta \bar{v}(h_a)$, but the reader can easily check that this assumption was not actually used in the derivation, and one can therefore simply substitute $\beta \bar{v}(h_a) \to \beta_a \bar{v}(h_a - \eta_a)$ in Eq. (4.56). Recall that we omit the factor β in front of the free energy, because the temperatures are now different, as in Eq. (4.15). We thus obtain

$$-f_{s+1}(\widehat{\varphi}_a, T_a, \hat{\Delta}) = -(s+1)d\log(\Lambda/\ell_g) - \log(\rho\ell_g^d) + \frac{d\,s}{2}\log\left(\frac{2\pi e}{d^2}\right)$$

$$+ \frac{d}{2}\log[2\det(-\hat{\Delta}/2)(-\underline{1}^{\mathrm{T}}\hat{\Delta}^{-1}\underline{1})] - \frac{d\widehat{\varphi}_g}{2}\mathcal{F}(\hat{\Delta}), \qquad (4.59)$$

$$\mathcal{F}(\hat{\Delta}) = -\int_{-\infty}^{\infty}dh\,e^h\left\{e^{-\frac{1}{2}\sum_{a,b=1}^n \Delta_{ab}\frac{\partial^2}{\partial h_a\partial h_b}}\left[e^{-\sum_{a=1}^n \beta_a\bar{v}(h_a-\eta_a)} - 1\right]\right\}_{h_a=h},$$

which should be minimised over all matrix elements Δ_{ab} with $a \neq 1$ and $b \neq 1$. Note that computing the restricted thermodynamic glass free energy, according to Eq. (4.10), requires an additional minimisation over Δ_r. The two minimisations can be performed concurrently. In order to compute Eq. (4.10), one thus has to minimise the free energy over the full matrix $\hat{\Delta}$.

The submatrix Δ_{ab} for $a \neq 1$ and $b \neq 1$ is an $s \times s$ matrix. In principle, we should determine it by minimising the free energy for each integer value of s and then perform an analytical continuation to noninteger s in order to take the $s \to 0$ limit and extract the glass free energy according to Eq. (4.15). In general this task is very complex, unless we consider a special prescription for $\hat{\Delta}$ that permits a simple analytical continuation. Such a prescription is discussed in [254] and is at the core of the replica approach to disordered systems. In Section 4.3, we discuss the simplest example of this construction and its physical consequences.

4.3 Replicated Free Energy and Replica Symmetry

Finding explicitly the matrix $\hat{\Delta}$ that minimises the replicated free energy in Eq. (4.59) is greatly simplified by the use of symmetry. Because we consider a system prepared at $(\widehat{\varphi}_g, T_g)$ and then followed to $(\widehat{\varphi}, T)$, we need the replicated free energy in which the first replica is at $(\widehat{\varphi}_g, T_g)$ and all the remaining $s = n - 1$ replicas are at $(\widehat{\varphi}, T)$, as discussed in Section 4.1.3. As a consequence, the s replicas at $(\widehat{\varphi}, T)$ are equivalent, and the free energy must be symmetric under their exchange. This symmetry is manifest in Eq. (4.12), where the s replicas give rise to the s equivalent factors $Z[\varphi, \beta; \underline{Y}, D_r]$. This permutation symmetry of the s replicas $a = 2, \ldots, s + 1$ is called 'replica symmetry' and can be exploited analytically to obtain constraints on observables derived from the replicated free energy [254].

4.3.1 Replica Symmetric Matrices

The permutations of k objects form a group P_k with a generic element acting as $\mathcal{P}(a_1, \ldots, a_k) \to (a_{\mathcal{P}(1)}, \ldots, a_{\mathcal{P}(k)})$. For the molecular liquid, the function $\rho(\underline{x})$

expresses the probability of finding a given molecule (composed of an atom from each of the n replicas) in position $\underline{\mathbf{x}}$. Because of the replica symmetry, the free energy must satisfy

$$F[\rho(\underline{\mathbf{x}})] = F[\rho(\mathcal{P}\underline{\mathbf{x}})], \quad \mathcal{P}\underline{\mathbf{x}} = \{\mathbf{x}_1, \mathbf{x}_{\mathcal{P}(2)}, \ldots, \mathbf{x}_{\mathcal{P}(n)}\}, \quad \mathcal{P} \in P_{n-1}. \tag{4.60}$$

The same symmetry is also present when the free energy is expressed in terms of $\pi(\mathbf{u})$. The mean square displacement matrix, defined in Eq. (4.31) as $\Delta_{ab} \propto \langle |\mathbf{u}_a - \mathbf{u}_b|^2 \rangle$, transforms under the action of a permutation of the replicas as $\Delta_{ab} = \Delta_{\mathcal{P}(a)\mathcal{P}(b)}$. As a consequence, when $\widehat{\varphi}_a = \widehat{\varphi}$ and $T_a = T, \forall a \geq 2$, the replicated free energy in Eq. (4.59) satisfies

$$f_{s+1}(\widehat{\varphi}_a, T_a, \Delta_{ab}) = f_{s+1}(\widehat{\varphi}_a, T_a, \Delta_{\mathcal{P}(a)\mathcal{P}(b)}), \quad \mathcal{P} \in P_{n-1}. \tag{4.61}$$

In other words, the free energy takes the same value on all matrices $\hat{\Delta}$ that are obtained by permuting lines and columns in the block of the s replicas with $a \geq 2$.

This situation is similar to that discussed in Chapter 1 for the homogeneous Ising model in absence of an external magnetic field, in which the free energy is a symmetric function of magnetisation, with $f(m) = f(-m)$. In the paramagnetic phase, in which there is a single equilibrium state, the free energy has a unique minimum that is invariant under the symmetry operation and, therefore, has $m = -m = 0$. Conversely, in presence of many equilibrium states, the symmetry is spontaneously broken, and the free energy has multiple minima that transform one into another under the action of the symmetry group. In the case of the Ising model, only two minima can be present, one for $m = m^*$ and the other for $m = -m^*$.

For the molecular liquid, when there is a single equilibrium state, the free energy must have a unique minimum. This corresponds to a matrix $\hat{\Delta}$ that is invariant under the action of the permutation group, $\Delta_{ab} = \Delta_{\mathcal{P}(a)\mathcal{P}(b)}$. Such a matrix is called 'replica symmetric' (RS) and necessarily has the form:

$$\Delta_{aa} = 0, \qquad\qquad a = 1, \ldots, n,$$
$$\Delta_{1a} = \Delta_{a1} = \Delta_r, \qquad a = 2, \ldots, n, \tag{4.62}$$
$$\Delta_{ab} = \Delta, \qquad\qquad a, b = 2, \ldots, n, \text{ and } a \neq b.$$

For example, when $n = 4$,

$$\hat{\Delta} = \begin{pmatrix} 0 & \Delta_r & \Delta_r & \Delta_r \\ \Delta_r & 0 & \Delta & \Delta \\ \Delta_r & \Delta & 0 & \Delta \\ \Delta_r & \Delta & \Delta & 0 \end{pmatrix}. \tag{4.63}$$

In Section 4.2.3 we also introduced an $s \times s$ matrix for $a, b = 2, \ldots, n = s + 1$:

$$\alpha_{ab} = (d/\ell_g^2) \langle \mathbf{u}_a \cdot \mathbf{u}_b \rangle = (d/\ell_g^2) \langle (\mathbf{x}_a - \mathbf{x}_1) \cdot (\mathbf{x}_b - \mathbf{x}_1) \rangle$$
$$= (\Delta_{a1} + \Delta_{1b} - \Delta_{ab})/2. \tag{4.64}$$

See Section 4.6.1 for a proof of the last equality. The replica symmetric matrix has the form $\alpha_{ab} = \Delta_r \delta_{ab} + (\Delta_r - \Delta/2)(1 - \delta_{ab})$. For example, when $n = 4$,

$$\hat{\alpha} = \begin{pmatrix} \Delta_r & \Delta_r - \Delta/2 & \Delta_r - \Delta/2 \\ \Delta_r - \Delta/2 & \Delta_r & \Delta_r - \Delta/2 \\ \Delta_r - \Delta/2 & \Delta_r - \Delta/2 & \Delta_r \end{pmatrix}. \tag{4.65}$$

In the rest of this chapter, we restrict our analysis to the replica symmetric case. In Chapter 5, we will introduce the notion of spontaneous replica symmetry breaking and discuss its consequences.

4.3.2 Replica Symmetric Free Energy

The replicated free energy can be evaluated explicitly if the matrices $\hat{\Delta}$ and $\hat{\alpha}$ have the replica symmetric form of Eqs. (4.63) and (4.65). The computation is slightly simpler for the expression as a function of $\hat{\alpha}$, Eq. (4.54). Here again, we neglect the kinetic term proportional to $\log \Lambda$ in the free energy, which comes from integrating over the momenta, because it is an irrelevant additive constant. Taking into account the temperature and potential difference for the different replicas – i.e., $\beta_1 = \beta_g$, $\eta_1 = 0$, $\beta_a = \beta$ and $\eta_a = \eta$, $\forall a \geq 2$ – gives

$$-f_{s+1}(\widehat{\varphi}_a, T_a, \hat{\alpha}) = -\log(\rho \ell_g^d) + \frac{ds}{2} \log\left(\frac{2\pi e}{d^2}\right) + \frac{d}{2} \log \det \hat{\alpha} - \frac{d\widehat{\varphi}_g}{2} \mathcal{F}(\hat{\alpha}), \tag{4.66}$$

$$\mathcal{F}(\hat{\alpha}) = -\int_{-\infty}^{\infty} dh \, e^h \left\{ e^{-\beta_g \bar{v}(h)} e^{\sum_{a,b=2}^n \alpha_{ab} \frac{\partial^2}{\partial h_a \partial h_b}} \left[e^{-\beta \sum_{a=2}^n \bar{v}(h_a + \alpha_{aa} - \eta)} \right] \Big|_{h_a = h} - 1 \right\}.$$

In order to obtain the replica symmetric free energy, we first need to compute $\det \hat{\alpha}$ for $\hat{\alpha}$ given in Eq. (4.65). The eigenvectors of $\hat{\alpha}$ are the vector $\underline{1} = \{1, \ldots, 1\}$, with associated eigenvalue $\lambda_1 = s\Delta_r - (s-1)\Delta/2$ and all the $s-1$ vectors orthogonal to $\underline{1}$, with degenerate eigenvalues $\lambda_2 = \Delta/2$. Therefore,

$$\det \hat{\alpha} = \left[s\Delta_r - (s-1)\frac{\Delta}{2} \right] \left(\frac{\Delta}{2}\right)^{s-1}. \tag{4.67}$$

The function $\mathcal{F}(\hat{\alpha})$ that enters in the excess term, for a replica symmetric matrix, gives

$$
\mathcal{F}(\hat{\alpha}) = - \int_{-\infty}^{\infty} dh\, e^h \left\{ e^{-\beta_g \bar{v}(h)} e^{(\Delta_r - \frac{\Delta}{2})\left(\sum_{a=2}^{n} \frac{\partial}{\partial h_a}\right)^2 + \frac{\Delta}{2} \sum_{a=2}^{n} \frac{\partial^2}{\partial h_a^2}} \right.
$$

$$
\left. \times \left[e^{-\beta \sum_{a=2}^{n} \bar{v}(h_a + \Delta_r - \eta)} \right]\Big|_{h_a = h} - 1 \right\}. \tag{4.68}
$$

In the following, we denote the Gaussian convolution of a generic (integrable) function $r(h)$ as

$$
\gamma_\Delta \star r(h) = e^{\frac{\Delta}{2} \frac{d^2}{dh^2}} r(h) = \int_{-\infty}^{\infty} \frac{dz}{\sqrt{2\pi\Delta}} e^{-\frac{z^2}{2\Delta}} r(h - z). \tag{4.69}
$$

See Section 4.6.2 for details. From this definition, it follows that

$$
\mathcal{F}(\hat{\alpha}) = - \int_{-\infty}^{\infty} dh\, e^h \left\{ e^{-\beta_g \bar{v}(h)} e^{(\Delta_r - \frac{\Delta}{2})\left(\sum_{a=2}^{n} \frac{\partial}{\partial h_a}\right)^2} \right.
$$

$$
\left. \times \prod_{a=2}^{n} g_{\text{RS}}(\Delta, \beta; h_a + \Delta_r - \eta)\Big|_{h_a = h} - 1 \right\}, \tag{4.70}
$$

$$
g_{\text{RS}}(\Delta, \beta; h) \equiv \gamma_\Delta \star e^{-\beta \bar{v}(h)}.
$$

Applying the generic identity for a function of k variables $T(x_1, \ldots, x_k)$,

$$
\frac{d}{dt} T(t, \ldots, t) = \sum_{i=1}^{k} \frac{\partial}{\partial x_i} T(x_1, \ldots, x_k)\Big|_{x_i = t}, \tag{4.71}
$$

to Eq. (4.70) gives

$$
\mathcal{F}(\hat{\alpha}) = - \int_{-\infty}^{\infty} dh\, e^h \left\{ e^{-\beta_g \bar{v}(h)} e^{(\Delta_r - \frac{\Delta}{2}) \frac{d^2}{dh^2}} g_{\text{RS}}(\Delta, \beta; h + \Delta_r - \eta)^s - 1 \right\}. \tag{4.72}
$$

Plugging these results into Eq. (4.66), one obtains an explicit expression in terms of s; hence, the analytical continuation to real s is straightforward. Expanding in powers of s up to linear order, one obtains the following expression for the glass free energy:

$$
-\beta f_g(\hat{\varphi}, T | \hat{\varphi}_g, T_g) = \frac{d}{2} \log\left(\frac{\pi e \Delta}{d^2}\right) + \frac{d}{2} \frac{2\Delta_r - \Delta}{\Delta}
$$

$$
+ \frac{d\hat{\varphi}_g}{2} \int_{-\infty}^{\infty} dh\, e^{h - \beta_g \bar{v}(h)} e^{(\Delta_r - \frac{\Delta}{2}) \frac{d^2}{dh^2}} \log g_{\text{RS}}(\Delta, \beta; h + \Delta_r - \eta). \tag{4.73}
$$

The analytical continuation to $s \to 0$ of the replica symmetric free energy can thus be performed explicitly, and the glass free energy be written as a function of

Δ and Δ_r. These last two parameters should be determined by minimising the free energy. It is important to stress, however, that because of the analytical continuation, the convexity property of the free energy in Eq. (4.73) with respect to Δ changes when $s < 1$. The thermodynamic value of Δ is the maximum, and not the minimum, of the free energy. This fact is well known in the context of spin glasses [254] and we will not discuss it further. Because the free energy must then be maximised with respect to Δ and minimised with respect to Δ_r, the thermodynamic values of Δ, Δ_r correspond to a saddle point of Eq. (4.73). The values of Δ, Δ_r should thus be determined by setting to zero the derivatives of f_g with respect to these parameters, and to avoid any confusion, we refer to this process as an 'extremisation' rather than a minimisation of the free energy.

4.3.3 Recipe for State Following within the Replica Symmetric Scheme

In this section, we provide in compact form all the formulae that are needed to compute the thermodynamic properties of glass states prepared at an initial state $(\widehat{\varphi}_g, T_g)$ and followed to a new state $(\widehat{\varphi}, T)$, for a general interaction potential $\bar{v}(h)$. The equations presented here can be used in full generality; examples for specific systems are presented in Section 4.4.

The free energy in Eq. (4.73) can be compactly written as follows. One can first integrate by parts the differential operator in Eq. (4.73) in such a way that it acts on the term at its left, then use the identity in Eq. (4.89) with $r(h) \rightarrow e^{-\beta_g \bar{v}(h)}$ and $\Delta \rightarrow 2\Delta_r - \Delta$ and finally shift the variable $h \rightarrow h - \Delta_r + \Delta/2$. By defining the function

$$q(\Delta, \beta; h) = g_{RS}(\Delta, \beta; h + \Delta/2) = \gamma_\Delta \star e^{-\beta \bar{v}(h + \Delta/2)}, \qquad (4.74)$$

one then obtains

$$
-\beta f_g(\widehat{\varphi}, T | \widehat{\varphi}_g, T_g) = \frac{d}{2} \log\left(\frac{\pi e \Delta}{d^2}\right) + \frac{d}{2} \frac{2\Delta_r - \Delta}{\Delta}
$$

$$
+ \frac{d\widehat{\varphi}_g}{2} \int_{-\infty}^{\infty} dh \, e^h \, q(2\Delta_r - \Delta, \beta_g; h) \log q(\Delta, \beta; h - \eta),
$$

$$(4.75)$$

which is the final expression of the replica symmetric free energy of the glass state with $\eta = \log(\widehat{\varphi}/\widehat{\varphi}_g)$. As discussed at the end of Section 4.3.2, Δ and Δ_r should be determined by extremisation of Eq. (4.75), which results in the two coupled equations:

$$
2\Delta_r = \Delta + \widehat{\varphi}_g \Delta^2 \int_{-\infty}^{\infty} dh \, e^h \frac{\partial}{\partial \Delta} \left[q(2\Delta_r - \Delta, \beta_g; h) \log q(\Delta, \beta; h - \eta) \right],
$$

$$(4.76)$$

$$
\frac{2}{\Delta} = -\widehat{\varphi}_g \int_{-\infty}^{\infty} dh \, e^h \left[\frac{\partial}{\partial \Delta_r} q(2\Delta_r - \Delta, \beta_g; h) \right] \log q(\Delta, \beta; h - \eta).
$$

In practice, Eqs. (4.76) can be solved efficiently by iteration. Start from a reasonable guess for Δ and Δ_r, and compute numerically the right-hand side to obtain new estimates of Δ and Δ_r. Upon iteration, the new estimates typically get closer to the previous ones; hence, the procedure converges. The derivatives of $q(\Delta, \beta; h)$ with respect to Δ are not given here explicitly but can be deduced easily from Eq. (4.74).

Once the equations for Δ and Δ_r are solved, plugging the results in Eq. (4.75) gives the thermodynamic free energy of the glass. Note that $(\widehat{\varphi}_g, T_g)$ specify the preparation of the glass, through the reference configuration $\underline{Y}$ that acts as a quenched disorder, as discussed in Section 4.1, while $(\widehat{\varphi}, T)$ control the thermodynamic state of the system. By taking derivatives of the free energy with respect to $(\widehat{\varphi}, T)$, one can thus obtain the averages of the thermodynamic observables.[12] For example, the average glass energy is the derivative of the free energy with respect to the inverse temperature,

$$e_g = \frac{\partial(\beta f_g)}{\partial \beta} = -\frac{d\widehat{\varphi}_g}{2} \int_{-\infty}^{\infty} dh \, e^h q(2\Delta_r - \Delta, \beta_g; h) \frac{\partial}{\partial \beta} \log q(\Delta, \beta; h - \eta),$$

(4.77)

and the entropy is $s_g = -\beta(f_g - e_g)$. The pressure can be obtained from Eq. (2.7); observing that $\rho \propto \widehat{\varphi}$, and recalling that $\eta = \log(\widehat{\varphi}/\widehat{\varphi}_g)$, the 'reduced pressure' or 'compressibility factor' is given by

$$p = \frac{\beta P}{\rho} = \rho \frac{\partial(\beta f)}{\partial \rho} = \widehat{\varphi} \frac{\partial(\beta f)}{\partial \widehat{\varphi}} = \frac{\partial(\beta f)}{\partial \eta}.$$

(4.78)

For the glass, one therefore obtains

$$p_g = \frac{\partial(\beta f_g)}{\partial \eta} = -\frac{d\widehat{\varphi}_g}{2} \int_{-\infty}^{\infty} dh \, e^h q(2\Delta_r - \Delta, \beta_g; h) \frac{\partial}{\partial \eta} \log q(\Delta, \beta; h - \eta).$$

(4.79)

The case in which a glass state is both prepared and studied at the same state point – i.e., $(\widehat{\varphi}, T) = (\widehat{\varphi}_g, T_g)$ – is special and displays an additional symmetry. Replica 1 is then equivalent to all the others and $\Delta = \Delta_r$, as discussed in

---

[12] When taking derivatives of the free energy with respect to a thermodynamic control parameter, one should keep in mind the following structure. The free energy depends on the state point $(\widehat{\varphi}, T)$ but also on Δ, Δ_r which are determined by setting the derivatives of the free energy to zero and, therefore, depend implicitly on $(\widehat{\varphi}, T)$. When taking the first derivatives, one has, for example,

$$\frac{df_g}{dT} = \frac{\partial f_g}{\partial T} + \frac{\partial f_g}{\partial \Delta} \frac{\partial \Delta}{\partial T} + \frac{\partial f_g}{\partial \Delta_r} \frac{\partial \Delta_r}{\partial T}.$$

Because the derivatives with respect to Δ, Δ_r are set to zero on the thermodynamic values of these parameters, only the first term remains and $\frac{df_g}{dT} = \frac{\partial f_g}{\partial T}$, which considerably simplifies the calculation. Note, however, that this is only true for first derivatives. When taking second derivatives (e.g., to compute the specific heat), the derivatives of Δ, Δ_r with respect to $(\widehat{\varphi}, T)$ appear explicitly.

Section 4.1. For $(\widehat{\varphi}, T) = (\widehat{\varphi}_g, T_g)$, one can indeed check that choosing $\Delta = \Delta_r$, both Eqs. (4.76) reduce to the same equation for Δ_r, which can be conveniently written as

$$\frac{1}{\widehat{\varphi}_g} = \mathcal{F}_1(\Delta_r; \beta_g) = -\Delta_r \int_{-\infty}^{\infty} dh\, e^h \log[q(\Delta_r, \beta_g; h)] \frac{\partial q(\Delta_r, \beta_g; h)}{\partial \Delta_r}. \tag{4.80}$$

Eq. (4.80) coincides with Eq. (3.65), derived in Chapter 3. This key result proves that the hypothesis that the state following construction can reproduce the long time limit of dynamical correlations in the arrested phase is indeed correct. Furthermore, for $(\widehat{\varphi}, T) = (\widehat{\varphi}_g, T_g)$ and $\Delta = \Delta_r$ the glass reduced pressure in Eq. (4.79) simplifies to

$$\begin{aligned}
p_g &= -\frac{d\widehat{\varphi}}{2} \int_{-\infty}^{\infty} dh\, e^h q(\Delta, \beta; h) \left. \frac{\partial}{\partial \eta} \log q(\Delta, \beta; h - \eta) \right|_{\eta=0} \\
&= -\frac{d\widehat{\varphi}}{2} \int_{-\infty}^{\infty} dh\, e^{h - \beta \bar{v}(h)} \beta \bar{v}'(h),
\end{aligned} \tag{4.81}$$

which coincides with the reduced pressure in the liquid phase given in Eq. (2.78). The pressure is thus continuous at the glass transition; the same also holds for the internal energy.

Examples of the Franz–Parisi potential, Eq. (4.75), are given in Figure 4.3. The corresponding function $\mathcal{F}_1(\Delta_r)$, defined in Eq. (4.80), is illustrated in Figure 3.2. The solution of Eq. (4.80), which can be easily found numerically, can be used as initial condition for the iteration of Eqs. (4.76) while one changes slowly $(\widehat{\varphi}, T)$ to follow the evolution of the state [299]. Note that there are usually at least two solutions of Eq. (4.80), and the correct solution corresponds to a local minimum of the Franz–Parisi potential f_g; in the example of Figure 4.3, this is the solution with smaller value of Δ_r.

As discussed in Section 3.3.2, one can show that the function $\mathcal{F}_1(\Delta_r; \beta_g)$ generally vanishes both for $\Delta_r \to 0$ and $\Delta_r \to \infty$ and has a maximum in between. Therefore, no solution to Eq. (4.80) can be found if $\widehat{\varphi}_g < \widehat{\varphi}_d(\beta_g)$, where

$$\frac{1}{\widehat{\varphi}_d(\beta_g)} = \max_{\Delta_r} \mathcal{F}_1(\Delta_r; \beta_g). \tag{4.82}$$

The absence of a solution means that the Franz–Parisi potential has no local minimum at finite Δ_r. No stable glass phase then exists, and the system is therefore a liquid. This condition allows one to determine the dynamical transition density $\widehat{\varphi}_d(\beta_g)$ for each temperature, consistently with the dynamical treatment of Chapter 3. For $\widehat{\varphi}_g > \widehat{\varphi}_d(\beta_g)$, stable glass states exist and can be followed at different state points.

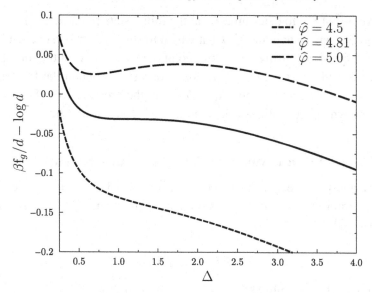

Figure 4.3 Franz–Parisi potential Eq. (4.75), multiplied by β and scaled to have a finite limit in $d \to \infty$, for hard spheres in equilibrium, corresponding to $\widehat{\varphi} = \widehat{\varphi}_g$ and $\Delta_r = \Delta$, at various densities, below, close to, and above the dynamical transition at $\widehat{\varphi}_d = 4.8067\ldots$ The corresponding function $\mathcal{F}_1(\Delta_r)$, defined by Eq. (4.80), is illustrated in Figure 3.2.

In addition, the Franz–Parisi potential develops an inflection point at the dynamical transition (Figure 4.3). It can be shown that the presence of this inflection point gives rise to the dynamical criticality discussed in Section 3.4.2 [146, 204]. A development of the potential around its inflection point then gives access to the dynamical critical exponents of mode-coupling theory [72, 288] and to the critical properties of the four-point dynamical correlations [37, 145], which have been discussed in Section 3.4.2.

Note that in the dynamically arrested phase, the Franz–Parisi potential is not convex, as shown in Figure 4.3. This non-convexity is a necessary condition to have a local minimum at finite Δ_r, and it is possible because, in the calculation of the potential, Δ_r was assumed to be spatially uniform. The Franz–Parisi potential is then a 'potential' similar to the function $v(m)$ discussed in Section 1.4. In the non-convex region, phase coexistence would lead to a non-uniform spatial profile $\Delta_r(\mathbf{x})$ and restore the convexity of the free energy [73, 257]. We do not further discuss this issue in this book; see Section 4.5.2 for additional references.

To summarise, the recipe to study the glass states of a given potential $\bar{v}(h)$ is the following [299]:

1. Solve Eq. (4.82) to determine the dynamical transition line $\widehat{\varphi}_d(\beta_g)$.
2. Choose an initial state $(\widehat{\varphi}_g, T_g)$ in the dynamically arrested region $\widehat{\varphi}_g > \widehat{\varphi}_d(\beta_g)$.

3. Solve Eq. (4.80) to determine Δ_r at the initial state point $(\widehat{\varphi}_g, T_g)$.
4. Choose a new state point $(\widehat{\varphi}, T)$ at which to study the glass state, and iteratively solve Eqs. (4.76), using as starting point the value of Δ_r at the initial state.
5. Once Δ and Δ_r at $(\widehat{\varphi}, T)$ are determined, one can obtain the free energy from Eq. (4.75), the energy from Eq. (4.77) and the pressure from Eq. (4.79). Other thermodynamic quantities can be computed along similar lines.

4.4 Replica Symmetric Phase Diagram of Simple Glasses

We now apply the recipe discussed in Section 4.3.3 to compute and discuss the phase diagram of two simple glass models: hard spheres [299] and soft repulsive spheres [315].

4.4.1 Hard Spheres

Hard spheres are the simplest model of glass forming liquid. The scaled infinite-dimensional potential $\bar{v}(h)$ is deduced from Eq. (2.47). It is infinite for $h < 0$ and vanishes for $h > 0$, and thus, the Mayer function $\overline{f}(h) = e^{-\beta \bar{v}(h)} - 1 = -\theta(-h)$. Because the potential has no energy scale (it is either zero or infinite), the Mayer function does not depend on β. As a consequence, the only energy scale in the problem is the temperature, which can thus be eliminated by expressing the thermodynamic quantities in units of T. The only remaining control parameter is thus the scaled packing fraction $\widehat{\varphi} = 2^d \varphi / d$. Conveniently, an explicit expression for the function $q(\Delta, \beta; h)$ (which is also independent of β) can be derived:

$$
q_{\mathrm{HS}}(\Delta; h) = \int_{-\infty}^{\infty} \frac{dz}{\sqrt{2\pi\Delta}} e^{-\frac{z^2}{2\Delta}} \theta(h - z + \Delta/2) = \Theta\left(\frac{h + \Delta/2}{\sqrt{2\Delta}}\right),
$$

$$
\Theta(x) = \frac{1}{2}(1 + \mathrm{erf}(x)) = \frac{1}{2}\mathrm{erfc}(-x).
$$
(4.83)

The recipe of Section 4.3.3 can be applied using this analytical expression, but the rest of the calculation has to be performed numerically [299]. The resulting phase diagram is reported in Figure 4.4.

According to the analysis of Chapter 2, the liquid equation of state is $p = 1 + d\widehat{\varphi}/2$. The glass region corresponds to finite values of $\widehat{\varphi}$ when $d \to \infty$, and in that case, the liquid reduced pressure is $p \sim d\widehat{\varphi}/2$; one can therefore define a scaled pressure $\widehat{p} = p/d$, which in the liquid phase is given by $\widehat{p} = \widehat{\varphi}/2$. As a first step, one has to determine the dynamical transition point. Because $q_{\mathrm{HS}}(\Delta; h)$ does not depend on β, neither does $\mathcal{F}_1(\Delta_r)$, whose maximum (Figure 3.2) defines $\widehat{\varphi}_d = 4.8067\ldots$ according to Eq. (4.82). The corresponding pressure is $\widehat{p}_d = \widehat{\varphi}_d/2 = 2.4033\ldots$ If $\widehat{\varphi} < \widehat{\varphi}_d$, the liquid phase is ergodic and the corresponding dynamics is diffusive, as discussed in Chapter 3.

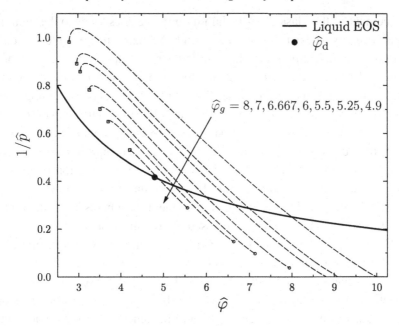

Figure 4.4 Following hard-sphere glasses in compression and decompression, within the replica symmetric ansatz [299]. The inverse of the scaled reduced pressure $\widehat{p} = p/d$ is plotted versus the scaled packing fraction $\widehat{\varphi} = 2^d \varphi/d$. The liquid equation of state (full line) is $\widehat{p} = \widehat{\varphi}/2$. The dynamical transition $\widehat{\varphi}_d$ is marked by a full circle. For $\widehat{\varphi}_g > \widehat{\varphi}_d$, the liquid is a collection of glasses. The equations of state of glasses for different $\widehat{\varphi}_g$ (dashed lines) intersect the liquid equation of state at $\widehat{\varphi}_g$. Upon compression, for low $\widehat{\varphi}_g$ they end at an unphysical spinodal point (open circle), while at high $\widehat{\varphi}_g$ they end at a jamming point at which pressure diverges. Upon decompression, the glass pressure falls below that of the liquid, and reaches a minimum before growing again until a physical spinodal point (open square) at which it melts into the liquid.

If $\widehat{\varphi} > \widehat{\varphi}_d$, the liquid dynamics is arrested. Glass states appear and each equilibrium liquid configuration at $\widehat{\varphi}_g > \widehat{\varphi}_d$ selects a glass. For each choice of $\widehat{\varphi}_g$, we can follow the recipe of Section 4.3.3 to solve the equations for Δ, Δ_r and compute the pressure of the corresponding glass. Recall that the energy is always zero for hard-sphere systems. In Figure 4.4, the equations of state of several glasses, corresponding to different choices of $\widehat{\varphi}_g$, are plotted. The pressure of the glass coincides with that of the liquid at $\widehat{\varphi}_g$, implying that the pressure is continuous for each glass, but the slope of the glass equation of state at $\widehat{\varphi}_g$ is different from that of the liquid equation of state. When the system becomes confined in the glass state selected at $\widehat{\varphi}_g$, the compressibility thus has a jump. Following glasses in compression, the pressure increases faster than in the liquid – i.e., the compressibility is smaller.

For larger values of $\widehat{\varphi}_g$, it is found that, upon compression, the pressure increases and eventually diverges at a finite 'jamming' density $\widehat{\varphi}_j(\widehat{\varphi}_g)$, where the mean square displacement of the glass $\Delta \to 0$. Jamming thus corresponds to the point at which the glass ceases to exist; beyond that point, the hard-sphere exclusion constraints cannot be satisfied while remaining within the same glass state. Because pressure is infinite and $\Delta = 0$, at jamming spheres touch their first neighbours. Jamming thus also corresponds to the 'close-packed' limit of a given glass. Jamming points have several interesting properties that will be carefully studied in Chapter 9. The glasses prepared at lower $\widehat{\varphi}_g$, by contrast, show a different behaviour upon compression. Above some density $\widehat{\varphi}_{\rm sp}^+(\widehat{\varphi}_g)$, the solution for Δ, Δ_r disappears, and the glass state cannot be followed anymore. The disappearance of the solution happens through a bifurcation as in a spinodal point. This spinodal point is, however, unphysical because it is reasonable to expect that once the hard spheres are confined to a solid phase, this solid phase can be compressed up to infinite pressure without becoming unstable. Compression should only stabilise the solid. We will confirm in Chapter 6 that this spinodal point is, in fact, an artefact of the replica symmetric ansatz. It disappears because replica symmetry is broken in that region.

Finally, a glass prepared at $\widehat{\varphi}_g$ can also be followed in decompression – i.e., for $\widehat{\varphi} < \widehat{\varphi}_g$. In this case, the glass pressure becomes lower than that of the liquid until, upon decreasing density, a spinodal point $\widehat{\varphi}_{\rm sp}(\widehat{\varphi}_g)$ is reached, at which the solution for Δ, Δ_r disappears through a bifurcation. This spinodal point is physical and corresponds to the glass becoming unstable and thus melting into the liquid. At the spinodal point, due to its bifurcation nature, all thermodynamic quantities display a square-root singularity – e.g., $\Delta \sim \sqrt{\widehat{\varphi}_{\rm sp} - \widehat{\varphi}}$, and similarly for Δ_r and the pressure. Note also that the region of densities between the pressure minimum and the spinodal is thermodynamically unstable because the compressibility is then negative. Such an unstable region is an artefact of the mean field treatment.

4.4.2 Soft Spheres

We now discuss the simplest example of soft-sphere potential [315]: the pure power-law repulsive potential, $v(r) = \varepsilon(\ell/r)^{vd}$, which corresponds to a scaled potential $\bar{v}(h) = \varepsilon e^{-vh}$ (Section 2.3.2). The specific choice $v = 4$ corresponds to the repulsive part of a Lennard-Jones potential in $d = 3$, generalised to arbitrary dimension. In Section 2.3.2, it was shown that the thermodynamics of this system is controlled by a single parameter $\Gamma = \rho/T^{1/4}$; hence, temperature and density are not independent. We can choose an arbitrary value of density and study the phase diagram using temperature as a control parameter. The liquid equation of state has also been studied in Section 2.3.2.

In $d \to \infty$, it is convenient to use $\widehat{\Gamma} = \widehat{\varphi}(\varepsilon/T)^{1/4}$ as control parameter. The dynamical transition point is $\widehat{\Gamma}_{\rm d} = 4.304\ldots$ It is then convenient to work at fixed density $\widehat{\varphi} = \widehat{\Gamma}_{\rm d}$, such that $T_{\rm d}/\varepsilon = 1$. The results are reported in Figure 4.5. The

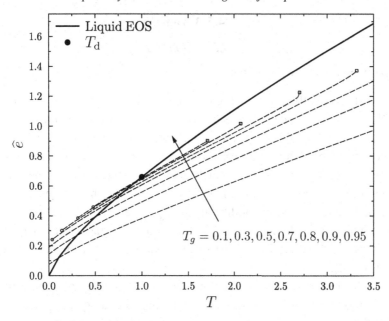

Figure 4.5 Following soft-sphere glasses with potential $\bar{v}(h) = \varepsilon e^{-4h}$, within the replica symmetric ansatz, at fixed density $\widehat{\varphi} = 4.304\ldots$ [315]. The evolution of the reduced energy $\widehat{e} = e/d$ with temperature T under heating and cooling is reported, both quantities being expressed in units of the interaction strength ε. On the liquid equation of state (solid line), dynamical arrest happens at the dynamical transition $T_d = 1$ (solid circle). For $T_g < T_d$, the liquid is a collection of glasses. The glass equations of state are reported for several choices of T_g (dashed lines) and intersect the liquid equation of state at T_g. Upon cooling, for high T_g, they end at an unphysical spinodal point (open circle), while for low T_g, they can be continued down to zero temperature. Upon heating, the glass energy falls below that of the liquid, until a physical spinodal point (open square) is reached, at which the glass melts into the liquid.

phase diagram is very similar to that of hard spheres, identifying temperature with inverse pressure and inverse density with energy. For $T > T_d$, the liquid dynamics is diffusive, while for $T < T_d$, it is arrested. Equilibrium liquid configurations at $T_g < T_d$ select a glass state that can be followed to both lower and higher temperatures. The specific heat (i.e., the derivative of the energy with respect to temperature) has a discontinuous jump at T_g, the specific heat of the liquid being larger than that of the glass. Upon cooling, the glasses with lower T_g can be followed down to $T = 0$, while glasses with higher T_g display an unphysical spinodal before $T = 0$ can be reached. Upon heating, the glass energy remains lower than that of the liquid, until a physical spinodal point $T_{sp}(T_g)$ is reached, at which the glass melts into the liquid.

4.5 Wrap-Up

4.5.1 Summary

In this chapter, we have seen that

- In the dynamically arrested region, equilibrium configurations remain stuck inside a glass state. The averages of dynamical observables, in the long time limit, reach a stationary value (Section 4.1.1). These long-time dynamical averages can be computed by defining a probability measure over phase space, restricted to a glass state. This procedure, which goes under the name of the Franz–Parisi construction (Section 4.1.2), allows one to compute the free energy of a glass state that was prepared in equilibrium at $(\widehat{\varphi}_g, T_g)$ and then followed to $(\widehat{\varphi}, T)$.

- The computation of the Franz–Parisi potential (or glass free energy) requires the introduction of replicas of the system to average over the quenched disorder (Section 4.1.3).

- The replicated free energy corresponds to the free energy of a molecular liquid. The methods of Chapter 2 can also be generalised to a liquid made of molecules, if the limit $d \to \infty$ is taken by imposing a strong localisation of the molecules, expressed by Eq. (4.29) with finite $\hat{\alpha}$ (Section 4.2.1). The order parameter is the molecular density $\pi(\mathbf{u})$ and the resulting free energy is the sum of an ideal (molecular) gas term plus a two-molecule interaction, Eq. (4.20).

- When $d \to \infty$, under the same hypothesis of strong molecule localisation, the dependence of the molecular free energy Eq. (4.20) on the molecular distribution $\pi(\mathbf{u})$ is greatly simplified. In fact, the free energy can be expressed as a function of the matrix $\alpha_{ab} \propto \langle \mathbf{u}^a \cdot \mathbf{u}^b \rangle$ or $\Delta_{ab} \propto \langle |\mathbf{u}^a - \mathbf{u}^b|^2 \rangle$ only, resulting in Eq. (4.54) or Eq. (4.56), respectively (Sections 4.2.2 and 4.2.3). The matrices $\hat{\alpha}$ or $\hat{\Delta}$ are then determined by minimising the free energy. The mean square displacement matrix $\hat{\Delta}$ corresponds to the dynamical mean square displacement in the long time regime.

- The number of replicas s should be sent to zero to extract the glass free energy from the replicated free energy. This requires an analytical continuation. To simplify the procedure, the symmetry of the free energy under permutations of the replicas – i.e., replica symmetry – is used (Section 4.3.1). Assuming that the replica symmetry is not broken, the matrix $\hat{\Delta}$ depends only on two parameters, Δ and Δ_r, which have a simple physical meaning. The expression of the glass free energy simplifies considerably, and the analytic continuation to $s \to 0$ can be performed explicitly (Section 4.3.2). Minimisation of the free energy becomes an extremisation in the $s \to 0$ limit, so the analytically continued free energy should then be extremised with respect to Δ and Δ_r.

- The main result of this chapter is the final recipe to compute the thermodynamical properties of the glass states of any potential $\bar{v}(h)$, under the assumption of unbroken replica symmetry (Section 4.3.3).

- Results are given in Section 4.4 for hard spheres and soft repulsive spheres. The phase diagrams, with the liquid and glass equations of state, are computed in both cases.

4.5.2 Further Reading

We provide here a list of references that can be consulted to further explore the subjects discussed in this chapter, selected according to the criteria discussed in Section 1.6.2.

The state following construction was applied to structural glasses in $d = 3$ shortly after it was formulated. The replica-HNC approximation used in early papers gives good results close to the equilibrium liquid line but unfortunately fails when glasses are brought to high densities or low temperatures. Better approximation schemes for replicated liquids in finite dimensions have thus been subsequently developed. The relevant papers include

- Mézard and Parisi, *A tentative replica study of the glass transition* [252]
- Cardenas, Franz and Parisi, *Constrained Boltzmann–Gibbs measures and effective potential for glasses in hypernetted chain approximation and numerical simulations* [77]
- Mézard and Parisi, *Glasses and replicas* [253]
- Parisi and Zamponi, *Mean-field theory of hard-sphere glasses and jamming* [292]

In $d \to \infty$, the method has been applied to variations of the square-well potential discussed in Section 2.3.2 that model attractive or patchy colloids. For these models, the function $\mathcal{F}_1(\Delta; \beta)$ defined in Eq. (3.65) can have multiple maxima, which leads to the Franz–Parisi potential having multiple minima that describe possible values of the plateau Δ of the mean square displacement. Details can be found in

- Sellitto and Zamponi, *A thermodynamic description of colloidal glasses* [325]
- Maimbourg, Sellitto, Semerjian et al., *Generating dense packings of hard spheres by soft interaction design* [240]
- Yoshino, *Translational and orientational glass transitions in the large-dimensional limit: A generalized replicated liquid theory and an application to patchy colloids* [367]

Similar phenomena have been previously found within MCT, where the coexistence of multiple glass solutions leads to non-trivial dynamical effects. An introductory review can be found in Sciortino and Tartaglia, *Glassy colloidal systems* [323].

The state following method has also been used in the context of random optimisation problems, of interest for computer science. The approach then provides various insights into the structure of their energy landscape and the performances of algorithms trying to find good minima. A selection of relevant references includes

- Krzakala and Kurchan, *Landscape analysis of constraint satisfaction problems* [215]
- Zdeborová and Krzakala, *Generalization of the cavity method for adiabatic evolution of Gibbs states* [371]
- Krzakala and Zdeborová, *Performance of simulated annealing in p-spin glasses* [216]

In the context of particle glasses in finite dimensions, as mentioned in Section 3.5.2, the dynamical transition is not sharp. The Franz–Parisi construction described in this chapter provides a simple way of understanding this observation. In finite dimensions, the Franz–Parisi potential must be convex. As discussed in Chapter 1, a non-convex region indicates phase coexistence, which is realised via nucleation of the stable liquid phase into the metastable glass phase. In mean field, the glass state corresponds to a local minimum of the potential, and its lifetime is infinite, while in finite dimensions, it is not. Numerical measurements of the Franz–Parisi potential support this scenario. A selection of relevant references is

- Mézard and Parisi, *Statistical physics of structural glasses* [257]
- Parisi, *Glasses, replicas and all that* [286]
- Cammarota, Cavagna, Giardina, et al., *Phase-separation perspective on dynamic heterogeneities in glass-forming liquids* [73]
- Berthier, *Overlap fluctuations in glass-forming liquids* [39]

In addition to the dynamical processes mentioned in Section 3.5.2 that destroy the sharp dynamical transition observed in mean field, the restricted thermodynamic analysis developed in this chapter suggests that nucleation could help escape from metastable glass states. The identification of this process initiated the development of the random first-order transition (RFOT) theory of the glass transition beyond the mean field limit. Details on the nucleation processes within the RFOT approach can be found in

- Kirkpatrick, Thirumalai and Wolynes, *Scaling concepts for the dynamics of viscous liquids near an ideal glassy state* [208]
- Cavagna, *Supercooled liquids for pedestrians* [80]
- Berthier and Biroli, *Theoretical perspective on the glass transition and amorphous materials* [40]
- Wolynes and Lubchenko (eds), *Structural glasses and supercooled liquids: Theory, experiment, and applications* [357]

Thanks to the various escape processes mentioned previously, finite-dimensional systems can leave the metastable glass minimum of the Franz–Parisi potential. The liquid phase can thus be equilibrated even for $T_g < T_d(\rho_g)$. Because of the different microscopic time scales, this process is most efficient in structural glasses, while it is much less efficient in colloids, emulsions and grains.

Recent algorithmic and experimental developments further accelerate this sampling to efficiently achieve equilibration for $T_g < T_d(\rho_g)$. In numerical simulations, 'swap' algorithms can be used to speed up the equilibration process, while in experiments, vapour deposition techniques achieve the same goal. Examples of these recent developments can be found in

- Swallen, Kearns, Mapes, et al., *Organic glasses with exceptional thermodynamic and kinetic stability* [338]
- Ninarello, Berthier and Coslovich, *Models and algorithms for the next generation of glass transition studies* [273]
- Fullerton and Berthier, *Density controls the kinetic stability of ultrastable glasses* [157]

These methods provide very well equilibrated liquid configurations, corresponding to glass basins that lie very deep in the free energy landscape. One can thus use these starting configurations to perform simulations or experiments on time scales much shorter than the relaxation time of the liquid, in such a way that the system is effectively confined into a glass basin. The equation of state of thermal glasses (Figure 4.5) can then be directly compared with the results obtained experimentally in [338, figure 1B]. The sharp peak in the specific heat reported in [338, figure 1A] during heating indeed mirrors the divergence of the slope of $\widehat{e}(T)$ at the melting spinodal in Figure 4.5. The equation of state of hard-sphere glasses (Figure 4.4) can be compared with the numerical results of [157, figure 1]; the mean field melting spinodal is hidden, in finite dimensions, by nucleation processes.

4.6 Appendix

4.6.1 Determinant of the Scalar Product Matrix

We consider an $n \times n$ symmetric matrix $\hat{\gamma}$ that satisfies $\gamma_{1a} = \gamma_{a1} = 0$ for all $a = 1, \ldots, n$, and a second symmetric matrix $\hat{\Delta}$ defined by the relation $\Delta_{ab} = \gamma_{aa} + \gamma_{bb} - 2\gamma_{ab}$. Note that this definition implies $\Delta_{aa} = 0$, $\forall a$. We want to express the determinant of the $(1, 1)$ cofactor $\hat{\gamma}^{(1,1)} = \hat{\alpha}$ – i.e., the $(n-1) \times (n-1)$ matrix obtained by removing from $\hat{\gamma}$ the first (vanishing) line and column, as a function of $\hat{\Delta}$. This determinant can be computed using different techniques, and we present here one method that is particularly simple.

As a first step, it is convenient to express $\hat{\gamma}$ as a function of $\hat{\Delta}$. From the definition of $\hat{\Delta}$, one has $\Delta_{a1} = \gamma_{aa}$, and therefore, $\gamma_{ab} = (\Delta_{a1} + \Delta_{1b} - \Delta_{ab})/2$, which holds for all $a, b = 1, \ldots, n$. Next, we consider a Gaussian integral representation of $\det \gamma^{(1,1)}$:

$$\frac{1}{\sqrt{\det(\hat{\gamma}^{(1,1)})}} = \int \frac{\mathrm{d}y_2 \cdots \mathrm{d}y_n}{\sqrt{(2\pi)^{n-1}}} e^{-\frac{1}{2}\sum_{a,b=2}^{n} \alpha_{ab} y_a y_b}$$

$$= \int \frac{\mathrm{d}y_2 \cdots \mathrm{d}y_n}{\sqrt{(2\pi)^{n-1}}} e^{-\frac{1}{4}\sum_{a,b=2}^{n}(\Delta_{a1} + \Delta_{1b} - \Delta_{ab}) y_a y_b}. \tag{4.84}$$

One can then add to the Gaussian integration an additional variable $y_1 = -\sum_{a=2}^{n} y_a$ and express the argument in the exponential by means of this additional variable (recall that $\Delta_{11} = 0$) by the identity

$$\frac{1}{\sqrt{\det(\hat{\gamma}^{(1,1)})}} = \int \frac{\mathrm{d}y_1 \cdots \mathrm{d}y_n}{\sqrt{(2\pi)^{n-1}}} \delta\left(\sum_{a=1}^{n} y_a\right) e^{\frac{1}{4}\sum_{a,b=1}^{n} \Delta_{ab} y_a y_b}$$

$$= \int \frac{\mathrm{d}\lambda}{2\pi} \int \frac{\mathrm{d}y_1 \cdots \mathrm{d}y_n}{\sqrt{(2\pi)^{n-1}}} e^{i\lambda \sum_{a=1}^{n} y_a + \frac{1}{4}\sum_{a,b=1}^{n} \Delta_{ab} y_a y_b}$$

$$= \frac{1}{\sqrt{\det(-\hat{\Delta}/2)}} \int \frac{\mathrm{d}\lambda}{\sqrt{2\pi}} e^{\lambda^2 \sum_{ab}(\hat{\Delta}^{-1})_{ab}} \tag{4.85}$$

$$= \frac{1}{\sqrt{2\det(-\hat{\Delta}/2)(-\underline{1}^T \hat{\Delta}^{-1} \underline{1})}},$$

where $\underline{1} = \{1, \ldots, 1\}$ is the vector of all ones. We therefore obtain

$$\det \hat{\alpha} = 2\det(-\hat{\Delta}/2)(-\underline{1}^T \hat{\Delta}^{-1} \underline{1}). \tag{4.86}$$

4.6.2 Gaussian Convolutions

We collect here some useful identities involving Gaussian integration and convolution that are used throughout this chapter. We begin by defining the Gaussian convolution as in Eq. (4.69),

$$\gamma_\Delta \star r(h) = e^{\frac{\Delta}{2}\frac{\mathrm{d}^2}{\mathrm{d}h^2}} r(h) = \int_{-\infty}^{\infty} \frac{\mathrm{d}z}{\sqrt{2\pi\Delta}} e^{-\frac{z^2}{2\Delta}} r(h-z), \tag{4.87}$$

where $r(h)$ is such that the expressions in Eq. (4.87) are well defined and otherwise arbitrary. The identity of the second and third expressions in Eq. (4.87) can be proven by expanding $e^{\frac{\Delta}{2}\frac{\mathrm{d}^2}{\mathrm{d}h^2}}$ in powers of Δ in the second expression and expanding $r(h-z)$ in powers of z in the third.

Eq. (4.87) can be generalised to the multidimensional case as

$$e^{\frac{1}{2}\sum_{a,b=1}^{k} M_{ab}\frac{\partial^2}{\partial h_a \partial h_b}} \prod_{a=1}^{k} r(h_a) = \int_{-\infty}^{\infty} \frac{\prod_{a=1}^{k} \mathrm{d}z_a}{\sqrt{(2\pi)^k \det \hat{M}}} e^{-\frac{1}{2}\sum_{a,b=1}^{k} z_a \hat{M}_{ab}^{-1} z_b} \prod_{a=1}^{k} r(h_a - z_a), \tag{4.88}$$

where $\hat{M}$ is a $k \times k$ symmetric and invertible matrix, and $r(h)$ is once again such that the expressions are well defined and otherwise arbitrary. The proof of Eq. (4.88) can be obtained similarly to Eq. (4.87), by expanding the exponential of the differential operator in a power series of $\hat{M}$ in the left-hand side of the identity and expanding $r(h_a - z_a)$ in a power series of z_a in its right-hand side.

Another useful identity is

$$
e^{\frac{\Delta}{2}\frac{d^2}{dh^2}}[e^h r(h)] = e^{h + \frac{\Delta}{2}} e^{\frac{\Delta}{2}\frac{d^2}{dh^2}} r(h + \Delta) , \tag{4.89}
$$

which can be proven either by series expansion in Δ or by an appropriate change of variable in the integral representation in Eq. (4.87).

Finally, by defining a matrix $\hat{\gamma}$ as in Section 4.6.1, we obtain the identity

$$
\int_{-\infty}^{\infty} dh\, e^h \left\{ e^{-\beta \bar{v}(h)} e^{\sum_{a,b=2}^n \alpha_{ab} \frac{\partial^2}{\partial h_a \partial h_b}} \left[e^{-\beta \sum_{a=2}^n \bar{v}(h_a + \alpha_{aa})} \right] \Big|_{h_a = h} - 1 \right\}
$$

$$
= \int_{-\infty}^{\infty} dh\, e^h \left\{ e^{\sum_{a,b=1}^n \gamma_{ab} \frac{\partial^2}{\partial h_a \partial h_b} + \sum_{a=1}^n \gamma_{aa} \frac{\partial}{\partial h_a}} \left[e^{-\beta \sum_{a=1}^n \bar{v}(h_a)} - 1 \right] \right\}_{h_a = h} \tag{4.90}
$$

$$
= \int_{-\infty}^{\infty} dh\, e^h \left\{ e^{-\frac{1}{2}\sum_{a,b=1}^n (\gamma_{aa} + \gamma_{bb} - 2\gamma_{ab}) \frac{\partial^2}{\partial h_a \partial h_b}} \left[e^{-\beta \sum_{a=1}^n \bar{v}(h_a)} - 1 \right] \right\}_{h_a = h} .
$$

To prove the last equality in Eq. (4.90), one can replace

$$
e^{-\frac{1}{2}\sum_{a,b=1}^n \gamma_{aa} \frac{\partial^2}{\partial h_a \partial h_b}} \rightarrow e^{-\frac{1}{2}\left(\sum_{a=1}^n \gamma_{aa} \frac{\partial}{\partial h_a}\right)\frac{d}{dh}} \tag{4.91}
$$

when these operators are acting inside the integral, which is an application of Eq. (4.71). Then one can expand the exponential in Taylor series and observe that $\left(\frac{d}{dh}\right)^k$ can be integrated by parts, acting on the term e^h, giving rise to a factor $(-1)^k$. Then one can replace

$$
e^{-\frac{1}{2}\left(\sum_{a=1}^n \gamma_{aa} \frac{\partial}{\partial h_a}\right)\frac{d}{dh}} \rightarrow e^{\frac{1}{2}\sum_{a=1}^n \gamma_{aa} \frac{\partial}{\partial h_a}} . \tag{4.92}
$$

4.6.3 Gaussian Ansatz for the Glass Free Energy

In this section, we briefly sketch the derivation of the glass free energy based on a Gaussian ansatz for the molecular density $\rho(\mathbf{x})$. This derivation is simpler than the one discussed in Section 4.2 but relies on an additional, and a priori unjustified, assumption.

We start from the free energy of the molecular liquid truncated at the second order in the virial expansion, as given by Eq. (4.20), but neglecting the kinetic contribution, which is but an irrelevant constant,

$$-\beta F[\rho] = \int d\underline{x}\rho(\underline{x})[1 - \log \rho(\underline{x})] + \frac{1}{2} \int d\underline{x}d\underline{y}\rho(\underline{x})\rho(\underline{y})f(\underline{x} - \underline{y}),$$

$$f(\underline{x} - \underline{y}) = e^{-\beta_g v_g(x_1 - y_1)} \prod_{a=2}^{s+1} e^{-\beta v(x_a - y_a)} - 1. \tag{4.93}$$

Here, as discussed in Section 4.2.4, the first replica has temperature T_g and a potential $v_g(r)$ which contains an interaction length scale ℓ_g, while the other replicas have temperature T and a potential $v(r)$ containing ℓ. As in Section 4.2.2, we then perform the change of variable $\mathbf{u}^a = \mathbf{x}^a - \mathbf{x}^1$, as given in Eq. (4.21), and introduce the normalised distribution $\pi(\underline{u}) = \rho(\underline{u})/\rho$. We assume that $\pi(\underline{u})$ is Gaussian, with average and covariance given by Eq. (4.29) – i.e.,

$$\pi_{\hat{A}}(\underline{u}) = \frac{1}{[(2\pi)^s \det \hat{A}]^{d/2}} e^{-\frac{1}{2}\sum_\mu \sum_{a,b=2}^{s+1} u_\mu^a A_{ab}^{-1} u_\mu^b}, \qquad \hat{A} = \frac{\ell^2}{d}\hat{\alpha}. \tag{4.94}$$

Under this assumption, the calculation of the ideal gas term is straightforward. Starting from Eq. (4.41), we obtain

$$-\beta f^{id} = \frac{1}{N} \int d\underline{x}\rho(\underline{x})[1 - \log \rho(\underline{x})] = 1 - \log \rho - \int d\underline{u}\pi_{\hat{A}}(\underline{u}) \log \pi_{\hat{A}}(\underline{u})$$

$$= 1 - \log \rho + \frac{d}{2} \log[(2\pi)^s \det \hat{A}] + \frac{d}{2} \sum_{a,b} A_{ab} A_{ab}^{-1} \tag{4.95}$$

$$= 1 - \log \rho + \frac{d}{2} \log \det \hat{A} + \frac{d\,s}{2} \log(2\pi e).$$

The excess term is also easily evaluated. Starting from Eq. (4.23), we get

$$-\beta f^{ex} = \frac{\rho}{2} \int d\mathbf{R}\,d\underline{u}d\underline{v}\,\pi_{\hat{A}}(\underline{u})\pi_{\hat{A}}(\underline{v})\,f(\underline{u} - \underline{v} + \mathbf{R}). \tag{4.96}$$

Given that $\underline{u}$ and $\underline{v}$ are independently and identically distributed according to $\pi_{\hat{A}}(\underline{u})$, their difference $\underline{w} = \underline{u} - \underline{v}$ is also Gaussian, with twice the covariance, such that

$$-\beta f^{ex} = \frac{\rho}{2} \int d\mathbf{R}\,d\underline{w}\pi_{2\hat{A}}(\underline{w})\,f(\underline{w} + \mathbf{R})$$

$$= \frac{\rho}{2} \int d\mathbf{R} \left[e^{-\beta_g v_g(\mathbf{R})} \int d\underline{w}\pi_{2\hat{A}}(\underline{w}) \prod_{a=2}^{s+1} e^{-\beta v(w_a + \mathbf{R})} - 1 \right] \tag{4.97}$$

$$= \frac{\rho}{2} \int d\mathbf{R} \left\{ e^{-\beta_g v_g(\mathbf{R})}\, e^{\sum_\mu \sum_{a,b=2}^{s+1} A_{ab} \frac{\partial^2}{\partial w_\mu^a \partial w_\mu^b}}\, e^{-\beta \sum_{a=2}^{s+1} v(w_a + \mathbf{R})} \bigg|_{\underline{w}=\underline{0}} - 1 \right\},$$

where in the last line we used the identity in Eq. (4.88). The reader can check that the total free energy $f = f^{id} + f^{ex}$ coincides with Eq. (4.54), if the limit $d \to \infty$ is taken with the appropriate scalings.

The replica symmetric matrix has the form $A_{ab} = D_r \, \delta_{ab} + (D_r - D/2)(1 - \delta_{ab})$. Plugging this form into the replicated free energy, the determinant and the Gaussian convolution can then be evaluated explicitly and the limit $s \to 0$ can be taken to extract the glass free energy, as in Section 4.3.2. For completeness, we give the final result:

$$
-\beta f_g(\varphi, T | \varphi_g, T_g) = \frac{d}{2} \log(\pi e \, \mathsf{D}) + \frac{d}{2} \frac{2\,\mathsf{D}_r - \mathsf{D}}{\mathsf{D}}
$$
$$
+ \frac{\rho}{2} \Omega_d \int dr \, r^{d-1} \, q_g(2\,\mathsf{D}_r - \mathsf{D}, \beta_g; r) \log q(\mathsf{D}, \beta; r),
$$
(4.98)

where $q(\mathsf{D}, \beta; r)$ is given in Eq. (3.61), and $q_g(2\,\mathsf{D}_r - \mathsf{D}, \beta_g; r)$ is defined with the potential $v_g(r)$. Details on how to take the limit $d \to \infty$ of this expression can be found in [292], and the result coincides with Eq. (4.75). Note that the equation for the mean square displacement plateau (and then for the dynamical transition) obtained from this expression coincides with Eq. (3.60) in any finite dimension d.

5

Replica Symmetry Breaking and Hierarchical
Free Energy Landscapes

In Chapter 4, we have illustrated how the replica method can be a natural tool to compute the properties of the glass states of simple systems, despite the absence of quenched disorder in the interaction potential. The Franz–Parisi construction provides a way to average over an ensemble of glass states selected by a reference equilibrium configuration. The replica symmetric glass phase diagram has then been computed. Both in the case of hard-sphere and soft-sphere potentials, the replica symmetric solution predicts some unphysical features – in particular, a spinodal instability of the glass state upon compression or cooling. Something might thus be wrong with the assumption of replica symmetry, which motivates a deeper investigation of its validity.

In this chapter, we clarify the physical meaning of the replica symmetry. In particular, we discuss how, as for ordinary symmetries, it can be spontaneously broken at phase transitions and what the ensuing consequences are. Replica symmetry breaking was first discovered in the context of spin glasses, and it has since been applied in physics and in research areas as distant as computer science and neural networks. Because the subject is already discussed in great detail in several books [10, 138, 251, 254, 274], we here only provide a compendium of the basic notions and tools needed for the rest of the book. The reader who is not interested in the details of replica symmetry breaking can skip this chapter in a first reading. Note that in this chapter, we provide general results about the replica formalism while, in Chapter 6, we apply them to the specific case of particle glasses in the limit $d \to \infty$.

5.1 An Introduction to Replica Symmetry Breaking

Replica symmetry may seem different from ordinary symmetries of physical systems. For example, in ferromagnetic systems without an external magnetic field (see Chapter 1), the spin inversion symmetry $S_i \to -S_i$ is a symmetry of the Hamiltonian. It then plays a crucial role in understanding the thermodynamic properties of the model – e.g., its thermodynamic pure states in the low-temperature phase, as discussed in Chapter 1. By contrast, in glassy systems, the Hamiltonian does not display any symmetry beyond the trivial translational and rotational symmetries. In particular, once the system is frozen into a glass state, particles interact in a random environment fixed by the initial configuration which breaks translational and rotational invariance, and therefore, the Hamiltonian is not invariant under any symmetry group. Replica symmetry only emerges as a consequence of the average over all possible glass states, which is encoded by the average over the initial configuration. One may therefore wonder if replica symmetry has any physical meaning or if it is just a technical artefact. Before going into the details of replica symmetry breaking, we thus provide here a short discussion of the main physical properties behind this phenomenon.

5.1.1 Trivial Replica Symmetry Breaking

We have seen in Chapter 1 that in the ferromagnetic Ising model, at low temperatures and zero magnetic field, there are two translationally invariant equilibrium states, corresponding to magnetisation $\pm m_{eq}$. There may also be non-translationally invariant equilibrium states, with a free energy that is higher by a factor proportional to L^{d-1}, where L is the linear size of the system.

The existence of more than one equilibrium state is common in physics. For instance, it occurs everywhere along the gas–liquid coexistence line. In order to generalise this notion, let us summarise the procedure to monitor the presence of two phases first described in Chapter 1. We consider a spatially dependent observable $\mathcal{O}(\mathbf{x})$ that takes a different value in the two phases, such as the magnetisation $m(\mathbf{x})$ in spin systems and the local density $\rho(\mathbf{x})$ near the gas–liquid transition, and add to the Hamiltonian a small field ϵ conjugated to this observable:

$$\Delta H = -\epsilon \int d\mathbf{x} \mathcal{O}(\mathbf{x}). \tag{5.1}$$

Denoting $\langle \bullet \rangle_\epsilon$ the Gibbs–Boltzmann average in the thermodynamic limit with the additional field, we can define $\langle \bullet \rangle_\pm = \lim_{\epsilon \to 0^\pm} \langle \bullet \rangle_\epsilon$. Phase coexistence is observed if $\langle \mathcal{O}(\mathbf{x}) \rangle_- \neq \langle \mathcal{O}(\mathbf{x}) \rangle_+$.

We next consider the slightly more complex problem of a liquid that crystallises. In this case, we have an infinite number of equilibrium states, which differ from each other by a global translation. If one does not know the structure of the crystal, it is difficult to identify an observable $\mathcal{O}(\mathbf{x})$ that takes a different value in each of the many possible phases, and the previous construction is not viable. A possible way out of this problem, introduced by Edwards and Anderson [136], consists in considering two replicas (or clones) $\underline{X}$ and $\underline{Y}$ of the same system and writing the total Hamiltonian as

$$H_\epsilon(\underline{X}, \underline{Y}) = H(\underline{X}) + H(\underline{Y}) + \epsilon N\, \mathsf{D}(\underline{X}, \underline{Y}), \tag{5.2}$$

where

$$\mathsf{D}(\underline{X}, \underline{Y}) = \frac{1}{N} \min_{\mathcal{P} \in P_N} \sum_{i=1}^{N} |\mathbf{x}_i - \mathbf{y}_{\mathcal{P}(i)}|^2 \tag{5.3}$$

is the mean square displacement between the two configurations. It is important to note that, with respect to the definition given in Eq. (4.1), we added here a minimisation over all possible permutations $\mathcal{P} \in P_N$ of the N particles. In Chapter 4, we were interested in using $\underline{Y}$ as a reference configuration to constrain $\underline{X}$ into a restricted equilibrium. Here, by contrast, we want to treat $\underline{X}$ and $\underline{Y}$ on equal footing in a full equilibrium, and therefore, a minimisation over particle permutations is required to define a meaningful distance. The Hamiltonian in Eq. (5.2) is invariant under the $\mathbb{Z}_2$ group, whose elements exchange the two replicas and which coincides with replica symmetry in this context.

In a crystal phase, the limits $\epsilon \to 0^\pm$ produce different results for many observables that involve two replicas, while single-replica observables have the same value in the two limits. In particular, for $\epsilon > 0$, the two configurations are in the same crystal, and the value of the mean square displacement $\mathsf{D}(\epsilon) = \langle \mathsf{D}(\underline{X}, \underline{Y}) \rangle_\epsilon$ is as small as possible; it would be $\mathsf{D}(\epsilon) = 0$ at zero temperature. By contrast, for $\epsilon < 0$, one configuration is at the maximum distance (of the order of half the lattice spacing) from the other, and $\mathsf{D}(\epsilon)$ takes the largest possible value. The point $\epsilon = 0$ is singular, thus indicating a phase transition.

Note that at $\epsilon = 0$, one recovers the original Hamiltonian in which the two replicas are independent, but the structure of equilibrium states indicates that replica symmetry is then spontaneously broken. In order to show this, we focus, for simplicity, on the case of a simple cubic lattice with lattice spacing equal to unity. Each replica then occupies a cubic lattice, and the two lattices are shifted by an arbitrary vector $\mathbf{a} \in [-1/2, 1/2]^d$, because there is no coupling between the replicas. The corresponding value of mean square displacement is $\mathsf{D}(\underline{X}, \underline{Y}) = |\mathbf{a}|^2$, and its probability distribution is

$$P(\mathsf{D}) = \int_{[-1/2, 1/2]^d} \mathbf{da}\, \delta(|\mathbf{a}|^2 - \mathsf{D}). \tag{5.4}$$

A distribution $P(\mathsf{D})$ different from a delta function then signals the presence of multiple states and replica symmetry breaking. Note, however, that in this case, as soon as $\epsilon \neq 0$, replica symmetry breaking disappears and $P(\mathsf{D})$ becomes a delta function.

5.1.2 Non-Trivial Replica Symmetry Breaking

In the example of crystallisation discussed in Section 5.1.1, replica symmetry breaking is a consequence of an underlying spontaneous breaking of an internal symmetry – i.e., translations. In this case, the effect disappears if we change the definition of the distance between two configurations in a way that respects the internal symmetry,

$$D(\underline{X}, \underline{Y}) = \frac{1}{N} \min_{\mathbf{a} \in \mathbb{R}^d} \min_{\mathcal{P} \in P_N} \sum_{i=1}^{N} |\mathbf{x}_i - \mathbf{y}_{\mathcal{P}(i)} + \mathbf{a}|^2, \tag{5.5}$$

that is, if we define the distance as the minimum with respect to all possible translations of one configuration with respect to the other. With this new definition, $P(D)$ is a delta function both in the liquid and in the crystal phases, and no replica symmetry breaking occurs. This treatment is easily generalised to any situation in which an internal symmetry is spontaneously broken.

Phase coexistence, which is the essence of replica symmetry breaking, is present at first-order transition points, such as the gas–liquid transition. In this case, only two phases are present. Three phases are present at the triple point: gas, liquid and solid [175]. The Gibbs phase rule states that in order to have coexistence of $K + 1$ phases, we need to tune K control parameters. To prove this, let us label phases by an index α. The probability w_α that the system is in phase α at equilibrium in a finite volume is given by

$$w_\alpha = \frac{e^{-\beta F_\alpha}}{Z}, \qquad Z = \sum_\alpha e^{-\beta F_\alpha}, \qquad \sum_\alpha w_\alpha = 1, \tag{5.6}$$

where F_α is the total free energy of phase α. Then, consider adding a perturbation ΔH to the Hamiltonian, which perturbs the free energies as

$$F_\alpha \rightarrow F_\alpha + \langle \Delta H \rangle_\alpha, \tag{5.7}$$

where $\langle \bullet \rangle_\alpha$ denotes the equilibrium average restricted to phase α (as discussed in Chapter 1). Because the different phases have different structures (consider, e.g., gas, liquid and solid phases), it is very unlikely that the $\langle \Delta H \rangle_\alpha$ are equal. The perturbed free energies are then also different. Because the difference is of order N, only one phase survives in the thermodynamic limit. The only way to remain at coexistence is to impose that all the $\langle \Delta H \rangle_\alpha$ be equal, which imposes K conditions on the control parameters. This proves the Gibbs phase rule. In other words, having $K + 1$ equal free energies is an unlikely event that is destroyed by a small perturbation.

The preceding argument, however, fails in disordered systems, because one can then construct many distinct phases that have statistically identical properties. One can think, for instance, of the many glasses that can be formed by a same system of interacting particles. Suppose that at a given state point, in the thermodynamic limit, there is an infinite number of coexisting states with free energy F_α, and a perturbation is added to the Hamiltonian. Because the states have statistically independent properties, $\langle \Delta H \rangle_\alpha$ is then self-averaging and thus independent of α. The shift of the extensive free energy, ΔF_α, is therefore the same for all states. More concretely, consider adding a small perturbation $\Delta v(r)$ to the pair interaction potential of a particle system. According to Eq. (2.29), the variation of the free energy of state α is

$$\Delta F_\alpha = \frac{N\rho}{2} \int \mathbf{dr}\, g_\alpha(\mathbf{r}) \Delta v(r). \qquad (5.8)$$

If the coexisting phases are structurally different, such as a gas and a crystal, then their $g_\alpha(\mathbf{r})$ are different, which results in different ΔF_α. By contrast, if the coexisting phases are microscopically different glasses, $g_\alpha(\mathbf{r})$ is the same, as is well known numerically and experimentally [40, 80, 120], resulting in identical ΔF_α. Thanks to disorder, infinitely many phases might thus coexist in a whole region of parameter space.

Note that the coexistence of infinitely many phases could also be described within a fully probabilistic framework, known as the 'cavity method' [254], thus avoiding the use of the replica formalism. We do not make this choice in this book for three reasons:

- The probabilistic arguments used in the cavity method are subtle and not easy to follow for a reader who is not already familiar with the structure of a replica symmetry broken phase.
- The cavity method is much easier to develop for lattice models. The absence of an underlying lattice for particle systems complicates the derivation.
- The replica approach is computationally much easier to handle, and new applications are often first developed using this method. This is certainly the case for particles in $d \to \infty$.

Therefore, here we use the very compact algebraic replica formalism, even if this may sometimes hide the underlying physical interpretation. Hopefully, the main elements of its physical interpretation provided in this chapter somewhat compensate for this choice.

5.1.3 Stability of the Replica Symmetric Solution

It has been shown in Chapter 4 that the thermodynamic properties of glass states can be derived, within the Franz–Parisi construction, by minimising the replicated free energy $f_{s+1}(\widehat{\varphi}_a, T_a, \hat{\Delta})$, defined by Eq. (4.59), with respect to the $n \times n$ mean square displacement matrix $\hat{\Delta}$. Then, one should take the analytic continuation of the minimum to real $s = n - 1$ and take the limit $s \to 0$. In order to perform this operation, in Chapter 4 it was assumed that the matrix $\hat{\Delta}$ is invariant under replica permutations – i.e., it has the 'replica symmetric' form, $\hat{\Delta}^{RS}$, given by Eq. (4.62) and illustrated by Eq. (4.63). The assumption of replica symmetry allows one to perform the analytic continuation straightforwardly. However, the replica symmetric form was taken as an assumption, and we did not check that the minimum of the free energy is really assumed on matrices of this form.

For the ferromagnetic model discussed in Chapter 1 the paramagnetic minimum at $m = 0$ can become unstable for two reasons. Either the curvature of the free energy $f''(m = 0)$ vanishes, leading to a linear instability of the minimum, identified with a second-order phase transition. Or another minimum appears at $m \neq 0$, and its free energy becomes smaller than $f(m = 0)$, thus leading to a discontinuous (first-order) transition.

By analogy, a minimal way to support the replica symmetric assumption is to check that $\hat{\Delta}^{RS}$ is a stable local minimum of the replicated free energy.[1] This check amounts to excluding the possibility of a second-order phase transition. Although this does not exclude the existence of another free energy minimum characterised by replica symmetry breaking – i.e., a first-order transition – it at least excludes the possibility that the replica symmetric solution is unstable to small fluctuations. To perform this check, we need to compute the stability operator (or Hessian) of the free energy, defined as

$$\mathcal{H}_{ab;cd} = \left. \frac{\partial^2 f_{s+1}}{\partial \Delta_{ab} \partial \Delta_{cd}} \right|_{\hat{\Delta} = \hat{\Delta}^{RS}}, \qquad a < b, \ c < d. \tag{5.9}$$

We define the replica symmetric solution to be locally stable if all the eigenvalues of $\mathcal{H}$ are positive. Recall that the matrix $\hat{\Delta}$ is by definition symmetric and has $\Delta_{aa} = 0$. Hence, only elements with $a < b$ can be varied independently, and the Hessian is a $n(n-1)/2 \times n(n-1)/2$ symmetric matrix. Note that a pair $a < b$ is here considered as a single index taking $n(n-1)/2$ values.

The positivity of the Hessian matrix given by Eq. (5.9) is a natural condition. The lowest-order corrections to mean field theory indeed contain terms that involve the matrix

$$(\mathcal{H}^{-1})_{ab;cd} = \chi_{ab;cd} = \langle \Delta_{ab} \Delta_{cd} \rangle - \langle \Delta_{ab} \rangle \langle \Delta_{cd} \rangle. \tag{5.10}$$

The matrix $\chi_{ab;cd}$ is positive definite by construction, analogously to the magnetic susceptibility defined in Eq. (1.9). Because it is the inverse of the Hessian, it

[1] Recall that in the analytical continuation to $s < 1$ a minimum may become a saddle point. The correct procedure is thus to check the stability for $s > 1$ and then analytically continue the eigenvalues to $s < 1$.

diverges when the Hessian develops a zero eigenvalue and becomes (unphysically) negative in the region where the Hessian has negative eigenvalues. The divergence of $\chi_{ab;cd}$ thus signals a phase transition, and the presence of negative eigenvalues in a positive definite matrix signals that the replica symmetric computation is inconsistent, because a phase transition has been ignored.

Replica symmetry strongly constrains the form of the Hessian. Its eigenvalues can thus be computed explicitly. We do not report this calculation here, but it can be found in several classic references [113, 254]. In many models, this computation reveals that at some point in parameter space, one of the eigenvalues of the stability operator vanishes, signalling the breaking of replica symmetry. The set of degenerate eigenvectors ϕ_{ab} that correspond to a vanishing eigenvalue are called 'replicon' modes and satisfy the condition $\sum_b \phi_{ab} = 0$. This phase transition was first discovered by de Almeida and Thouless in the context of the equilibrium thermodynamics of the Sherrington–Kirkpatrick model [113], a mean field spin glass model. In that case, the phase transition separates the high-temperature paramagnetic phase from the low-temperature spin glass phase. In other models, the same transition was also shown to happen inside a glass state, by Gross, Kanter and Sompolinsky [171] and by Gardner [160], who first computed exactly the transition point in a given model. In this context, the phase transition is known as a Gardner transition. We will describe more precisely this transition in Chapter 6.

In all these models, it has been shown that the correct structure of the solution in the replica symmetry broken phase is given by 'hierarchical replica matrices' [254]. The structure of these matrices allows one to perform a simple analytic continuation to real s and then take the limit $s \rightarrow 0$, as in the replica symmetric case. It has been proven mathematically that this solution gives the correct free energy and that it correctly describes the underlying ultrametric structure of pure states, which we discuss in Section 5.3.3. In the context of infinite dimensional structural glass models, there is no rigorous proof that the hierarchical construction gives the correct free energy. Nevertheless, we will make this assumption because it provides physically consistent results. The consequences of this assumption will be discussed in Chapter 6.

In the rest of this chapter, we show how to compute the transition point based on the hierarchical matrix construction, without explicitly obtaining the eigenvalues of the stability operator defined by Eq. (5.9). We also show how to compute the properties of the low-temperature phase, first in an expansion around the transition point and then by solving the replica symmetry broken equations.

5.2 The Algebra of Hierarchical Matrices

In this section, we discuss some of the mathematical preliminaries needed for the replica formalism. In particular, because in all known cases the replica symmetry broken phase displays a hierarchical structure in replica space, we introduce and discuss the algebraic properties of this structure. Note that within the Franz–Parisi construction, one has $n = 1 + s$ replicas. The first one is special, and the other s are all equivalent. Considerations of replica symmetry are thus restricted to the $s \times s$ sub-block of the matrices with indices $a, b = 2, \ldots, s + 1$. Here we consider a simplified setting, in which there are only s equivalent replicas – i.e., we momentarily discard the special replica. To generalise the discussion to the Franz–Parisi construction, it will suffice to add back the special replica with $a = 1$ at the end (see Section 5.3).

5.2.1 k-Step Replica Symmetry Breaking

In the following, we construct the algebra of hierarchical matrices step by step. We consider $s \times s$ real symmetric matrices $\hat{q}$, whose elements are denoted by q_{ab}, and for notational simplicity, we use indices $a, b = 1, \ldots, s$.

Replica Symmetric (RS) Algebra

The replica symmetric algebra is composed of matrices of the form

$$q_{aa} = q_d, \qquad q_{a \neq b} = q. \tag{5.11}$$

It is easy to check that these matrices form a closed commutative algebra, in the sense that the product of two RS matrices is still a RS matrix, and the order of the matrices in the product is irrelevant.

One-Step Replica Symmetry Breaking (1RSB) Algebra

Consider an integer m_0, such that s is divisible by m_0 – i.e., $s \bmod m_0 = 0$. We divide the replica indices in s/m_0 groups, and define $\mathrm{ceil}(x) = \lceil x \rceil$, a function that returns the smallest integer larger than or equal to x. If $a \in \{1, \ldots, s\}$, then replica a belongs to the group $l_a^{(1)} = \left\lceil \frac{a}{m_0} \right\rceil$, with $l_a^{(1)} \in [1, \ldots, s/m_0]$. The 1RSB parametrisation of $\hat{q}$ is then given by

$$q_{ab} = \begin{cases} q_d & \text{if } a = b, \\ q_1 & \text{if } a \neq b \text{ but } l_a^{(1)} = l_b^{(1)}, \\ q_0 & \text{if } l_a^{(1)} \neq l_b^{(1)}. \end{cases} \tag{5.12}$$

In other words, the diagonal is set to q_d, and the off-diagonal elements are equal to q_1 if a, b are in the same group and to q_0 if they are in a different group. A graphical representation of this construction is given in Figure 5.1.

Two-Step Replica Symmetry Breaking (2RSB) Algebra

The 2RSB parametrisation can be obtained starting from the 1RSB one. Consider an integer m_1, such that $m_0 \bmod m_1 = 0$. We divide each group $l_a^{(1)} = 1, \ldots, s/m_0$

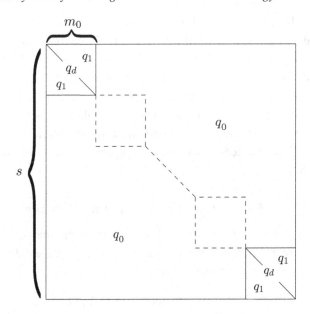

Figure 5.1 Illustration of a matrix $\hat{q}$ of the 1RSB form.

in m_0/m_1 subgroups of replicas, labelled by $l_a^{(2)} = \lceil \frac{a}{m_1} \rceil$, each group containing m_1 replicas. If two replicas a and b belong to different groups, $l_a^{(1)} \neq l_b^{(1)}$, the matrix element $q_{ab} = q_0$ is unchanged. Let us now consider the case in which a and b belong to the same group, $l_a^{(1)} = l_b^{(1)}$. On the diagonal, we still have $q_{aa} = q_d$. If a and b belong to the same subgroup, $l_a^{(2)} = l_b^{(2)}$, the matrix element is $q_{ab} = q_2$, while it remains $q_{ab} = q_1$ for $l_a^{(2)} \neq l_b^{(2)}$. The graphical representation of this construction is given in Figure 5.2.

From kRSB to $(k+1)$RSB: Iterative Construction

This construction can be iterated an arbitrary number of times, provided s is large enough to find integers that divide the matrix in subgroups. Let us suppose that a kRSB parametrisation has been constructed. We denote its parameters

$$\{s = m_{-1}, m_0, \ldots, m_{k-1}, m_k = 1\} \rightarrow \{q_0, \ldots, q_k; q_d\}, \qquad (5.13)$$

where we defined $m_k = 1$ and $m_{-1} = s$ for later convenience. We also have $s \geq m_0 \geq \ldots \geq m_{k-1} \geq m_k = 1$. The RS case corresponds to $k = 0$, and we have already discussed the cases $k = 1, 2$.

To move from kRSB to $(k+1)$RSB, we change the value of $m_k = 1$ to a number $m_k > 1$, such that $m_{k-1} \bmod m_k = 0$, and we add a new $m_{k+1} = 1$. Then, we consider the innermost blocks of size m_{k-1} and divide the replicas that belong to each of these blocks in m_{k-1}/m_k groups of m_k replicas. We keep all matrix elements as before, except if a pair of replicas a and b belong to the same subgroup (but are

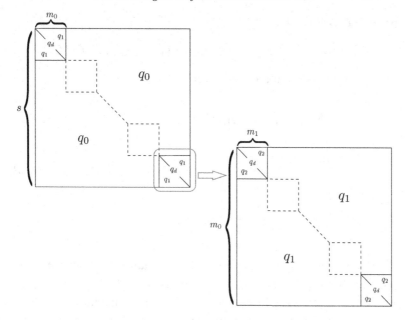

Figure 5.2 Illustration of a matrix $\hat{q}$ of the 2RSB form.

not identical), in which case, we replace the matrix element $q_{ab} = q_k$ with a new $q_{ab} = q_{k+1}$.

Properties of Hierarchical Matrices

An important property of hierarchical matrices is that each line is a permutation of the other lines. Hence, for any function $f(x)$,

$$\sum_b f(q_{ab}) = \sum_b f(q_{cb}), \qquad \forall a,c. \tag{5.14}$$

In other words, hierarchical matrices break the replica symmetry in a way that does not differentiate the single replicas but only induces correlations among replicas. One-replica observables thus remain symmetric under the action of the permutation group. This property also implies that the limit $\lim_{s\to 0} \frac{1}{s} \sum_{ab} f(q_{ab}) = \sum_b f(q_{ab})$, which appears in the computation of the replicated free energy, is finite. Other schemes of replica symmetry breaking, however, do not guarantee the existence of this limit.

Another important property is that, at any level k of RSB, the hierarchical matrices with fixed $\{m_i\}$ form a closed commutative algebra, because for any two kRSB matrices $\hat{A}$ and $\hat{B}$, their product is a kRSB matrix $\hat{C} = \hat{A}\hat{B} = \hat{B}\hat{A}$. If the matrices $\hat{A}$ and $\hat{B}$ are parametrised by

$$\{s = m_{-1}, m_0, \ldots, m_{k-1}, m_k = 1\} \to \{A_0, \ldots, A_k; A_d\},$$
$$\{s = m_{-1}, m_0, \ldots, m_{k-1}, m_k = 1\} \to \{B_0, \ldots, B_k; B_d\}, \tag{5.15}$$

their product $\hat{C}$ is given by

$$\{s = m_{-1}, m_0, \ldots, m_{k-1}, m_k = 1\} \rightarrow \{C_0, \ldots, C_k; C_d\}, \tag{5.16}$$

where the parameters C_d and C_i are given by

$$C_d = A_d B_d + \sum_{i=0}^{k} (m_{i-1} - m_i) A_i B_i,$$

$$C_i = \left(A_d + \sum_{l=0}^{k} A_l (m_{l-1} - m_l) \right) B_i + \left(B_d + \sum_{l=0}^{k} B_l (m_{l-1} - m_l) \right) A_i \tag{5.17}$$

$$+ \sum_{l=0}^{i} (m_{l-1} - m_l)(A_l - A_i)(B_l - B_i).$$

Using these equations, one can compute, for instance, the inverse of a matrix $\hat{A}$, by imposing that $C_d = 1$ and $C_i = 0$ for all $i = 0, \ldots, k$. This property thus guarantees that, once a RSB structure with a given k is chosen, all the terms that appear in the free energy preserve this same structure. We come back to this point in Section 5.2.2, after having discussed the analytic continuation to $s \rightarrow 0$.

5.2.2 Full Replica Symmetry Breaking

The kRSB matrices, as defined in Section 5.2.1, can only exist for large enough values of s. The smallest size corresponds to choosing only two groups of replicas at each step – i.e., $s = 2^{k+1}$. Because s cannot be smaller than 2^{k+1}, one might wonder how to perform the analytic continuation to the non-integer and small values of s and, in particular, to the limit $s \rightarrow 0$ in which we are ultimately interested. The continuation to real values of s is possible, however, because a kRSB matrix is fully parametrised by $\{m_i\}$, $\{q_i\}$, as in Eq. (5.13), and specifying a proper analytic continuation of these parameters is indeed straightforward. Here, we are specifically interested in the case where s is continued to a real positive value[2] $s \geq 0$.

Let us first consider the analytic continuation of the parameters m_i. The simplest case of 1RSB has $s \geq m_0 \geq m_1 = 1$. If $m_0 = s$ and $m_0 = 1$, the 1RSB matrix reduces to a RS matrix. In particular, when $s = 1$, one necessarily has $s = m_0 = m_1 = 1$, and the matrix $\hat{q}$ then becomes a single number q_d. Because m_0 is bounded between 1 and s and is necessarily equal to 1 for $s = 1$, any meaningful continuation to real values of s should be such that for $s < 1$ the inequality is

---

[2] Negative values of s have also been considered in order to compute large deviations of the free energy. See for example [131].

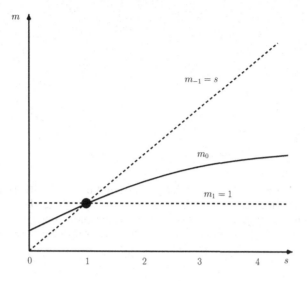

Figure 5.3 Analytic continuation of a 1RSB matrix to $s < 1$. The inequality $s = m_{-1} \geq m_0 \geq m_1 = 1$ is naturally reversed to $s \leq m_0 \leq 1$ in the analytical continuation to $s < 1$.

reversed. One then has $s \leq m_0 \leq 1$, as illustrated in Figure 5.3. Generalising this reasoning to kRSB, all the inequalities between the $\{m_i\}$ are reversed:

$$s \geq m_0 \geq m_1 \ldots \geq m_k = 1 \quad \rightarrow \quad s \leq m_0 \leq m_1 \ldots \leq m_k = 1. \tag{5.18}$$

One can then collect the $\{m_i\}$ and the off-diagonal parameters $\{q_i\}$ into a piecewise constant function, as illustrated in Figure 5.4. Once represented this way, a kRSB matrix can be analytically continued to any value of s. The product of two kRSB matrices can also be written as in Eq. (5.17), which remains well defined even for non-integer $\{m_i\}$ and s. As we discuss in more detail in Section 5.4, for generic forms of the free energy, if one considers only kRSB matrices, then the free energy can be analytically expressed as a function of $\{m_i\}$ and $\{q_i\}$, i.e., the piecewise constant function $q(x)$. It can then also be continued to real s. In summary, the algebra of kRSB matrices remains formally well defined for any continuous value of m_i for $s \geq 0$, provided Eq. (5.18) is respected. Inserting this structure in the free energy gives a function of $\{m_i\}$ and $\{q_i\}$, which upon extremisation determines the $\{m_i\}$ and $\{q_i\}$.

Full replica symmetry breaking (fullRSB) is reached when the number k of replica symmetry breaking steps is sent to infinity. If the kRSB algebra is represented by a piecewise function $q(x)$, then the limit $k \rightarrow \infty$ is well defined, provided $q(x)$ converges to a continuous function. This process, which is illustrated

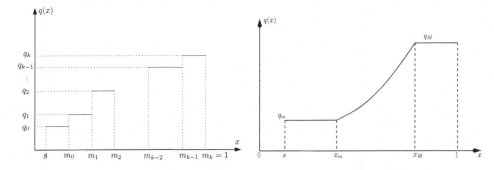

Figure 5.4 Parametrisation of a kRSB (left) and of a fullRSB (right) matrix for $s < 1$ via a piecewise constant function $q(x)$, which encodes the off-diagonal matrix elements of $\hat{q}$. The function $q(x)$ is always monotonous; depending on the definition of q_{ab}, it can either be an increasing or a decreasing function of the parameter x.

in Figure 5.4, can only happen if $\{m_i\}$ converges to a dense set on the interval $[0, 1]$. The difference between two successive m_i must thus become infinitesimal, i.e., $m_{i+1} - m_i = \mathrm{d}x$. In this continuum limit, sums over replica indices can be transformed into integrals. An explicit and useful example is (recall the convention $m_{-1} = s$ and $m_k = 1$)

$$\sum_{a,b}^{1,s} q_{ab}^p = s \left[q_d^p + \sum_{i=0}^{k} (m_{i-1} - m_i) q_i^p \right] = s \left[q_d^p - \int_0^1 \mathrm{d}x\, q(x)^p \right]. \qquad (5.19)$$

5.2.3 The Algebra of fullRSB Matrices

We now give explicit expressions for the product of two fullRSB matrices, and from it, we derive an expression for the inverse and for the determinant of a fullRSB matrix, following [255]. These results are especially useful because these quantities usually appear in the replicated free energy of interesting models.

Product

A fullRSB version of Eq. (5.17) can be obtained in the continuum limit. Given two fullRSB matrices $\hat{A}$, $\hat{B}$ parametrised by

$$\hat{A} \to \{A(x); A_d\}, \qquad \hat{B} \to \{B(x); B_d\}, \qquad (5.20)$$

their product

$$\hat{C} = \hat{A}\hat{B} = \hat{B}\hat{A} \to \{C(x); C_d\} \qquad (5.21)$$

is expressed by taking the continuum limit of Eq. (5.17) as

$$C_d = A_d B_b - \int_s^1 dx\, A(x) B(x),$$

$$C(x) = A(x)\, (B_d - \langle B \rangle) + B(x)\, (A_d - \langle A \rangle) \qquad (5.22)$$

$$- \int_s^x dy\, (A(y) - A(x))\, (B(y) - B(x)),$$

where we have defined

$$\langle A \rangle = \int_s^1 dx\, A(x). \qquad (5.23)$$

Hence, fullRSB matrices also form a closed commutative algebra. Remarkably, the fullRSB algebra contains any finite-kRSB algebra as a particular case. Inserting piecewise constant functions $A(x)$ and $B(x)$ in Eq. (5.22), as illustrated in Figure 5.4, one indeed recovers Eq. (5.17).

Fourier Transform

For simplicity we now restrict to the case $s \to 0$, which is the interesting limit in most applications, and discuss some additional properties of fullRSB matrices. To each hierarchical matrix $\hat{A} \to \{A(x); A_d\}$, we can associate its 'Fourier transform' [106, 116, 289] $\hat{\tilde{A}} \to \{\tilde{A}(x); \tilde{A}_d\}$, defined by the linear transformation

$$\tilde{A}(x) = A_d - x A(x) - \int_x^1 dy\, A(y), \qquad \tilde{A}_d = A_d - \int_0^1 dx\, A(x). \qquad (5.24)$$

The inverse Fourier transform is then given by[3]

$$A(x) = A(1) - \int_x^1 \frac{dy}{y} \frac{d\tilde{A}(y)}{dy}, \qquad A_d = \tilde{A}(1) + A(1). \qquad (5.25)$$

These relations can also be expressed in a differential form,

$$\frac{d\tilde{A}(x)}{dx} = -x \frac{dA(x)}{dx}, \qquad \tilde{A}(1) = A_d - A(1), \qquad (5.26)$$

which is equivalent to the integral relations in Eqs. (5.24) and (5.25).

The reader can verify that the multiplication of two matrices, as defined by Eq. (5.22), becomes a simple multiplication in terms of the Fourier transformed matrices:

$$\tilde{C}(x) = \tilde{A}(x) \tilde{B}(x), \qquad \tilde{C}_d = \tilde{A}_d \tilde{B}_d. \qquad (5.27)$$

---

[3] Note that the transformation $\{A(x); A_d\} \to \{A(x) + C; A_d + C\}$ leaves the Fourier transform invariant for any constant C. As a consequence, the inversion of the Fourier transform leaves an undetermined constant in $\{A(x); A_d\}$. This is why $A(1)$ appears in the right-hand side of Eqs. (5.25).

In other words, the Fourier transform reduces the matrix multiplication to a simple product and diagonalises the algebra of fullRSB matrices.[4]

Determinant and Inverse

The expressions for the determinant and the inverse of a fullRSB matrix $\hat{q} \rightarrow \{q(x); q_d\}$ can be derived in compact form in terms of the Fourier transform [255]. The standard notation for the Fourier transform of $\hat{q}$ is $\hat{\lambda}$,

$$\lambda(x) = q_d - xq(x) - \int_x^1 dy\, q(y), \qquad \dot{\lambda}(x) = -x\dot{q}(x), \qquad (5.28)$$

where here and in the rest of this chapter, the dot denotes a derivative with respect to the argument of a fullRSB function. One then obtains [255]

$$\lim_{s \to 0} \frac{d}{ds} \log \det \hat{q} = \log(\lambda(0)) + \frac{q(0)}{\lambda(0)} - \int_0^1 \frac{dx}{x^2} \log\left(\frac{\lambda(x)}{\lambda(0)}\right)$$

$$= \log(\lambda(1)) + \frac{q(0)}{\lambda(0)} + \int_0^1 dx\, \frac{\dot{q}(x)}{\lambda(x)}. \qquad (5.29)$$

The inverse of $\hat{q}$ can be parametrised by $\hat{q}^{-1} \rightarrow \{[q^{-1}](x); [q^{-1}]_d\}$, and in the $s \to 0$ limit, one obtains

$$[q^{-1}]_d = \frac{1}{\lambda(0)}\left(1 - \int_0^1 \frac{dy}{y^2}\frac{\lambda(0) - \lambda(y)}{\lambda(y)} - \frac{q(0)}{\lambda(0)}\right),$$

$$[q^{-1}](x) = -\frac{1}{\lambda(0)}\left[\frac{q(0)}{\lambda(0)} + \frac{\lambda(0) - \lambda(x)}{x\lambda(x)} + \int_0^x \frac{dy}{y^2}\frac{\lambda(0) - \lambda(y)}{\lambda(y)}\right]. \qquad (5.30)$$

Note that from Eq. (5.30) one also obtains a useful relation,

$$\lim_{s \to 0} \frac{1}{s}\sum_{a,b}^{1,s} (q^{-1})_{ab} = [q^{-1}]_d - \int_0^1 dx\, [q^{-1}](x) = \frac{1}{\lambda(0)}. \qquad (5.31)$$

5.3 Probability Distribution of the Mean Square Displacement

We now examine a first physical consequence of the hierarchical structure of the replica matrices by considering the probability distribution of the mean square displacement (MSD) between replicas. In Section 5.2.1, we discussed the algebra of hierarchical matrices in a general setting, but here we revisit the setting of Chapter 4, in which one follows a glass state prepared at a given initial temperature and density to a different state point. The free energy, in the Franz–Parisi construction, is then expressed in terms of the MSD matrix $\hat{D}$ of $1 + s$ replicas,

---

[4] This result is not surprising. It is well known by the Gelfand representation theorem that any closed commutative algebra can be diagonalised [104].

where the first replica encodes the initial state, and the block of the remaining s replicas is symmetric under permutations. We assume, in the following, that this block is described by a hierarchical matrix of the type discussed in Section 5.2.1.

5.3.1 Replicas and Hierarchical Matrices

As discussed in Section 4.1.2, in the glass phase the long time dynamics of the system is described by a restricted Gibbs–Boltzmann measure, given by Eq. (4.7). Consider two particle configurations $\underline{X}_1$ and $\underline{X}_2$, independently extracted according to Eq. (4.7) with identical parameters. Their relative distance, or MSD, $D = D\left(\underline{X}_1, \underline{X}_2\right)$, is a random variable[5] with probability distribution

$$P_{\underline{Y}, D_r}(D) = \int d\underline{X}_1 d\underline{X}_2 P(\underline{X}_1, \varphi, \beta | \underline{Y}, D_r) P(\underline{X}_2, \varphi, \beta | \underline{Y}, D_r)$$

$$\times \delta\left(D - D(\underline{X}_1, \underline{X}_2)\right). \tag{5.32}$$

Because $\underline{Y}$ is a random configuration extracted from the Gibbs–Boltzmann equilibrium measure, $P_{\underline{Y}, D_r}(D)$ is itself a random object and can be averaged over $\underline{Y}$:

$$\overline{P_{\underline{Y}, D_r}(D)} = \int \frac{d\underline{Y} d\underline{X}_1 d\underline{X}_2}{Z[\varphi_g, \beta_g]} e^{-\beta_g V[\underline{Y}, \ell_g]}$$

$$\times P(\underline{X}_1, \varphi, \beta | \underline{Y}, D_r) P(\underline{X}_2, \varphi, \beta | \underline{Y}, D_r) \delta\left(D - D(\underline{X}_1, \underline{X}_2)\right). \tag{5.33}$$

Following the reasoning and notations of Section 4.1.3, one can introduce replicas and write Eq. (5.33) as[6]

$$\overline{P_{\underline{Y}, D_r}(D)} = \lim_{s \to 0} \frac{1}{Z_{s+1}[\varphi_a, \beta_a, D_r]} \int \left(\prod_{a=1}^{s+1} d\underline{X}^a e^{-\beta_a V[\underline{X}^a, \ell_a]}\right)$$

$$\times \left(\prod_{a=2}^{s+1} \delta\left(D_r - D(\underline{X}^a, \underline{X}^1)\right)\right) \left[\frac{2}{s(s-1)} \sum_{a \neq b}^{2, s+1} \delta\left(D - D(\underline{X}^a, \underline{X}^b)\right)\right]$$

$$= \left\langle \frac{2}{s(s-1)} \sum_{a \neq b}^{2, s+1} \delta\left(D - D(\underline{X}^a, \underline{X}^b)\right)\right\rangle \equiv \mathcal{P}(D), \tag{5.34}$$

where the average is over the replicated liquid defined in Section 4.1.3.

---

[5] Note that in Chapter 4, D denoted the average of this random variable. The motivation for this change of notation will be given shortly.

[6] In short, the proof is as follows. By replica symmetry, one can eliminate the average over pairs $a \neq b$ in the last term of the second line of Eq. (5.34) and choose two arbitrary replicas of the s block, say $a = 2$ and $b = 3$. The other $s - 2$ replicas can then be integrated, giving rise to a factor $Z[\varphi, \beta | \underline{Y}, D_r]^{s-2}$. In the limit $s \to 0$, this factor provides the denominator of the two identical copies of $P(\underline{X}, \varphi, \beta | \underline{Y}, D_r)$ that appear in Eq. (5.33). The result follows after recalling that $Z_{s+1}[\varphi_a, \beta_a, D_r] \to Z[\varphi_g, \beta_g]$ when $s \to 0$.

In the $d \to \infty$ limit, the free energy can be expressed as a function of the rescaled MSD matrix $\hat{\Delta} = (d/\ell_g^2)\hat{D}$, via the closed Eq. (4.59), as discussed in Section 4.2.4. It follows from the general theory of thermodynamic fluctuations that the probability distribution of $\hat{\Delta}$ is given by[7]

$$P(\hat{\Delta}|\widehat{\varphi}_a, T_a) \propto e^{-N f_{s+1}(\widehat{\varphi}_a, T_a, \hat{\Delta})}. \tag{5.35}$$

Therefore, in the thermodynamic limit, the probability distribution of $\hat{\Delta}$ becomes extremely sharply peaked around the value that minimises the free energy. Rescaling the argument of $P(D)$ as $\Delta = (d/\ell_g^2)\, D$, the average in Eq. (5.34) simplifies to [254]

$$\mathcal{P}(\Delta) = \lim_{s \to 0} \frac{1}{s(s-1)} \sum_{a \neq b}^{2,s+1} \delta(\Delta - \Delta_{ab}), \tag{5.36}$$

where Δ_{ab} are the elements of the matrix $\hat{\Delta}$ that minimises (for $s > 1$) or extremises (for $s < 1$) the free energy $f_{s+1}(\widehat{\varphi}_a, T_a, \hat{\Delta})$. Eq. (5.36) is a general relation between this matrix and the equilibrium probability distribution $\mathcal{P}(\Delta)$ of the MSD between two identical copies, extracted from the distribution of the same glass state – i.e., generated by the same reference configuration $\underline{Y}$ – and followed to a target state point.

The replica symmetric structure for Δ_{ab}, as discussed in Chapter 4, Eq. (4.62), reads[8]

$$\begin{aligned}
\Delta_{aa} &= \Delta_d = 0, & a &= 1, \ldots, n, \\
\Delta_{1a} &= \Delta_{a1} = \Delta_r, & a &= 2, \ldots, n, \\
\Delta_{ab} &= \Delta_0, & a,b &= 2, \ldots, n, \text{ and } a \neq b,
\end{aligned} \tag{5.37}$$

where Δ_0 and Δ_r are the solutions of Eqs. (4.76). Plugging this structure in Eq. (5.36), we get

$$\mathcal{P}(\Delta) = \delta(\Delta - \Delta_0). \tag{5.38}$$

Therefore, within the replica symmetric assumption, the average probability distribution of the mean square displacement of two copies of the system sampled from the same glass state is peaked around the mean value Δ_0. In phase space, each glass state thus corresponds to a restricted portion of the set of configurations, which is sampled ergodically by the dynamics.

---

[7] Remember that $f_{s+1}(\widehat{\varphi}_a, T_a, \hat{\Delta})$ incorporates a factor β, and it is therefore adimensional. See Section 4.2.4.
[8] Note that while in Chapter 4, Δ denoted the off-diagonal elements of Δ_{ab}, here we denote them by Δ_0 to be consistent with the notation of Section 5.2.1. Instead, Δ denotes the argument of the probability distribution $\mathcal{P}(\Delta)$.

5.3.2 The Replica Symmetry Broken Phase

We now show that Eq. (5.36) is the basis for a physical interpretation of replica symmetry breaking. We consider a RSB parametrisation for the matrix $\hat{\Delta}$ in the block of the s identical replicas labelled by $a = 2, \ldots, n$, and we derive the average distribution of the MSD between two replicas that belong to a same glass state. Let us consider first a 1RSB structure – that is,

$$
\begin{aligned}
&\Delta_{aa} = 0, & a &= 1, \ldots, n, \\
&\Delta_{1a} = \Delta_{a1} = \Delta_r, & a &= 2, \ldots, n, \\
&\Delta_{ab} = \begin{cases} \Delta_1 & \text{if } a, b = 2, \ldots, n \text{ in same subgroup and } a \neq b, \\ \Delta_0 & \text{if } a, b = 2, \ldots, n \text{ in different subgroups}, \end{cases}
\end{aligned}
\tag{5.39}
$$

where m_0 is the size of the 1RSB subgroups, as illustrated in Figure 5.5. The MSD probability distribution, from Eq. (5.36), is then

$$
\mathcal{P}(\Delta) = m_0 \delta(\Delta - \Delta_0) + (1 - m_0)\delta(\Delta - \Delta_1). \tag{5.40}
$$

In the 1RSB case, two identical glass configurations $\underline{X}_1$ and $\underline{X}_2$ can typically be found either at a distance Δ_0, with probability m_0, or at distance Δ_1, with probability $1 - m_0$. This result can be interpreted in terms of the phase space structure. In the replica symmetric case, a glass state is a unique cluster of particle configurations. Two typical configurations visited by the dynamics, separated by a long time, are always at a distance Δ_0. Within the 1RSB ansatz, by contrast, what was a glass state

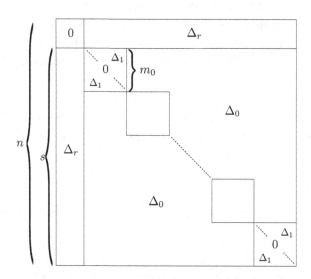

Figure 5.5 Illustration of the 1RSB parametrisation for the state following mean square displacement matrix, within the Franz–Parisi construction.

in the RS construction becomes a 'metabasin' that contains several glass substates. Configurations that belong to the same substate are typically at distance Δ_1 apart, while configurations that belong to different substates within the same metabasin are typically at distance Δ_0 apart. For this reason, one physically expects that $\Delta_0 > \Delta_1$. This reasoning is a first illustration of the physical consequences of the hierarchical structure of the MSD matrix in replica space.

Consider next a kRSB or fullRSB parametrisation of the form

$$\Delta_{aa} = 0, \qquad\qquad a = 1,\ldots,n,$$

$$\Delta_{1a} = \Delta_{a1} = \Delta_r, \qquad a = 2,\ldots,n, \qquad (5.41)$$

$$\Delta_{ab}|_{a,b=2,\ldots,n,\ a\neq b} \rightarrow \{\Delta(x);0\},$$

where the function $\Delta(x)$ encodes the off-diagonal elements of the block of the s identical replicas. As in the 1RSB case, values of $\Delta(x)$ for smaller x encode replicas that are farther away, and therefore, we expect $\Delta(x)$ to be a monotonically decreasing function of x, as illustrated[9] in Figure 5.6. From Eq. (5.36), one can show that

$$P(\Delta) = \left| \frac{dx(\Delta)}{d\Delta} \right|, \qquad (5.42)$$

where $x(\Delta)$ is the inverse function of $\Delta(x)$ over the interval $x \in [0,1]$. This inverse is well defined and monotonically decreasing, because $\Delta(x)$ is assumed to be monotonically decreasing. The RSB profile $\Delta(x)$, therefore, defines the average probability distribution $P(\Delta)$ of the MSD between two typical configurations that belong to a same metabasin. Note that Eq. (5.42) can also be applied to a piecewise constant function. For example, in the RS case where $\Delta(x) = \Delta_0$, the inverse function is formally $x(\Delta) = \theta(\Delta_0 - \Delta)$, which is consistent with $P(\Delta) = \delta(\Delta - \Delta_0)$. The 1RSB result in Eq. (5.40) is also reproduced by this general expression.

A typical profile for $\Delta(x)$ is shown in Figure 5.6. It displays two flat parts, $\Delta(x) = \Delta_m$ for $x \in [0, x_m]$ and $\Delta(x) = \Delta_M$ for $x \in [x_M, 1]$, with a continuously changing part $\Delta_c(x)$ in between. Using Eq. (5.42) one obtains that the average probability distribution of the overlap is given by

$$P(\Delta) = x_M \delta(\Delta - \Delta_M) + x_m \delta(\Delta - \Delta_m) + P_c(\Delta), \qquad (5.43)$$

where $P_c(\Delta)$ denotes the regular part of the probability distribution.

---

[9] In the spin glass and structural glass literatures, an overlap $q_{ab} \in [0,1]$ is often used as a measure of distance, such that $q_{ab} = 1$ corresponds to identical configurations, and decreasing q_{ab} corresponds to increasingly different configurations. In that context $q(x)$ is an increasing function of x, as illustrated in Figure 5.4.

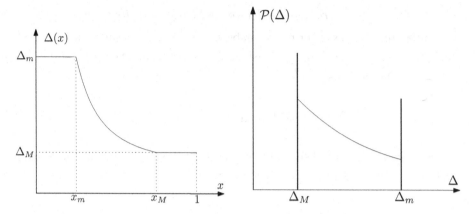

Figure 5.6 (Left) Illustration of a typical profile of the function $\Delta(x)$. The two points x_m and x_M separate the flat and continuous regions of $\Delta(x)$. These points are often called 'breaking points'. (Right) Illustration of the corresponding $\mathcal{P}(\Delta)$. The two flat parts of $\Delta(x)$ are responsible for the two Dirac delta functions appearing at Δ_M and Δ_m. Their weights are respectively given by x_M and x_m.

5.3.3 The Ultrametric Structure of States

The replica symmetric assumption implies that a glass state is a restricted portion of phase space ergodically sampled by the dynamics. Two typical configurations that belong to this portion of phase space have a MSD that satisfies the replica symmetric saddle point equations. At this level, a glass is a simple thermodynamic state.

At the 1RSB level, a glass state is not a simple state anymore. It is a metabasin that contains many individual substates: a pair of typical configurations within this metabasin can have two different MSDs depending on whether they belong to the same substate or not.

Upon iterating the RSB construction – i.e., moving to higher number of replica symmetry breaking steps – the glass metabasin splits into sub-basins, each containing other sub-basins, and so on. This hierarchy terminates when one obtains individual states. As a consequence, the MSD of two typical configurations that belong to the same glass metabasin can take a set of k values, corresponding to the configurations being separated at different levels in the hierarchical structure of substates. This set becomes continuous for $k \to \infty$ and is thus encoded in the profile $\Delta(x)$. For completeness, in this section, we discuss briefly a few additional properties of the organisation of glass states in phase space that are important in some applications. We refer the reader to [254, chapter IV] for a more detailed discussion.

Triplet Distribution and Ultrametricity

The hierarchical states description can be refined by computing additional geometrical properties of phase space. Consider, for instance, three configurations $\underline{X}_1$, $\underline{X}_2$ and $\underline{X}_3$, with mutual MSDs[10]

$$\check{\Delta}_{ab} = \frac{d}{\ell_g^2} \, \mathsf{D}\left(\underline{X}_a, \underline{X}_b\right), \qquad a, b = 1, \dots, 3. \tag{5.44}$$

The average probability distribution of the three MSDs can be computed using an approach along the same lines as Section 5.3.1,

$$\mathcal{P}(\check{\Delta}_{12}, \check{\Delta}_{13}, \check{\Delta}_{23}) = \lim_{s \to 0} \frac{1}{s(s-1)(s-2)}$$
$$\times \sum_{a \neq b \neq c (\neq a)}^{2, s+1} \delta(\check{\Delta}_{12} - \Delta_{ab}) \delta(\check{\Delta}_{13} - \Delta_{ac}) \delta(\check{\Delta}_{23} - \Delta_{bc}). \tag{5.45}$$

Assuming a hierarchical structure for Δ_{ab}, one then obtains [254]

$$\mathcal{P}(\check{\Delta}_{12}, \check{\Delta}_{13}, \check{\Delta}_{23}) = \frac{1}{2} \mathcal{P}(\check{\Delta}_{12}) x(\check{\Delta}_{12}) \delta(\check{\Delta}_{12} - \check{\Delta}_{13}) \delta(\check{\Delta}_{12} - \check{\Delta}_{23})$$
$$+ \frac{1}{2} \big[\mathcal{P}(\check{\Delta}_{12}) \mathcal{P}(\check{\Delta}_{13}) \theta(\check{\Delta}_{13} - \check{\Delta}_{12}) \delta(\check{\Delta}_{13} - \check{\Delta}_{23})$$
$$+ \mathcal{P}(\check{\Delta}_{13}) \mathcal{P}(\check{\Delta}_{23}) \theta(\check{\Delta}_{23} - \check{\Delta}_{13}) \delta(\check{\Delta}_{23} - \check{\Delta}_{12})$$
$$+ \mathcal{P}(\check{\Delta}_{23}) \mathcal{P}(\check{\Delta}_{12}) \theta(\check{\Delta}_{12} - \check{\Delta}_{23}) \delta(\check{\Delta}_{12} - \check{\Delta}_{13}) \big]. \tag{5.46}$$

Therefore, this probability is non-vanishing only if two MSDs are equal and the third is equal or smaller. In other words, for every triplet of configurations extracted according to the Boltzmann–Gibbs measure restricted to one glass metabasin, the triangle formed by their mutual distances is either equilateral or isosceles, with the two equal edges longer than the third. This structure, which follows directly from the hierarchical structure of $\hat{\Delta}$, is called 'ultrametric' in mathematics [300] and is also common in taxonomy; see, e.g., [17]. In the framework of disordered systems, it was first discovered in spin glasses [254].

The replica symmetry breaking structure thus corresponds to a hierarchical organisation of states in phase space. The states form an ultrametric tree, as illustrated in Figure 5.7 for a 3RSB landscape. We refer the reader to [254, chapter IV and reprint 16] and [300] for more details.

---

[10] In this section, we denote by $\check{\Delta}_{ab}$ the MSDs that are used as arguments of probability distributions to distinguish them from the thermodynamic matrix Δ_{ab} that extremises the free energy.

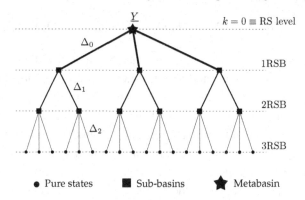

$$k = 0 \equiv \text{RS level}$$
$$\text{1RSB}$$
$$\text{2RSB}$$
$$\text{3RSB}$$

● Pure states　■ Sub-basins　★ Metabasin

Figure 5.7 An example of a 3RSB ultrametric structure of states. Individual states are the leaves of the tree. They contain configurations whose typical MSD is Δ_3. The states are organised in metabasins with a hierarchical structure: configurations belonging to states at distance 1 in the tree have typical MSD Δ_2, those at distance 2 have Δ_1, and those at distance 3 have Δ_0.

Fluctuations of the MSD Distribution

Another important property that follows from the hierarchical matrix structure can be obtained from the fluctuations of $P_{\underline{Y}, D_r}(D)$ from metabasin to metabasin. This probability density, defined by Eq. (5.32), is a random object because it depends on the configuration $\underline{Y}$ that selects one of the possible glass metabasins. One can thus study its probability distribution over the realisations of $\underline{Y}$. Consider four configurations $\{\underline{X}_i\}_{i=1,\dots,4}$ with mutual distances $\check{\Delta}_{ab}$ defined by Eq. (5.44). Following the same strategy as in Section 5.3.1, the probability distribution of $\check{\Delta}_{12}$ and $\check{\Delta}_{34}$ is [254]

$$\overline{P_{\underline{Y}}(\check{\Delta}_{12}, \check{\Delta}_{34})}^{\underline{Y}} = \frac{1}{3}\overline{P_{\underline{Y}}(\check{\Delta}_{12})}^{\underline{Y}}\delta(\check{\Delta}_{12} - \check{\Delta}_{34}) + \frac{2}{3}\overline{P_{\underline{Y}}(\check{\Delta}_{12})}^{\underline{Y}}\overline{P_{\underline{Y}}(\check{\Delta}_{34})}^{\underline{Y}}. \quad (5.47)$$

This result shows that

$$\overline{P_{\underline{Y}}(\check{\Delta}_{12}, \check{\Delta}_{34})}^{\underline{Y}} \neq \overline{P_{\underline{Y}}(\check{\Delta}_{12})}^{\underline{Y}}\overline{P_{\underline{Y}}(\check{\Delta}_{34})}^{\underline{Y}} \quad (5.48)$$

and, thus, that $P_{\underline{Y}}(\Delta)$ is not self-averaging. This quantity does not converge in probability to $\mathcal{P}(\Delta)$, and finite fluctuations persist even in the thermodynamic limit. The structure of the fluctuations of $P_{\underline{Y}}(\Delta)$ is also entirely fixed by the hierarchical ansatz for the MSD matrix. We refer the reader to [254, reprint 16] for more details.

Generation of the Ultrametric Tree of States

We conclude this section by discussing the statistical properties of the free energies of glass metabasins, refining the analysis of Section 5.1. Recall that the total partition function of a given metabasin, selected by a configuration $\underline{Y}$, is given by Eq. (4.7),

$$Z[\varphi, \beta | \underline{Y}, D_r] = \int d\underline{X} e^{-\beta V[\underline{X}, \ell]} \delta(D_r - D(\underline{X}, \underline{Y})). \tag{5.49}$$

When replica symmetry is broken, the metabasin is a collection of sub-basins, each broken into sub-basins, and so on, down to the lowest level in the hierarchy of Figure 5.7 that corresponds to pure states. Each pure state is a minimum of the free energy and corresponds to a cluster of configurations. These clusters are disjoint, and the probability that the system is found far away from one of them is exponentially small in N. The partition function is therefore the sum of the partition functions of the pure states.[11] For notational simplicity, here we consider the parameters φ, β, D_r as fixed and focus on the dependence on $\underline{Y}$. The partition function can then be written as

$$Z(\underline{Y}) = \sum_\alpha Z_\alpha(\underline{Y}) = \sum_\alpha e^{-\beta N f_\alpha(\underline{Y})}, \qquad f_\alpha(\underline{Y}) = -\frac{T}{N} \log Z_\alpha(\underline{Y}), \tag{5.50}$$

where α labels the pure states that compose the metabasin selected by $\underline{Y}$, and $f_\alpha(\underline{Y})$ is the free energy of α. The weight of state α is therefore

$$w_\alpha(\underline{Y}) = \frac{e^{-\beta N f_\alpha(\underline{Y})}}{\sum_\gamma e^{-\beta N f_\gamma(\underline{Y})}}, \qquad \sum_\alpha w_\alpha(\underline{Y}) = 1, \tag{5.51}$$

and given an observable $\mathcal{O}$, one can write its equilibrium average as

$$\langle \mathcal{O} \rangle_{\underline{Y}} = \sum_\alpha w_\alpha(\underline{Y}) \langle \mathcal{O} \rangle_\alpha, \tag{5.52}$$

where $\langle \mathcal{O} \rangle_\alpha$ is the average of $\mathcal{O}$ restricted to state α.

All the pure states that contribute to the equilibrium measure must have the same (and lowest possible) free energy per particle,[12] $f_\alpha = f_{min}$, as discussed in Section 5.1. States with $f_\alpha > f_{min}$ are unstable to nucleation, as discussed in Section 1.5.3. We therefore assume that, if equilibrium can be attained within the glass metabasin selected by $\underline{Y}$, the relevant equilibrium pure states have $N f_\alpha = N f_{min} + \delta f_\alpha$, where δf_α is a finite correction in the thermodynamic limit.[13] Their weight is then finite and given by

---

[11] In the case of a ferromagnet, this statement can be verified based on the discussion of Chapter 1. For $T < T_c$ and zero external field, the two minima of the free energy correspond to positive and negative magnetisation $m = \pm m_{eq}$. Excluding the possibility of phase coexistence – i.e., focusing on homogeneous magnetisation profiles – the probability that the system has magnetisation $m \neq m_{eq}$ is exponentially small in N, and the fluctuations around $\pm m_{eq}$ within each state are of the order of $N^{-1/2}$. The partition function is therefore dominated by configurations with $m = \pm m_{eq} + O(N^{-1/2})$ and can be written as the sum of two partition functions, corresponding to disjoint sets of configurations with positive or negative magnetisation.

[12] We will discuss in Chapter 7 how to extend this reasoning to take into account metastable states.

[13] This is certainly the case for all states in fully connected models such as the Curie–Weiss model introduced in Section 1.5.2, for which mean field theory is exact. In this case, the free energy can be expressed as a series in $1/N$, as can be shown from the high-temperature expansion discussed in Chapter 1. In finite dimensions, this is only true for pure states. For instance, in presence of interfaces, we have $\delta f_\alpha \propto L^{d-1} = N^{(d-1)/d}$, where L is the linear size of the system.

$$w_\alpha(\underline{Y}) = \frac{e^{-\beta\delta f_\alpha(\underline{Y})}}{\sum_\gamma e^{-\beta\delta f_\gamma(\underline{Y})}}. \tag{5.53}$$

We wish to characterise the statistical properties of δf_α, in order to provide a statistical procedure to generate the ultrametric tree of states.

We first discuss these statistical properties for a 1RSB hierarchical structure, in which the glass metabasin is partitioned into independent pure states. Because of the randomness of $\underline{Y}$, the pure states, their partition functions and their weights are themselves random variables. It turns out that the random free energies δf_α have an extremely simple distribution [254]. They are extracted independently from a Poisson point process with intensity $\rho_{m_0}(\delta f) = e^{\beta m_0 \delta f}$, where m_0 is the point at which $\Delta(x)$ is discontinuous. The number of states with free energy $\delta f_\alpha \in [\delta f, \delta f + d\delta f]$ is then given by

$$d\rho_{m_0}(\delta f) = e^{\beta m_0 \delta f} d\delta f. \tag{5.54}$$

We now consider the case in which the glass metabasin has a 2RSB structure. Individual pure states are then organised into sub-basins. We assign to each sub-basin a label $\alpha^{(0)}$, and we label $(\alpha^{(0)}, \alpha^{(1)})$ the individual states within this basin. As for individual states, we can assign to each sub-basin a free energy, which is the logarithm of the sum of the partition functions of the individual states that belong to that basin. We denote $Nf_{min} + \delta f_{\alpha^{(0)}}$ the reference free energy for the sub-basins, and $Nf_{min} + \delta f_{\alpha^{(0)}} + \delta f_{\alpha^{(0)},\alpha^{(1)}}$ the free energy of the individual states. It turns out that the free energies of the sub-basins are here again extracted according to a Poisson point process. The number of free energies $\delta f_{\alpha^{(0)}}$ in the interval $[\delta f_0, \delta f_0 + d\delta f_0]$ is thus

$$d\rho_{m_0}(\delta f_0) = e^{\beta m_0 \delta f_0} d\delta f_0. \tag{5.55}$$

The free energies of individual states within a sub-basin of free energy $\delta f_{\alpha^{(0)}}$ are extracted according to another point process, for which the density of free energies in the interval $[\delta f_1, \delta f_1 + d\delta f_1]$ is given by

$$d\rho_{m_1}(\delta f_1) = e^{\beta m_1 \delta f_1} d\delta f_1. \tag{5.56}$$

This construction can be repeated an arbitrary number of times in order to generate the free energies of states and basins on arbitrary ultrametric trees. This construction is particularly helpful to describe the evolution of the ultrametric tree under an external perturbation in order to characterise, for example, the equilibrium statistical distribution of avalanches [148, 225]. We refer the reader to [254, reprint 18] for more details.

It is interesting to discuss the behaviour of the weights when a random perturbation is added to the original Hamiltonian, as discussed in Section 5.1. The free energies are modified, and the new ones, denoted Nf'_α, are

$$Nf'_\alpha = Nf'_{min} + \delta f'_\alpha, \qquad f'_{min} = f_{min} + \Delta f, \qquad \delta f'_\alpha = \delta f_\alpha + \epsilon_\alpha, \tag{5.57}$$

where the shift of the extensive part of the free energy, Δf, is the same for all states, as discussed in Section 5.1. We assume once again that the perturbations ϵ_α are finite in the thermodynamic limit; otherwise, one state dominates as soon as the perturbation is added. Let us also assume that ϵ_α are random Gaussian variables with zero mean and variance σ^2, i.e., with a probability distribution $\gamma_{\sigma^2}(\epsilon)$ $= e^{-\epsilon^2/(2\sigma^2)}/\sqrt{2\pi\sigma^2}$. The new density of free energy states is then

$$\rho'(\delta f') = \int d\epsilon \, \gamma_{\sigma^2}(\epsilon)\rho(\delta f' - \epsilon). \tag{5.58}$$

If the moments of δf are well defined, we have

$$[\delta f^2]_c = [\delta f^2]_c + \sigma^2, \qquad [\delta f^2]_c = [\delta f^2] - [\delta f]^2, \tag{5.59}$$

where $[\bullet]$ here denotes an average over $\rho(\delta f)$. The distribution $\rho(\delta f)$ thus widens when the perturbation is added. The situation is completely different if $\rho(\delta f) = e^{k\delta f}$. We then have

$$\rho'(\delta f') = e^{k(\delta f' - c)} = \rho(\delta f' - c), \qquad c = -\frac{\sigma^2}{2k}. \tag{5.60}$$

In this case, the shape of the distribution of weights is left invariant by the perturbation. This invariance property of the exponential weight distribution characterises ultrametric trees, as given by Eq. (5.55).

Let us stress one final point. Although in the fullRSB limit, the ultrametric tree has an infinite number of branching points, and at each branching point, the number of branches is infinite, the distribution of weights is such that only states with positive weight contribute.[14] More precisely, if we restrict our analysis to states such that $w_\alpha > \epsilon$, we have for small ϵ

$$\sum_\alpha w_\alpha \theta(w_\alpha - \epsilon) = 1 - O(\epsilon^\omega), \tag{5.61}$$

with a positive exponent $\omega = 1 - x_M$ [254].

5.4 Replicated Free Energies and Hierarchical Matrices

The replicated free energies of many mean field models have a common structure, and their evaluation on hierarchical matrices can be done using the same approach. In this section, we consider the free energy $f(\hat{q})$ to be a function of a replica matrix $\hat{q}$, of a generic form, and we provide some mathematical results that are useful to

---

[14] A contrarian could argue that a very large number of states with vanishing weight gives a finite contribution. This possibility is excluded by Eq. (5.61).

evaluate this free energy on a hierarchical RSB matrix $\hat{q}$. The expressions we obtain can be applied to many different problems, and in Chapter 6, we consider the case of particle systems in the limit $d \to \infty$ more specifically.

Consider a generic free energy that depends on a $s \times s$ matrix, $\hat{q}$, written as the sum of two terms,

$$f(\hat{q}) = f^a(\hat{q}) + f^d(\hat{q}), \qquad (5.62)$$

the first term being a simple 'algebraic' function of $\hat{q}$ and the second being a 'differential' term to be specified later. Under the assumption that $\hat{q} \to \{q(x); q_d\}$ is a hierarchical matrix, our aim is to compute the limit $s \to 0$ of the free energy

$$f^{a,d}[q(x); q_d] = \lim_{s \to 0} \frac{d}{ds} f^{a,d}(\hat{q}). \qquad (5.63)$$

The term $f^a(\hat{q})$ is assumed to be a simple explicit function of $\hat{q}$, for which the evaluation on the hierarchical matrix and the limit $s \to 0$ can be taken easily. Two common examples are the sum of an arbitrary power p of matrix elements,

$$f^a(\hat{q}) = \sum_{a,b}^{1,s} q_{ab}^p \qquad \Rightarrow \qquad f^a[q(x); q_d] = q_d^p - \int_0^1 dx\, q(x)^p, \qquad (5.64)$$

where we used Eq. (5.19), and $f^a(\hat{q}) = \log \det \hat{q}$, which – using Eq. (5.29) – leads to

$$f^a[q(x); q_d] = \log(\lambda(1)) + \frac{q(0)}{\lambda(0)} + \int_0^1 dx\, \frac{\dot{q}(x)}{\lambda(x)}. \qquad (5.65)$$

The term $f^d(\hat{q})$ is expressed in terms of a differential operator and a function $r(h)$ as

$$f^d(\hat{q}) = e^{\frac{1}{2} \sum_{a,b=1}^s q_{ab} \frac{\partial^2}{\partial h_a \partial h_b}} \prod_{c=1}^s r(h_c) \Bigg|_{\{h_c = H\}}. \qquad (5.66)$$

The aim of this section is to show how the differential term can be treated when $\hat{q}$ has a hierarchical structure and the limit $s \to 0$ is taken. We will give a general recipe for computing this term, and deduce equations for $q_d, q(x)$ from extremising the free energy.

In the context of spin glasses, for instance, the free energy of the Sherrington–Kirkpatrick model, as well as many of its generalisations, has this form [254]. In the context of glasses, $\hat{q}$ corresponds to the MSD matrix, and the replicated free energy given in Eq. (4.59) is written in a similar form, with some small complications that will be discussed in Chapter 6. Hence, the results of this section are broadly applicable to mean field models.

5.4.1 The Differential Free Energy Term

We now describe how to compute the differential term $f^d(\hat{q})$ using the iterative algorithm developed in [133]. In the rest of this section, to simplify the notation, we do not explicitly indicate the ranges of indices $a, b = 1, \ldots, s$ in the sums and products. For the new indices we introduce, the range is specified only at the moment they are defined.

The Replica Symmetric Case

Let us start by the simplest case, in which $\hat{q}$ has a replica symmetric structure, i.e., $q(x) = q_0$. This case has already been discussed in Section 4.3.2 in the specific setting of particles in the limit $d \to \infty$. The differential operator appearing in Eq. (5.66) can then be written as

$$\frac{1}{2} \sum_{a,b} q_{ab} \frac{\partial^2}{\partial h_a \partial h_b} = \frac{q_d - q_0}{2} \sum_a \frac{\partial^2}{\partial h_a^2} + \frac{q_0}{2} \left(\sum_a \frac{\partial}{\partial h_a} \right)^2. \tag{5.67}$$

Applying the identity in Eq. (4.69) to Eq. (5.67) gives

$$e^{\frac{q_d - q_0}{2} \sum_a \frac{\partial^2}{\partial h_a^2} + \frac{q_0}{2} \left(\sum_a \frac{\partial}{\partial h_a} \right)^2} \prod_c r(h_c) = e^{\frac{q_0}{2} \left(\sum_a \frac{\partial}{\partial h_a} \right)^2} \prod_c \gamma_{q_d - q_0} \star r(h_c). \tag{5.68}$$

Next, we can apply the identity in Eq. (4.71) to express the derivatives with respect to the variables $\{h_a\}$ evaluated at a point where they all take the same value $\{h_a = H\}$,

$$e^{\frac{q_0}{2} \left(\sum_a \frac{\partial}{\partial h_a} \right)^2} \prod_c \gamma_{q_d - q_0} \star r(h_c) \Bigg|_{\{h_c = H\}} = e^{\frac{q_0}{2} \frac{\partial^2}{\partial h^2}} \left[\gamma_{q_d - q_0} \star r(h) \right]^s \Bigg|_{h = H} \tag{5.69}$$

$$= \gamma_{q_0} \star \left[\gamma_{q_d - q_0} \star r(H) \right]^s.$$

Taking the $s \to 0$ limit, we then obtain

$$f^d[q(x); q_d] = \lim_{s \to 0} \frac{d}{ds} \gamma_{q_0} \star \left[\gamma_{q_d - q_0} \star r(H) \right]^s = \gamma_{q_0} \star \log \left[\gamma_{q_d - q_0} \star r(H) \right]. \tag{5.70}$$

The 1RSB Case

We now repeat the same calculation within a 1RSB parametrisation. The function $q(x)$ is now a step function, $q(x) = q_0$ for $x < m_0$ and $q(x) = q_1$ for $x > m_0$. For finite s, this corresponds to dividing the s replicas in s/m_0 blocks that we label with an index $l \in \{1, \ldots, s/m_0\}$. We write $a \in l$ if replica a is one of the m_0 replicas that belongs to block l. The differential operator can then be written as

$$\frac{1}{2} \sum_{a,b} q_{ab} \frac{\partial^2}{\partial h_a \partial h_b} = \frac{q_d - q_1}{2} \sum_a \frac{\partial^2}{\partial h_a^2} + \frac{q_1 - q_0}{2} \sum_l \left(\sum_{a \in l} \frac{\partial}{\partial h_a} \right)^2$$
$$+ \frac{q_0}{2} \left(\sum_a \frac{\partial}{\partial h_a} \right)^2 .$$

(5.71)

The action of this operator on the product $\prod_a r(h_a)$ can be evaluated in three steps:

1. The operator $\frac{q_d - q_1}{2} \sum_a \frac{\partial^2}{\partial h_a^2}$ acts independently on each replica, via Eq. (4.69), and thus leads to the replacement

$$r(h_a) \rightarrow \gamma_{q_d - q_1} \star r(h_a) = g(1, h_a),$$

(5.72)

for each replica. From this point on, there is no need to consider different fields h_a in replicas of the same block because the remaining operators act on each group of replicas via Eq. (4.71). Therefore, we can consider that $h_a = h_l$, if $a \in l$, and

$$\prod_a g(1, h_a) \rightarrow \prod_l g(1, h_l)^{m_0}.$$

(5.73)

2. Replacing the operator $\frac{q_1 - q_0}{2} \sum_l \left(\sum_{a \in l} \frac{\partial}{\partial h_a} \right)^2$ with $\frac{q_1 - q_0}{2} \sum_l \frac{\partial^2}{\partial h_l^2}$ via Eq. (4.71) makes it act independently on each block of replicas, leading to the replacement

$$g(1, h_l)^{m_0} \rightarrow \gamma_{q_1 - q_0} \star g(1, h_l)^{m_0} = g(m_0, h_l)$$

(5.74)

in each block. From this point on, there is no need to consider different fields h_l in different blocks because the remaining operator acts on all replicas via Eq. (4.71). Therefore, we can consider that $h_l = h$, and

$$\prod_l g(m_0, h_l) \rightarrow g(m_0, h)^{s/m_0}.$$

(5.75)

3. Replacing the operator $\frac{q_0}{2} \left(\sum_a \frac{\partial}{\partial h_a} \right)^2$ with $\frac{q_0}{2} \frac{\partial^2}{\partial h^2}$ via Eq. (4.71) gives

$$f^d(\hat{q}) = e^{\frac{q_0}{2} \frac{\partial^2}{\partial h^2}} g(m_0, h)^{s/m_0} = \gamma_{q_0} \star g(m_0, h)^{s/m_0},$$

(5.76)

which should be computed in $h = H$.

Collecting the results, we obtain

$$f^d(\hat{q}) = \gamma_{q_0} \star \left[\gamma_{q_1 - q_0} \star \left[\gamma_{q_d - q_1} \star r(H) \right]^{m_0} \right]^{s/m_0}, \tag{5.77}$$

and

$$f^d[q(x); q_d] = \lim_{s \to 0} \frac{d}{ds} f^d(\hat{q}) = \frac{1}{m_0} \gamma_{q_0} \star \log \left[\gamma_{q_1 - q_0} \star \left[\gamma_{q_d - q_1} \star r(H) \right]^{m_0} \right]. \tag{5.78}$$

The kRSB Case

The 1RSB construction we just discussed can be immediately generalised to a kRSB structure with finite k. Using the conventions

$$q_{-1} = 0, \qquad q_{k+1} = q_d, \qquad m_{-1} = s, \qquad m_k = 1, \tag{5.79}$$

we can introduce an index, $l^{(i)} \in \{1, \ldots, s/m_i\}$, to label the blocks at level i of the hierarchical structure. Note that $l^{(k)} \in \{1, \ldots, s\}$ coincides with the original replica index a, while $l^{(-1)} \in \{1\}$ can take a single value corresponding to the block of all replicas. As for the 1RSB case, we write $a \in l^{(i)}$ if replica a belongs to block $l^{(i)}$. With those conventions, the differential operator in Eq. (5.66) can be written as

$$\frac{1}{2} \sum_{a,b} q_{ab} \frac{\partial^2}{\partial h_a \partial h_b} = \sum_{i=-1}^{k} \frac{(q_{i+1} - q_i)}{2} \sum_{l^{(i)}} \left(\sum_{a \in l^{(i)}} \frac{\partial}{\partial h_a} \right)^2. \tag{5.80}$$

As in the 1RSB case, the action of this operator on the product $\prod_a r(h_a)$ can then be evaluated in $k + 2$ steps. Because they are very similar to the 1RSB case, we describe them succinctly, the main purpose being the introduction of the notational convention.

1. The operator $\frac{q_d - q_k}{2} \sum_a \frac{\partial^2}{\partial h_a^2}$ acts independently on each replica, leading to

$$r(h) \to \gamma_{q_d - q_k} \star r(h) = g(m_k, h),$$

$$\prod_a r(h_a) \to \prod_{l^{(k-1)}} g(m_k, h_{l^{(k-1)}})^{m_{k-1}/m_k}, \tag{5.81}$$

recalling that $m_k = 1$ and that in each of the $l^{(k-1)}$ blocks there are m_{k-1} replicas.

2. The operator $\frac{q_k - q_{k-1}}{2} \sum_{l^{(k-1)}} \frac{\partial^2}{\partial h_{l^{(k-1)}}^2}$ acts independently on each $l^{(k-1)}$ block of replicas, leading to the replacement

$$g(m_k, h)^{m_{k-1}/m_k} \to \gamma_{q_k - q_{k-1}} \star g(m_k, h)^{m_{k-1}/m_k} = g(m_{k-1}, h) \tag{5.82}$$

in each block. All the $g(m_{k-1}, h)$ that pertain to the same $l^{(k-2)}$ block can now be grouped together. There are m_{k-2}/m_{k-1} sub-blocks in the same block, hence

$$\prod_{l^{(k-1)}} g(m_k, h_{l^{(k-1)}})^{m_{k-1}/m_k} \rightarrow \prod_{l^{(k-2)}} g(m_{k-1}, h_{l^{(k-2)}})^{m_{k-2}/m_{k-1}}. \qquad (5.83)$$

3. The process is iterated for all i, leading to the recursion relation

$$g(m_{i-1}, h) = e^{\frac{q_i - q_{i-1}}{2} \frac{\partial^2}{\partial h^2}} g(m_i, h)^{m_{i-1}/m_i} = \gamma_{q_i - q_{i-1}} \star g(m_i, h)^{m_{i-1}/m_i}. \qquad (5.84)$$

4. After all steps are completed, one obtains

$$f^d(\hat{q}) = e^{\frac{q_0}{2} \frac{\partial^2}{\partial h^2}} g(m_0, h)^{s/m_0} = \gamma_{q_0} \star g(m_0, h)^{s/m_0}, \qquad (5.85)$$

which should be evaluated in $h = H$.

In summary, one obtains

$$\begin{aligned}
g(m_k, h) &= \gamma_{q_d - q_k} \star r(h), \\
g(m_{i-1}, h) &= \gamma_{q_i - q_{i-1}} \star g(m_i, h)^{m_{i-1}/m_i}, \qquad i = k, \dots, 1, \\
f^d[q(x); q_d] &= \lim_{s \to 0} \frac{d}{ds} f^d(\hat{q}) = \frac{1}{m_0} \gamma_{q_0} \star \log g(m_0, H).
\end{aligned} \qquad (5.86)$$

The fullRSB Case

In the limit $k \to \infty$, assuming that $q(x)$ has the form illustrated in Figure 5.4, one can take a continuum limit of Eq. (5.86). At $m_k = 1$, we have $g(1, h) = \gamma_{q_d - q(1)} \star r(h)$. First, we note that for $x \in [x_M, 1]$, $q(x)$ is constant, $q(x) = q(1) = q_M$. Any discretisation of x would lead, using Eq. (5.84), to

$$g(x, h) = g(1, h)^x, \qquad x \in [x_M, 1]. \qquad (5.87)$$

We then consider the generic i-iteration step given by Eq. (5.84). In the region $x \in [x_m, x_M]$, $q(x)$ has a continuum limit with $\dot{q}(x) \neq 0$; one thus has

$$q_i - q_{i-1} \sim \dot{q}(x) dx, \qquad (5.88)$$

where $x = m_i$, $dx = m_i - m_{i-1}$, and the iteration Eq. (5.84) becomes

$$g(x - dx, h) = e^{\frac{1}{2} \dot{q}(x) dx \frac{\partial^2}{\partial h^2}} g(x, h)^{1 - dx/x}. \qquad (5.89)$$

By developing at linear order in dx we get the differential equation

$$\frac{\partial g(x, h)}{\partial x} = -\frac{1}{2} \dot{q}(x) \frac{\partial^2 g(x, h)}{\partial h^2} + \frac{1}{x} g(x, h) \log g(x, h), \qquad (5.90)$$

and defining

$$f(x,h) = \frac{1}{x} \log g(x,h) \tag{5.91}$$

gives a differential equation for $f(x,h)$,

$$\frac{\partial f(x,h)}{\partial x} = -\frac{1}{2}\dot{q}(x) \left[\frac{\partial^2 f(x,h)}{\partial h^2} + x \left(\frac{\partial f(x,h)}{\partial h} \right)^2 \right]. \tag{5.92}$$

Note that both Eq. (5.87) and Eq. (5.92) imply that when $\dot{q}(x) = 0$ – i.e., $q(x)$ is constant – the function $f(x,h)$ is independent of x. The initial condition for $f(x,h)$ can then be equivalently specified in $x = 1$ or in $x = x_M$ and is given by

$$f(1,h) = f(x_M,h) = \log g(1,h) = \log \gamma_{q_d - q(1)} \star r(h). \tag{5.93}$$

This initial condition should be evolved down to $x = x_m$ using Eq. (5.92). According to Eq. (5.86), the final value of the free energy is given by $f^d[q(x);q_d] = \gamma_{q(0)} \star f(m_0, H)$. The specific value of m_0 is irrelevant as long as $m_0 < x_m$, because $f(x,h)$ is independent of x for $x < x_m$. One can thus choose $m_0 = 0$. In summary, the free energy is given by the solution of

$$f(1,h) = \log \gamma_{q_d - q(1)} \star r(h),$$

$$\dot{f}(x,h) = -\frac{1}{2}\dot{q}(x) \left(f''(x,h) + xf'(x,h)^2 \right), \qquad 0 < x < 1, \tag{5.94}$$

$$f^d[q(x);q_d] = \lim_{s \to 0} \frac{d}{ds} f^d(\hat{q}) = \gamma_{q(0)} \star f(0,H),$$

recalling that all the functions are constant in the regions where $\dot{q}(x) = 0$, and using primes to denote derivatives with respect to h. The differential term defined by Eq. (5.66) can thus be computed implicitly by solving the partial differential Eq. (5.92). Note that Eqs. (5.94), which were first introduced in [285], are very general. They only depend on the differential structure of the term $f^d(\hat{q})$ and on the hierarchical structure of the matrix $\hat{q}$.

5.4.2 Variational Equations

We found that, for the fullRSB case, a generic replicated free energy, in the limit $s \to 0$, is given by

$$f[q(x);q_d] = f^a[q(x);q_d] + \gamma_{q(0)} \star f(0,H), \tag{5.95}$$

where $f^a[q(x);q_d]$ is a simple functional of $q(x)$ and q_d, and the second term depends on the function $f(x,h)$ that satisfies Eq. (5.94). The values of q_d and $q(x)$ have to be obtained as stationary points of the replicated free energy. This operation is difficult because unlike $f^a[q(x);q_d]$, for which the explicit dependence on $q(x)$ and q_d (and therefore the derivatives) are known, the term $f^d[q(x);q_d]$ is given implicitly in terms of the solution of Eq. (5.94). A standard way to solve this problem is to introduce Lagrange multipliers that enforce both the partial differential equation in (5.92) and its initial condition [333]. One can therefore define a variational replicated free energy as

$$f^L[q(x);q_d] = f^a[q(x);q_d] + \gamma_{q(0)} \star f(0,H)$$

$$- \int_{-\infty}^{\infty} dh\, P(1,h)\left[f(1,h) - \log \gamma_{q_d - q(1)} \star r(h)\right] \qquad (5.96)$$

$$+ \int_0^1 dx \int_{-\infty}^{\infty} dh\, P(x,h)\left[\dot{f}(x,h) + \frac{1}{2}\dot{q}(x)\left(f''(x,h) + xf'(x,h)^2\right)\right].$$

The Lagrange multipliers $P(x,h)$ and $P(1,h)$ actually have a physical meaning. For example, in the Sherrington–Kirkpatrick spin glass model, they encode the distribution of local magnetic fields within metabasins at level x in the hierarchical structure of states [254]. Having introduced the Lagrange multipliers $P(x,h)$ and $P(1,h)$, the variations must be taken in the enlarged space of all independent functions. These are the two functions $P(x,h)$ and $f(x,h)$, their initial conditions $P(1,h)$ and $f(1,h)$, the function $q(x)$ and q_d. Note that the initial condition $f(1,h)$ can also be replaced by the final condition $f(0,h)$ because the equation for $f(x,h)$ can be solved in both directions in x. This substitution has the advantage that $f(0,h)$ appears explicitly in the first line of Eq. (5.96). Differentiating the variational free energy with respect to $P(1,h)$, $P(x,h)$, $f(x,h)$ and $f(0,h)$ gives

$$f(1,h) = \log \gamma_{q_d - q(1)} \star r(h),$$

$$\dot{f}(x,h) = -\frac{1}{2}\dot{q}(x)\left(f''(x,h) + xf'(x,h)^2\right),$$

$$\dot{P}(x,h) = \frac{1}{2}\dot{q}(x)\left[P''(x,h) - 2x\left(P(x,h)f'(x,h)\right)'\right], \qquad (5.97)$$

$$P(0,h) = \frac{1}{\sqrt{2\pi q(0)}}e^{-\frac{(h-H)^2}{2q(0)}}.$$

The last two equations are obtained by integrating by parts the last line of Eq. (5.96) in order to make explicit the dependence on $f(x,h)$ and $f(0,h)$. Note that these equations are very general because they follow only by the form of the differential term $f^d[q(x);q_d]$ and its expression in terms of Eq. (5.94).

Finally, taking the derivatives with respect to q_d and $q(x)$ gives

$$\frac{df^a[q(x);q_d]}{dq_d} = -\frac{1}{2}\frac{\gamma_{q_d - q(1)} \star r''(h)}{\gamma_{q_d - q(1)} \star r(h)},$$

$$\frac{\delta f^a[q(x);q_d]}{\delta q(x)} = \frac{1}{2}\int dh\, P(x,h)f'(x,h)^2. \qquad (5.98)$$

Eqs. (5.97) and (5.98) form a complete set of equations that can be solved numerically to fully determine $f(x, h)$, $P(x, h)$, $q(x)$ and q_d. Once this is done, the free energy is obtained from Eq. (5.95).

5.5 De Almeida–Thouless Transition and Marginal Stability

The fullRSB variational equations obtained in Section 5.4.2 contain the kRSB ansatz, for any finite k, as a particular case. Because the kRSB matrices form a closed algebra, a kRSB ansatz for $q(x)$ always provides a consistent solution of the variational equations. This is true, in particular, for the case $k = 0$, which corresponds to a constant $q(x) = q_0$, and, hence, a RS solution always exists. One can ask, however, whether this solution is the correct one. As discussed in Section 5.1.3, the RS solution sometimes has (at least) one negative eigenvalue of the stability matrix, and it then corresponds to a saddle point in the space of $q(x)$, which makes it linearly unstable towards RSB. In this section, we discuss, using Eqs. (5.97) and (5.98), how this instability can occur via a continuous phase transition called the de Almeida–Thouless (dAT) transition [113, 254]. In order to discuss the stability of the RS solution, instead of computing the whole spectrum of the stability matrix, we take a shortcut due to Sommers [331, 332].

Our derivation relies on the assumption that the derivatives of $f^a[q(x); q_d]$ have a simple structure. An example, which we use in this section, is

$$\frac{d}{dx} \frac{\delta f^a[q(x); q_d]}{\delta q(x)} = f_a^{(2)}[q(x); q_d] \dot{q}(x),$$

$$\frac{d}{dx} f_a^{(n)}[q(x); q_d] = f_a^{(n+1)}[q(x); q_d] \dot{q}(x), \qquad n \geq 2. \tag{5.99}$$

One can check that Eq. (5.99) is true for the examples of $f^a[q(x); q_d]$ discussed in Section 5.4. In other examples – in particular, the one discussed in Chapter 6 – the term $\dot{q}(x)$ in Eq. (5.99) is replaced by the Fourier transform, $\dot{\lambda}(x) = -x\dot{q}(x)$, introduced in Eq. (5.28). This change only affects certain details of the formulae we derive in this section, but the overall procedure is generic.

5.5.1 Instability of the Replica Symmetric Solution

The second of Eqs. (5.98) holds for all $x \in [0, 1]$. Its derivative with respect to x, under the assumption of Eq. (5.99), gives

$$f_a^{(2)}[q(x); q_d] \dot{q}(x) = \frac{1}{2} \frac{d}{dx} \int dh\, P(x, h) f'(x, h)^2. \tag{5.100}$$

Using Eqs. (5.97), one can show[15] that

$$\frac{d}{dx} \int dh\, P(x,h) f'(x,h)^2 = \dot{q}(x) \int dh\, P(x,h) f''(x,h)^2, \tag{5.101}$$

and, therefore, Eq. (5.100) becomes

$$f_a^{(2)}[q(x); q_d]\, \dot{q}(x) = \frac{\dot{q}(x)}{2} \int dh\, P(x,h) f''(x,h)^2. \tag{5.102}$$

If there exist an interval for $x \in [x_m, x_M] \subseteq [0,1]$ over which $\dot{q}(x) \neq 0$, we can simplify the factor $\dot{q}(x)$ in Eq. (5.102) and obtain the relation

$$f_a^{(2)}[q(x); q_d] = \frac{1}{2} \int dh\, P(x,h) f''(x,h)^2, \qquad x \in [x_m, x_M]. \tag{5.103}$$

Let us consider a model for which replica symmetry is continuously broken at a dAT phase transition. In the replica symmetry broken region, close to the phase transition, we have a finite region $x \in [x_m, x_M]$ in which Eq. (5.103) holds. Upon approaching the phase transition point, $x_M - x_m \to 0$ so that the replica symmetric solution is recovered, and $q(x)$ becomes constant.[16] Yet, by continuity, at the transition point, Eq. (5.103) is satisfied at $x = x_m = x_M$. Because $q(x) = q_0$ is constant at the transition point, we have, in particular, $q(x_m) = q(x_M) = q_0$, and $P(x,h)$ and $f(x,h)$ are independent of x. We then obtain the condition

$$f_a^{(2)}[q(x); q_d]|_{q(x)=q_0} = \frac{1}{2} \int dh\, P(0,h) f''(1,h)^2, \tag{5.104}$$

where $f(1,h)$ and $P(0,h)$ are given by Eqs. (5.97) with $q(0) = q(1) = q_0$. Starting from the fullRSB equations, one can therefore obtain a simple condition for the instability of the replica symmetric solution. Eq. (5.104) is equivalent to the vanishing of one of the eigenvalues of the stability matrix.

5.5.2 Perturbative Expansion Close to the dAT Transition

In Section 5.5.1, we have shown that the instability of the RS solution is obtained from Eq. (5.104). Next, we can compute the properties of the function $q(x)$ close to that instability, in order to understand what happens when replica symmetry breaks down. In this section, we first derive an equation for the 'breaking point' x^* where the solution $q(x)$ develops a deviation from the constant profile – i.e., $\dot{q}(x^*) \neq 0$ – and we then derive an equation for the value of $\dot{q}(x^*)$. As we show next, these two numbers characterise the RSB solution in the vicinity of the dAT instability.

---

[15] The proof is based on using Eqs. (5.97) to replace $\dot{P}(x,h)$ and $\dot{f}(x,h)$ by derivatives over h and then integrating by parts on h. Note that the boundary terms at $h \to \pm\infty$ in the integrations by parts vanish because of the asymptotic properties of $P(x,h)$ and $f(x,h)$.

[16] A concrete example will be given in Figure 6.2 in a slightly different context.

The Breaking Point

Taking the derivative of Eq. (5.103) with respect to x, using Eqs. (5.99) and the relation

$$\frac{d}{dx} \int dh\, P(x,h) f''(x,h)^2 =$$
$$= \dot{q}(x) \left[\int dh\, P(x,h) f'''(x,h)^2 - 2x \int dh\, P(x,h) f''(x,h)^3 \right], \tag{5.105}$$

which follows from the variational Eqs. (5.97), one obtains

$$2f_a^{(3)}[q(x); q_d] = \int dh\, P(x,h) f'''(x,h)^2 - 2x \int dh\, P(x,h) f''(x,h)^3. \tag{5.106}$$

This equation can be inverted to obtain

$$x = \frac{\int dh\, P(x,h) f'''(x,h)^2 - 2f_a^{(3)}[q(x); q_d]}{2 \int dh\, P(x,h) f''(x,h)^3}. \tag{5.107}$$

Evaluating this last equation on the replica symmetric solution, at the dAT phase transition when the marginal stability condition in Eq (5.104) is satisfied, one obtains an expression for the breaking point x^*. There are then two possibilities:

1. $x^* \in [0,1]$: the instability of the replica symmetric phase gives rise to a continuous transition to a genuine RSB phase. The broken phase could be fullRSB or kRSB with finite k.
2. $x^* \notin [0,1]$: the dAT transition is unphysical and must be preceded by another instability, typically a discontinuous RSB transition.

The Nature of the Broken Phase

We now assume that $x^* \in [0,1]$ and that $q(x)$ is analytic in the vicinity of the transition point and of $x = x^*$. Taking the derivative of Eq. (5.106) with respect to x, and using Eqs. (5.99) and the relation

$$\frac{d}{dx} \left[\int dh\, P(x,h) f'''(x,h)^2 - 2x \int dh\, P(x,h) f''(x,h)^3 \right] =$$
$$= \dot{q}(x) \int dh\, P(x,h) q_4(x,h) - 2 \int dh\, P(x,h) f''(x,h)^3, \tag{5.108}$$

$$q_4(x,h) = f''''(x,h)^2 - 12x f''(x,h) f'''(x,h)^2 + 6x^2 f''(x,h)^4,$$

which again follows from Eqs. (5.97), we obtain

$$\dot{q}(x) = \frac{2 \int dh\, P(x,h) f''(x,h)^3}{\int dh\, P(x,h) q_4(x,h) - 2f_a^{(4)}[q(x); q_d]}. \tag{5.109}$$

Evaluating this equation on the replica symmetric solution in x^* has two possible outcomes:

1. $\dot{q}(x^*) > 0$: the solution is of the continuous fullRSB type, and the profile $q(x)$ is continuous in the vicinity of the transition point.
2. $\dot{q}(x^*) < 0$: the fullRSB solution cannot be accepted, because the term $f^d[q(x); q_d]$ is well defined only for monotonously increasing functions, as follows from Eq. (5.86), which is not well defined if $q_i < q_{i-1}$. Usually, the transition point is then a continuous transition towards a 1RSB phase, in which the profile $q(x)$ has a discontinuous jump at $x = x^*$.

Going on with the Expansion

When the dAT instability is a transition towards a continuously broken phase, one can iterate this approach to compute higher derivatives of $q(x)$ at the breaking point x^*. In this way, one can construct a series expansion of $q(x)$ around the replica symmetric solution near the phase transition point. This strategy was carried out extensively in the study of the Sherrington–Kirkpatrick model close to the transition temperature [107]. This analysis shows that the transition is continuous and thermodynamically of third order. In other words, the derivatives of the free energy up to the second order are continuous, but from the third order on, they display a jump with no divergence.

5.5.3 Marginal Stability

The existence of multiple statistically equivalent equilibrium states is the distinctive hallmark of replica symmetry breaking. The way in which these states are distributed in phase space defines two very different scenarios.

1. States are scattered in phase space in a nearly random way and stay at a minimum distance from one another, and the free energy barriers separating them are very high. In this situation, if the system is inside one state, it does not feel the existence of the other equilibrium states. This scenario physically corresponds to stable glasses and mathematically corresponds to a finite (and typically small) number of replica symmetry breaking steps.
2. States are not randomly distributed in phase space, but each state is surrounded by a large number of other states that are arbitrarily close. The barriers that separate one state from another might then be very small. This situation mathematically corresponds to full replica symmetry breaking. Physically, if the system is inside a state, it can move in phase space in many directions (typically corresponding to the directions of nearby states) without the free energy increasing too much. In other words, there are nearly flat directions in

the free energy function, exactly as at a second-order phase transition point. The spectrum of small oscillations within an equilibrium state then displays an excess of low-frequency modes that in some cases dominate over those coming from more conventional sources (e.g., phonons or magnons) [150, 254]. In the limit $s \to 0$, the permutation symmetry indeed becomes similar to a continuous symmetry, and one can thus derive Ward identities and the existence of Goldstone modes [117]. The precise form of this anomalous low-frequency spectrum and the localisation properties of the eigenvalues depend on the details of the model, but marginally stable glasses of this kind are all self-organised critical systems.

The two scenarios described previously differ in many physically observable ways:

- Correlation functions of non-conserved local quantities, in both space and time, decay exponentially in stable glasses, while they decay as power laws in marginally stable glasses. This is a consequence of the stability matrix in Eq. (5.9) having zero modes.
- In marginally stable glasses, there is a divergent linear susceptibility, given by a four-point correlation $\chi_{ab;cd} = \langle q_{ab}q_{cd} \rangle - \langle q_{ab} \rangle \langle q_{cd} \rangle$, which is the inverse of the stability matrix. The physical meaning of $\chi_{ab;cd}$ depends on the model, but it is generally the response of $\langle q_{ab} \rangle$ to a field conjugated to q_{ab}. In spin glass models, this can be written as the so-called spin glass susceptibility [254]. In addition, non-linear susceptibilities are finite in stable glasses, while they are often infinite in marginally stable glasses [64].
- If we consider a marginally stable glass on which an infinitesimal perturbation is applied, typically, the system changes state, leading to a rearrangement (an 'avalanche') of the order of the size of the system. These avalanches are power law distributed at zero temperature, with an exponent related to the shape of the function $x(q)$ [148, 225]. For both stable and marginally stable glasses, the application of a finite external perturbation may force the system to jump from one state to another, quite different, state. This behaviour is said to be 'chaotic' because the set of equilibrium states has a chaotic dependence on the parameters of the Hamiltonian [68].

5.6 Wrap-Up

5.6.1 Summary

In this chapter, we have provided a compendium of mathematical and physical results on replica symmetry breaking. In particular, we have seen that

- Replica symmetry breaking corresponds to the coexistence of a large number of equivalent thermodynamic states. In disordered systems, the Hamiltonian has no explicit symmetry, and the coexistence of many states results from the disorder. The states are statistically equivalent and can thus coexist over a full region of the phase diagram (Section 5.1).

- In all known mean field models, replica symmetry breaking is mathematically described by a particular, hierarchical, structure of the mean square displacement matrix Δ_{ab} (or overlap matrix q_{ab}). Hierarchical matrices are defined, for finite number s of replicas, by a block structure with k levels, called a kRSB structure. These matrices have many interesting properties: (1) at fixed k, they form a closed commutative algebra; (2) they break the replica symmetry by inducing correlations between replicas but preserving the statistical equivalence of each replica; (3) a kRSB hierarchical matrix $\hat{q}$ is fully specified by a finite set of $2k+2$ parameters $\{m_i\}, \{q_i\}, q_d$ for any s and, thus, an explicit analytical continuation to any real s and in particular to $s \to 0$ is possible (Section 5.2.1).

- When the number of RSB steps k goes to infinity, the $\{m_i\}$ accumulate and become a continuous parameter $x \in [0,1]$ (Section 5.2.2). The resulting fullRSB matrices are then parametrised by $\{q(x); q_d\}$, where q_d is the diagonal element and $q(x)$ is a continuous function of $x \in [0,1]$, which encodes the off-diagonal elements. FullRSB matrices also form a closed commutative algebra and allow an explicit analytical continuation to real s. The algebra of fullRSB matrices has interesting mathematical properties. In particular, it is diagonalised by a simple Fourier transform, which permits the easy computation of the inverse and determinant of such matrices (Section 5.2.3).

- In a replica symmetry broken phase, the coexisting thermodynamic states have a non-trivial distribution of mean square displacements, $\mathcal{P}(\Delta)$, which can be expressed in terms of the matrix Δ_{ab} (Section 5.3.1). For the replica symmetric case ($k = 0$), $\mathcal{P}(\Delta)$ is a delta function; for kRSB, $\mathcal{P}(\Delta)$ has $k + 1$ delta peaks; for the fullRSB limit, $\mathcal{P}(\Delta)$ has a finite support on an interval $[\Delta_M, \Delta_m]$ (Section 5.3.2). From the distribution of distances between three states, one can further infer that the states are organised in an ultrametric tree, which can be generated by a simple Poisson process (Section 5.3.3).

- In many mean field models, the replicated free energy is expressed as a function of the overlap matrix $\hat{q}$, by the sum of two terms: a simple 'algebraic' term and a 'differential' term of a particular form. Thanks to this structure, if $\hat{q}$ is a hierarchical matrix, the free energy can be compactly written in the limit $s \to 0$ as a functional of $\{q(x); q_d\}$, in terms of a differential equation for a function $f(x,h)$. In addition, the variational equations for $\{q(x); q_d\}$ can be written explicitly by introducing a Lagrange multiplier $P(x,h)$ conjugated to $f(x,h)$. One ends up with two differential equations that must be solved to obtain the thermodynamic free energy (Section 5.4).

- From the differential equations, one can investigate the stability of the replica symmetric solution via a simple argument and derive a condition for the instability point, which defines the de Almeida–Thouless (dAT) phase transition (Section 5.5.1). A perturbative expansion for the non-constant part of $q(x)$ can be developed around the dAT transition in order to investigate the nature of the replica symmetry broken phase in the vicinity of the transition (Section 5.5.2). A fullRSB phase is characterised by marginal stability, which leads to peculiar mathematical and physical properties (Section 5.5.3).

5.6.2 Further Reading

We provide here a list of references that can be consulted to further explore the subjects discussed in this chapter, selected according to the criteria discussed in Section 1.6.2.

As mentioned in the introduction to this chapter, replica symmetry breaking has been applied to a variety of problems, giving rise to a very large literature. Several books and reviews provide solid access points to this literature:

- Binder and Young, *Spin glasses: Experimental facts, theoretical concepts, and open questions* [51]
- Mézard, Parisi and Virasoro, *Spin glass theory and beyond* [254]
- Fischer and Hertz, *Spin Glasses* [140]
- Nishimori, *Statistical physics of spin glasses and information processing: An introduction* [274]
- Mézard and Montanari, *Information, Physics and Computation* [251]
- Talagrand, *Spin glasses: A challenge for mathematicians; Cavity and mean field models* [340]
- Talagrand, *Mean field models for spin glasses: Volume I; Basic examples* [341]
- Panchenko, *The Sherrington–Kirkpatrick model* [281]

The reader can consult the first four references [51, 140, 254, 274] for a basic introduction to the subject, including applications to spin glasses, neural networks, and information theory. The fifth one [251] provides a more recent and complete introduction to the application to information theory, including mathematically rigorous results. The last three [281, 340, 341] provide introductions to the subject for more probabilistically oriented readers.

The existence of replica symmetry broken phases is well established in mean field models. Their existence in finite dimensional models is, however, the subject of a hot debate. A variety of renormalisation group methods have been developed to investigate this question. From the field theory point of view, an introductory book and a few more recent papers are

- De Dominicis and Giardina, *Random fields and spin glasses: A field theory approach* [115]
- Moore and Bray, *Disappearance of the de Almeida–Thouless line in six dimensions* [264]
- Parisi and Temesvári, *Replica symmetry breaking in and around six dimensions* [290]
- Castellana and Parisi, *Non-perturbative effects in spin glasses* [78]
- Charbonneau, Hu, Raju et al., *Morphology of renormalization-group flow for the de Almeida–Thouless–Gardner universality class* [95]

Real space scaling arguments and renormalisation group techniques have also been used, a few examples being

- Fisher and Huse, *Equilibrium behavior of the spin-glass ordered phase* [141]
- Yeo and Moore, *Renormalization group analysis of the $M - p$-spin glass model with $p = 3$ and $M = 3$* [364]
- Angelini, Parisi and Ricci-Tersenghi, *Ensemble renormalization group for disordered systems* [15]
- Angelini and Biroli, *Spin glass in a field: A new zero-temperature fixed point in finite dimensions* [14]

Detailed numerical simulations have also been carried out in spin glass models, and their accuracy improves in step with the quick growth of standard computer performances. The use of dedicated supercomputers within the Janus collaboration has now matched the numerically and experimentally investigated time scales and has greatly improved the system sizes studied in equilibrium. A sample of results can be found in

- Marinari, Parisi, Ricci-Tersenghi et al., *Replica symmetry breaking in short-range spin glasses: Theoretical foundations and numerical evidences* [244]
- Leuzzi, Parisi, Ricci-Tersenghi et al., *Dilute one-dimensional spin glasses with power law decaying interactions* [229]
- Larson, Katzgraber, Moore et al., *Spin glasses in a field: Three and four dimensions as seen from one space dimension* [224]
- Wang, Machta, Munoz-Bauza et al., *Number of thermodynamic states in the three-dimensional Edwards-Anderson spin glass* [355]
- Janus collaboration, *An in-depth view of the microscopic dynamics of Ising spin glasses at fixed temperature* [32]
- Janus collaboration, *Nature of the spin-glass phase at experimental length scales* [26]
- Janus collaboration, *Critical parameters of the three-dimensional Ising spin glass* [23]

6

The Gardner Transition

In Chapter 4, we discussed the Franz–Parisi, or state following, construction and presented results for two typical potentials within the replica symmetric ansatz. In Chapter 5, we then saw that this replica symmetric ansatz can become unstable, which leads to a phase transition characterised by spontaneous replica symmetry breaking. In this chapter, we apply the general tools presented in Chapter 5 to the state following construction of Chapter 4. We show that deep into the glass a phase transition, at which replica symmetry is spontaneously broken, can happen. The resulting transition is known as a Gardner transition. We then discuss the impact of this phase transition on the phase diagram of hard and soft spheres, as well as general properties of the replica symmetry broken phase.

6.1 State Following in the Replica Symmetry Broken Phase

We begin by recalling some results of Chapter 4 and deriving from them the expression of the glass free energy in the replica symmetry broken (RSB) phase. As discussed in Section 4.1, we are interested in the average free energy $f_g(\varphi, T | \varphi_g, T_g)$ of glass states prepared at some initial equilibrium state point (φ_g, T_g) and then followed adiabatically to a new state point (φ, T). As shown in Section 4.2, this free energy is expressed, in the limit $d \to \infty$, as the extremum of a function of a $s \times s$ matrix $\hat{\alpha}$ or, equivalently, of a $n \times n$ (with $n = s + 1$) matrix $\hat{\Delta}$ in the limit $s \to 0$. The matrices $\hat{\alpha}$ and $\hat{\Delta}$ encode the phase space distances between the n replicas introduced to average over glass states. Replica 1 is special because it encodes the initial state at (φ_g, T_g), but all the other replicas, $a = 2, \ldots, n$ are equivalent. As a consequence, the mean square displacement, $\Delta_{a1} = \Delta_{1a} = \Delta_r$, is independent of $a > 1$ (recall that $\Delta_{11} = 0$). For $a, b = 2, \ldots, n$ we have, following Eq. (4.64):

$$\alpha_{ab} = (d/\ell_g^2) \langle (\mathbf{x}_a - \mathbf{x}_1) \cdot (\mathbf{x}_b - \mathbf{x}_1) \rangle = \Delta_r - \Delta_{ab}/2, \qquad (6.1)$$

where ℓ_g is the typical interaction scale of the potential in the initial state. One should keep in mind that replica symmetry only holds for the s replicas – i.e., $a = 2, \ldots, n$ – and, therefore, spontaneous RSB can only occur in that same sector. Replica 1 always remains distinct. In Section 4.3, we used the simplest assumption of replica symmetry for this sector and derived from it the replica symmetric (RS) expression of the glass free energy. From this, in Section 4.4, we obtained the corresponding phase diagram of glassy states. We now discuss the stability of this ansatz against RSB.

6.1.1 The Glass Free Energy

A convenient starting point is the expression of the free energy as a function of $\hat{\alpha}$, given by Eq. (4.54). This result should be adapted to the state following formalism, as discussed in Section 4.2.4. In particular, we consider $n = s + 1$ replicas, omit the factor β and the kinetic part of the ideal gas term of the free energy and keep replica 1 in state point $(\widehat{\varphi}_g, T_g)$ while the others are in state point $(\widehat{\varphi}, T)$, with $\widehat{\varphi} = \widehat{\varphi}_g e^{\eta}$, as defined in Eq. (4.57). We then obtain

$$
\begin{aligned}
-\mathrm{f}_{s+1}(\hat{\alpha}) = & -\log(\rho\ell_g^d) + \frac{d}{2}\frac{s}{2}\log\left(\frac{2\pi e}{d^2}\right) + \frac{d}{2}\log\det\hat{\alpha} \\
& + \frac{d\widehat{\varphi}_g}{2}\int_{-\infty}^{\infty} \mathrm{d}h\, e^h \left\{ e^{-\beta_g\bar{v}(h)}\,\mathrm{f}^\mathrm{d}(\hat{\alpha},h) - 1 \right\},
\end{aligned}
$$

$$
\mathrm{f}^\mathrm{d}(\hat{\alpha},h) = e^{\sum_{a,b=2}^n \alpha_{ab}\frac{\partial^2}{\partial h_a \partial h_b}} \left[e^{-\beta\sum_{a=2}^n \bar{v}(h_a-\eta+\alpha_{aa})} \right]\Big|_{h_a=h}.
$$

$$(6.2)$$

After extremising over $\hat{\alpha}$, the glass free energy is recovered taking the derivative with respect to s and setting $s = 0$, as in Eq. (4.16).

We now assume that the matrix $\hat{\Delta}$ has a hierarchical structure, as discussed in Chapter 5, and follow the procedure detailed in Section 5.4 to obtain the fullRSB expression of the free energy. Because the diagonal term $\Delta_{aa} = 0$, the hierarchical matrix is fully encoded by a function $\Delta(x)$. Correspondingly, the matrix $\hat{\alpha}$ is encoded, according to Eq. (6.1), by

$$
\hat{\alpha} \to \{\alpha(x) = \Delta_r - \Delta(x)/2; \alpha_d = \Delta_r\}. \tag{6.3}
$$

Our aim is to express the free energy as a function of Δ_r and $\Delta(x)$. The determinant of $\hat{\alpha}$ can then be obtained by substituting Eq. (6.3) into the general expression given in Eq. (5.29):

$$
\lim_{s\to 0}\frac{d}{ds}\log\det\hat{\alpha} = \log\left(\frac{\lambda(1)}{2}\right) + \frac{2\Delta_r - \Delta(0)}{\lambda(0)} - \int_0^1 \mathrm{d}x\,\frac{\dot{\Delta}(x)}{\lambda(x)}, \tag{6.4}
$$

where[1]

$$\lambda(x) = x\Delta(x) + \int_x^1 dy\,\Delta(y),$$

$$\dot{\lambda}(x) = x\dot{\Delta}(x), \qquad \lambda(1) = \Delta(1), \tag{6.5}$$

$$\Delta(x) = \lambda(1) + \int_x^1 dy\,\dot{\lambda}(y)/y.$$

The differential term $f^d(\hat{\alpha}, h)$ that appears in Eq. (6.2) has the same structure as that discussed in Section 5.4.1. Similarly to Eq. (5.80), but taking into account that $\alpha_d = \Delta_r$, $\Delta_d = 0$ and $\alpha_i = \Delta_r - \Delta_i/2$, we thus have (all the sums are over $a, b = 2, \ldots, n$)

$$\sum_{a,b} \alpha_{ab} \frac{\partial^2}{\partial h_a \partial h_b} = \left(\Delta_r - \frac{\Delta_0}{2}\right)\left(\sum_a \frac{\partial}{\partial h_a}\right)^2$$

$$+ \sum_{i=0}^{k-1} \frac{(\Delta_i - \Delta_{i+1})}{2} \sum_{l^{(i)}} \left(\sum_{a \in l^{(i)}} \frac{\partial}{\partial h_a}\right)^2 + \frac{\Delta_k}{2} \sum_a \left(\frac{\partial}{\partial h_a}\right)^2. \tag{6.6}$$

Following the same steps as in Section 5.4.1, we thus obtain

$$g(m_k, h) = \gamma_{\Delta_k} \star e^{-\beta\bar{v}(h - \eta + \Delta_r)},$$

$$g(m_{i-1}, h) = \gamma_{\Delta_{i-1} - \Delta_i} \star g(m_i, h)^{m_{i-1}/m_i}, \qquad i = k, \ldots, 1, \tag{6.7}$$

$$f^d(\hat{\alpha}, h) = \gamma_{2\Delta_r - \Delta_0} \star g(m_0, h)^{s/m_0}.$$

Plugging these results into Eq. (6.2), taking the derivative with respect to s and the limit $s \to 0$, one obtains in the continuous RSB limit

$$-\beta f_g(\widehat{\varphi}, T | \widehat{\varphi}_g, T_g) = \frac{d}{2}\left[\log\left(\frac{\pi e \lambda(1)}{d^2}\right) + \frac{2\Delta_r - \Delta(0)}{\lambda(0)} - \int_0^1 dx \frac{\dot{\Delta}(x)}{\lambda(x)}\right]$$

$$+ \frac{d\widehat{\varphi}_g}{2} \int_{-\infty}^{\infty} dh\, e^{h - \beta_g \bar{v}(h)} \gamma_{2\Delta_r - \Delta(0)} \star f(0, h), \tag{6.8}$$

with

$$f(1, h) = \log \gamma_{\Delta(1)} \star e^{-\beta\bar{v}(h - \eta + \Delta_r)},$$

$$\dot{f}(x, h) = \frac{1}{2}\dot{\Delta}(x)\left(f''(x, h) + xf'(x, h)^2\right), \qquad 0 < x < 1. \tag{6.9}$$

One can check that in the RS case, where $\Delta(x) = \lambda(x) = \Delta$ are constant and $f(x, h)$ is independent of x, the results of Chapter 4, Eq. (4.73), are recovered.

---

[1] Note that in order to compensate for the factor $1/2$ appearing in Eq. (6.3), $\lambda(x)$ has here been multiplied by 2 with respect to the definition given in Eq. (5.28).

6.1.2 Variational Equations

The variational equations for Δ_r and $\Delta(x)$ can be obtained following the approach described in Section 5.4.2. Adding to the free energy the Lagrange multipliers $P(x,h)$ and $P(1,h)$ gives

$$
-\beta f^L = \frac{d}{2}\left[\log\left(\frac{\pi e\lambda(1)}{d^2}\right) + \frac{2\Delta_r - \Delta(0)}{\lambda(0)} - \int_0^1 dx\,\frac{\dot{\Delta}(x)}{\lambda(x)}\right]
$$

$$
+ \frac{d\widehat{\varphi}_g}{2}\int_{-\infty}^{\infty} dh\, e^{h-\beta_g\bar{v}(h)}\gamma_{2\Delta_r-\Delta(0)}\star f(0,h) \tag{6.10}
$$

$$
- \frac{d}{2}\int_{-\infty}^{\infty} dh\, P(1,h)\left[f(1,h) - \log\gamma_{\Delta(1)}\star e^{-\beta\bar{v}(h-\eta+\Delta_r)}\right]
$$

$$
+ \frac{d}{2}\int_0^1 dx\int_{-\infty}^{\infty} dh\, P(x,h)\left[\dot{f}(x,h) - \frac{\dot{\Delta}(x)}{2}\left(f''(x,h) + xf'(x,h)^2\right)\right],
$$

and differentiating with respect to $f(x,h)$ and $f(0,h)$ leads to the differential equation

$$
P(0,h) = \widehat{\varphi}_g\,\gamma_{2\Delta_r-\Delta(0)}\star e^{h-\beta_g\bar{v}(h)}, \tag{6.11}
$$

$$
\dot{P}(x,h) = -\frac{1}{2}\dot{\Delta}(x)\left[P''(x,h) - 2x\left(P(x,h)f'(x,h)\right)'\right], \quad 0 < x < 1.
$$

Note that the expression for $P(0,h)$ can also be rewritten using the identity in Eq. (4.89). Equations for Δ_r and $\Delta(x)$ are then obtained by differentiating Eq. (6.10) with respect to these two quantities. Using the identity

$$
\frac{d}{dx}\int_{-\infty}^{\infty} dh\, P(x,h)f'(x,h) = 0, \tag{6.12}
$$

which can be proven using Eqs. (6.9) and (6.11) and a few additional manipulations, one obtains

$$
\frac{1}{\lambda(0)} = -\frac{1}{2}\int_{-\infty}^{\infty} dh\, P(0,h)[f''(0,h) + f'(0,h)],
$$

$$
\frac{2\Delta_r - \Delta(0)}{\lambda(0)^2} - \int_0^x dy\,\frac{\dot{\Delta}(y)}{\lambda(y)^2} = \frac{1}{2}\int_{-\infty}^{\infty} dh\, P(x,h)f'(x,h)^2. \tag{6.13}
$$

Eqs. (6.9), (6.11) and (6.13) constitute a set of closed equations for $f(x,h)$, $P(x,h)$, $\Delta(x)$ and Δ_r. In solving these equations, it is convenient to use $\lambda(x)$ instead of $\Delta(x)$; once $\lambda(x)$ is determined, one can obtain $\Delta(x)$ using Eq. (6.5). The solution can then be obtained iteratively using the following numerical procedure.

1. Start from a reasonable guess for Δ_r and $\lambda(x)$.
2. From $\lambda(x)$, compute $\Delta(x)$ by Eq. (6.5).

3. Solve Eq. (6.9), starting from the initial condition in $x = 1$ and evolving towards $x = 0$ to obtain $f(x, h)$ for all x.

4. Solve Eq. (6.11), starting from the initial condition in $x = 0$ and evolving towards $x = 1$ to obtain $P(x, h)$ for all x.

5. Compute $K(x) = \frac{1}{2} \int_{-\infty}^{\infty} dh \, P(x, h) f'(x, h)^2$.

6. Obtain a new estimate of $\lambda(x)$ by first computing a new estimate of $\lambda(0)$ from the first Eq. (6.13) and then integrating the relation[2]

$$\dot{K}(x) = -\frac{\dot{\lambda}(x)}{x\lambda(x)^2} \qquad \Rightarrow \qquad \frac{1}{\lambda(x)} = \frac{1}{\lambda(0)} + \int_0^x dy \, y \, \dot{K}(y). \qquad (6.14)$$

7. Compute a new estimate of Δ_r by using the second Eq. (6.13) computed in $x = 0$,

$$\Delta_r = \frac{\Delta(0) + \lambda(0)^2 K(0)}{2}. \qquad (6.15)$$

8. Repeat steps 2–7 until convergence of Δ_r and $\lambda(x)$ to a fixed point is reached.

6.1.3 Observables

Taking derivatives of the free energy in Eq. (6.10), we can obtain expressions for several interesting observables. For example, the energy is

$$
\begin{aligned}
e_g &= \frac{\partial(\beta f_g)}{\partial \beta} = -\frac{d}{2} \int_{-\infty}^{\infty} dh \, P(1, h) \frac{\partial}{\partial \beta} f(1, h) \\
&= \frac{d}{2} \int_{-\infty}^{\infty} dh \, P(1, h) \frac{\gamma_{\Delta(1)} \star [e^{-\beta \bar{v}(h - \eta + \Delta_r)} \bar{v}(h - \eta + \Delta_r)]}{\gamma_{\Delta(1)} \star e^{-\beta \bar{v}(h - \eta + \Delta_r)}},
\end{aligned} \qquad (6.16)
$$

and the reduced pressure is

$$
\begin{aligned}
p_g &= \frac{\partial(\beta f_g)}{\partial \eta} = \frac{d}{2} \int_{-\infty}^{\infty} dh \, P(1, h) f'(1, h) \\
&= -\frac{d}{2} \int_{-\infty}^{\infty} dh \, P(1, h) \frac{\gamma_{\Delta(1)} \star [e^{-\beta \bar{v}(h - \eta + \Delta_r)} \beta \bar{v}'(h - \eta + \Delta_r)]}{\gamma_{\Delta(1)} \star e^{-\beta \bar{v}(h - \eta + \Delta_r)}}.
\end{aligned} \qquad (6.17)
$$

More generally, Eq. (2.29) can be expressed in terms of $\bar{v}(h)$ to define a distribution function $\bar{g}(h)$ of interparticle gaps $h = d(r/\ell_g - 1)$. Starting from Eq. (2.29), using rotational invariance and changing variables from r to h, the functional derivative of the free energy with respect to the potential is

---

[2] Eq. (6.14) is obtained by taking a derivative with respect to x of the second Eq. (6.13) and recalling that $\dot{\lambda}(x) = x\dot{\Delta}(x)$.

$$\delta f_g = \frac{\rho}{2} \int d\mathbf{r} g(\mathbf{r}) \delta v(\mathbf{r}) = \frac{d\widehat{\varphi}_g}{2} \int_{-\infty}^{\infty} dh e^h \overline{g}(h) \delta \overline{v}(h)$$

$$\Rightarrow \quad \frac{\partial f_g}{\partial \overline{v}(h)} = \frac{d\widehat{\varphi}_g}{2} e^h \overline{g}(h). \tag{6.18}$$

Hence, the radial distribution function of the glass, $\overline{g}(h)$, can be computed by taking the variation of the free energy in Eq. (6.10) with respect to the potential. It is important to stress, however, that in the state following construction, all the thermodynamic and structural properties of the glass are expressed in terms of the s replicas with $a \geq 2$, while the replica 1 is used only to select one among all possible glasses. Therefore, one should formally introduce two distinct potentials, $\overline{v}_g(h)$ for replica $a = 1$ and $\overline{v}(h)$ for those with $a \geq 2$, and the variation in Eq. (6.18) should be taken only with respect to $\overline{v}(h)$ at constant $\overline{v}_g(h)$. From Eq. (6.10), one then obtains[3]

$$\widehat{\varphi}_g e^h \overline{g}(h) = e^{-\beta \overline{v}(h)} \gamma_{\Delta(1)} \star \left[\frac{P(1, h + \eta - \Delta_r)}{\gamma_{\Delta(1)} \star e^{-\beta \overline{v}(h)}} \right]. \tag{6.19}$$

This result can be used to express the average of any two-body observable that depends on the gaps, $\mathcal{O}[\underline{X}] = \frac{1}{2N} \sum_{i \neq j} \overline{\mathcal{O}}[d(|\mathbf{r}_i - \mathbf{r}_j|/\ell_g - 1)]$, as

$$\langle \mathcal{O} \rangle = \frac{d\widehat{\varphi}_g}{2} \int_{-\infty}^{\infty} dh e^h \overline{g}(h) \overline{\mathcal{O}}(h)$$

$$= \frac{d}{2} \int_{-\infty}^{\infty} dh P(1, h) \frac{\gamma_{\Delta(1)} \star [e^{-\beta \overline{v}(h - \eta + \Delta_r)} \overline{\mathcal{O}}(h - \eta + \Delta_r)]}{\gamma_{\Delta(1)} \star e^{-\beta \overline{v}(h - \eta + \Delta_r)}}. \tag{6.20}$$

Note that a particular case of this relation is the energy expression in Eq. (6.16).

6.2 Gardner Transition and Replica Symmetry Breaking

In this section, we obtain explicit expressions for the quantities that characterise the instability of the RS solution at the Gardner transition and then describe a general recipe to study RSB effects within the state following construction.

6.2.1 Instability of the Replica Symmetric Solution

Following the same procedure as in Section 5.5, we now write the equations that characterise the instability of the replica symmetric solution at the Gardner transition. Taking the derivative with respect to x of the second Eq. (6.13) gives that for $\dot{\Delta}(x) \neq 0$,

---

[3] Because $\overline{v}_g(h)$ (corresponding to replica $a = 1$) appears in the second line of Eq. (6.10), while $\overline{v}(h)$ (corresponding to replicas $a \geq 2$) appears in the third line, to compute $\overline{g}(h)$, one should take the variation of the third line of Eq. (6.10) only.

$$\frac{1}{\lambda(x)^2} = \frac{1}{2} \int_{-\infty}^{\infty} dh \, P(x,h) f''(x,h)^2. \tag{6.21}$$

This condition must be satisfied on the continuous part of the fullRSB solution. Taking the RS limit, in which $\lambda(x) = \Delta$, we then obtain[4] that

$$\lambda_R = 1 - \frac{\Delta^2}{2} \int_{-\infty}^{\infty} dh \, P(0,h) f''(1,h)^2 \tag{6.22}$$

$$= 1 - \frac{\widehat{\varphi}_g}{2} \Delta^2 \int_{-\infty}^{\infty} dh \, e^h q(2\Delta_r - \Delta, \beta_g; h) \left(\frac{\partial^2}{\partial h^2} \log q(\Delta, \beta; h - \eta) \right)^2$$

must vanish at the point where the replica symmetric solution becomes unstable. Hence, λ_R is proportional to the replicon eigenvalue.

Following Section 5.5, we take one additional derivative of Eq. (6.21) with respect to x using Eq. (5.105) and obtain another condition,

$$x = \frac{\int_{-\infty}^{\infty} dh \, P(x,h) f'''(x,h)^2}{\frac{4}{\lambda(x)^3} + 2\int_{-\infty}^{\infty} dh \, P(x,h) f''(x,h)^3}, \tag{6.23}$$

which holds in the continuous region of the fullRSB solution. Evaluating this equation on the RS solution at the instability point gives the breaking point

$$x^* = \frac{\int_{-\infty}^{\infty} dh \, P(0,h) f'''(1,h)^2}{\frac{4}{\Delta^3} + 2\int_{-\infty}^{\infty} dh \, P(0,h) f''(1,h)^3}$$

$$= \frac{\widehat{\varphi}_g \int_{-\infty}^{\infty} dh \, e^h q(2\Delta_r - \Delta, \beta_g; h) \left(\frac{\partial^3}{\partial h^3} \log q(\Delta, \beta; h - \eta) \right)^2}{\frac{4}{\Delta^3} + 2\widehat{\varphi}_g \int_{-\infty}^{\infty} dh \, e^h q(2\Delta_r - \Delta, \beta_g; h) \left(\frac{\partial^2}{\partial h^2} \log q(\Delta, \beta; h - \eta) \right)^3}. \tag{6.24}$$

Taking another derivative with respect to x using Eq. (5.108), one further obtains

$$\dot{\Delta}(x) = \frac{\frac{4}{\lambda(x)^3} + 2\int_{-\infty}^{\infty} dh \, P(x,h) f''(x,h)^3}{\frac{12x^2}{\lambda(x)^4} - \int_{-\infty}^{\infty} dh \, P(x,h) A(x,h)}, \tag{6.25}$$

$$A(x,h) = f''''(x,h)^2 - 12x f''(x,h) f'''(x,h)^2 + 6x^2 f''(x,h)^4.$$

Evaluating this expression at the instability point of the RS solution formally amounts (after some manipulation) to replacing, as before, $\lambda(x) \to \Delta$, $x \to x^*$, $P(x,h) \to \widehat{\varphi}_g e^h q(2\Delta_r - \Delta, \beta_g; h)$, $f(x,h) \to \log q(\Delta, \beta; h - \eta)$.

Given λ_R, x^* and $\dot{\Delta}(x^*)$, we can study the stability of the RS solution following the approach presented in Section 5.5.

---

[4] Note that the second line of Eq. (6.22) is obtained from the first by using the RS expressions of $P(0,h)$ and $f(1,h)$, applying the identity in Eq. (4.89) and using the definition of $q(\Delta, \beta; h)$ in Eq. (4.74).

6.2.2 Recipe for Replica Symmetry Breaking in State Following

Based on the results of Sections 5.5 and 6.1, we can outline a recipe for studying RSB effects in the state following construction for particle glasses in $d \to \infty$. For a given initial state $(\widehat{\varphi}_g, T_g)$ that falls into the dynamically arrested region (see Section 4.3.3), the procedure is as follows.

1. Use the recipe of Section 4.3.3 to follow the evolution of the initial state at any state point $(\widehat{\varphi}, T)$ within the RS ansatz. The RS values of Δ and Δ_r are then obtained.

2. At each $(\widehat{\varphi}, T)$, check the stability of the RS ansatz by computing the replicon λ_R according to Eq. (6.22). If $\lambda_R \geq 0$, the RS ansatz is stable. It is generically found that the RS ansatz is stable on the equilibrium line, i.e., for $(\widehat{\varphi}, T)$ $= (\widehat{\varphi}_g, T_g)$ [303]. Usually, upon heating or decompressing, the RS ansatz remains stable, while upon cooling or compressing, it may become unstable.

3. Identify the Gardner transition point by the condition $\lambda_R = 0$. Beyond this point, the RS solution is unstable. Note that the replicon vanishes at the dynamical transition; therefore, a glass state prepared on the dynamical transition line is immediately unstable and cannot be followed at the RS level (as discussed in Chapter 4).

4. Compute the breaking point x^* using the RS solution at the Gardner point, according to Eq. (6.24). If $x^* \in [0, 1]$, then the transition is continuous. Otherwise, there is an inconsistency, which signals that the RS ansatz has already become unstable via another mechanism (usually a discontinuous transition).

5. Compute the slope $\dot{\Delta}(x^*)$ at the Gardner point according to Eq. (6.25). If $\dot{\Delta}(x^*) < 0$, a perturbative fullRSB solution exists around the Gardner point, which suggests that the solution is fullRSB throughout the unstable phase. If instead $\dot{\Delta}(x^*) > 0$, a 1RSB solution describes the unstable phase in the vicinity of the instability.

6. Once the nature of the solution has been determined, solve numerically the RSB Eqs. (6.9), (6.11) and (6.13), as described in Section 6.1.2 (see also Section 6.6). From this solution, compute the observables (energy, pressure, etc.), as described in Section 6.1.3.

This procedure can be applied to any interaction potential. Specific examples are considered in Section 6.3.

6.3 Gardner Transition of Simple Glasses

In this section, we apply the procedure outlined in Section 6.2.2 to two representative potentials, hard spheres and the inverse power-law potential (soft spheres), already studied at the RS level in Chapter 4.

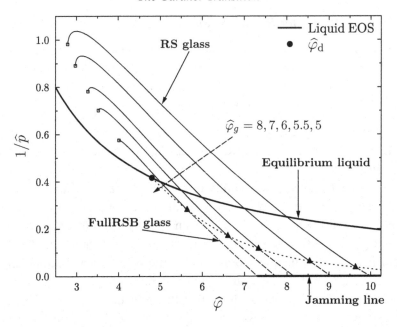

Figure 6.1 Following hard-sphere glasses in compression and decompression, within the full replica symmetric breaking ansatz [298]. The inverse of the reduced pressure $\widehat{p} = p/d$ is plotted versus packing fraction $\widehat{\varphi} = 2^d \varphi / d$. As in Figure 4.4, the liquid equation of state is plotted as a full line, and the dynamical transition $\widehat{\varphi}_d$ is marked by a full circle. The equations of state of glasses prepared at $\widehat{\varphi}_g > \widehat{\varphi}_d$ are reported here as full lines in the region where the replica symmetric ansatz is stable. Upon compression, all glasses undergo a Gardner transition at a density $\widehat{\varphi}_G(\widehat{\varphi}_g)$, marked by a triangle. The 'Gardner line' obtained by connecting the Gardner transitions of different glasses is plotted as a dotted line. Only data for $\widehat{\varphi}_g > \widehat{\varphi}_g^\dagger$ are reported, for which glasses are described by a fullRSB ansatz beyond the Gardner transition (dashed lines). The glass equations of state end at the jamming point, at which the pressure diverges[5].

6.3.1 Hard Spheres

The replica symmetric phase diagram of hard spheres has been discussed in Section 4.4.1. Replica symmetry breaking effects for this model have been studied in [88, 89, 221], and, in particular, state following results have been obtained in [298, 299]. See [93] for a review. Results obtained[5] by solving the kRSB equations with $k = 99$, following the procedure discussed in Section 6.6, are reported in Figure 6.1 and Table 6.1. These results remain stable upon further increasing k [298]. Upon decompression, the RS ansatz always remains stable, and

---

[5] Some of the results reported in Figure 6.1 and in Table 6.1 are taken from [298], but the numerical solution has been improved and new results have been produced for this book. Despite these improvements, for $(\widehat{\varphi}_g = 5, \widehat{p} \gtrsim 5.85)$, $(\widehat{\varphi}_g = 5.5, \widehat{p} \gtrsim 21.4)$ and $(\widehat{\varphi}_g = 6, \widehat{p} \gtrsim 324)$, the code is unstable, and convergence could not be reached. We believe that this is just a numerical instability of the code, and in Figure 6.1, the equations of state have thus been extended to infinite pressure by a linear fit. By contrast, for $\widehat{\varphi}_g \geq 7$, the code converges up to infinite pressure.

Table 6.1 *Gardner transition density* $\widehat{\varphi}_G$, *order parameters* Δ *and* Δ_r, *breaking point* x^*, *slope at the breaking point* $\dot{\Delta}(x^*)$ *and dynamical critical exponents* γ *and* a *for several initial densities* $\widehat{\varphi}_g$ *in hard spheres. The first line corresponds to* $\widehat{\varphi}_g = \widehat{\varphi}_d$ *and is reported with higher precision because the calculation is numerically simpler.*

$\widehat{\varphi}_g$	$\widehat{\varphi}_G$	Δ_r	Δ	x^*	$\dot{\Delta}(x^*)$	$\gamma = 1/a$
4.80677	4.80677	1.15336	1.15336	0.70698	19.357	3.08627
5	5.64	0.599	0.436	0.512	−6.950	2.547
5.5	6.61	0.333	0.169	0.397	−1.929	2.363
6	7.33	0.088	0.228	0.340	−0.990	2.295
7	8.54	0.133	0.033	0.280	−0.402	2.228
8	9.63	0.0888	0.0156	0.248	−0.204	2.194
9	10.67	0.0643	0.00847	0.228	−0.119	2.176

the results of Chapter 4 are unchanged. The glass states then melt into the liquid at a spinodal point. Upon compression, by contrast, there is always a Gardner transition, reached at a finite pressure, before jamming. For low-density states, this Gardner transition occurs before the unphysical RS spinodal (see Figure 4.4) is reached. Beyond the Gardner instability, a RSB solution appears and remains stable until reaching jamming, at infinite pressure.[5] Hence, no spinodal is observed upon compression once RSB is taken into account.

The values of the Gardner transition density are given in Table 6.1 for several values of initial density $\widehat{\varphi}_g$. By joining all the Gardner transitions corresponding to different $\widehat{\varphi}_g$ one obtains a 'Gardner transition line' in the $(\widehat{\varphi}, 1/\widehat{p})$ phase diagram (Figure 6.1). When the initial density tends to the dynamical transition point – i.e., for $\widehat{\varphi}_d \rightarrow \widehat{\varphi}_d^+$ – the Gardner transition approaches $\widehat{\varphi}_g$, and ultimately, $\widehat{\varphi}_G(\widehat{\varphi}_d) = \widehat{\varphi}_d$. This feature is generic of mean field models because it can be shown that the replicon eigenvalue vanishes at $\widehat{\varphi}_d$. As a consequence, glass states prepared at $\widehat{\varphi}_g = \widehat{\varphi}_d$ are always marginally stable. The Gardner line then originates from the dynamical transition and moves to higher pressures upon increasing $\widehat{\varphi}_g$.

The breaking point x^* and the corresponding slope $\dot{\Delta}(x^*)$ are also given in Table 6.1. Note that x^* always belong to the interval $[0, 1]$, which confirms that the transition is continuous. The inverse slope $1/\dot{\Delta}(x^*)$ vanishes linearly at $\widehat{\varphi}_g^\dagger \approx 4.85$ – i.e., the slope changes sign at $\widehat{\varphi}_g^\dagger$. For $\widehat{\varphi}_g > \widehat{\varphi}_g^\dagger$, the slope is negative, which is consistent with a continuous transition towards a fullRSB solution. In Figure 6.2, we report explicit examples of $\Delta(x)$ for $\widehat{\varphi}_g = 7$, in the vicinity of the Gardner transition. The fullRSB part of $\Delta(x)$ emerges continuously around the breaking point with the correct slope. In these cases, the solution remains fullRSB up to

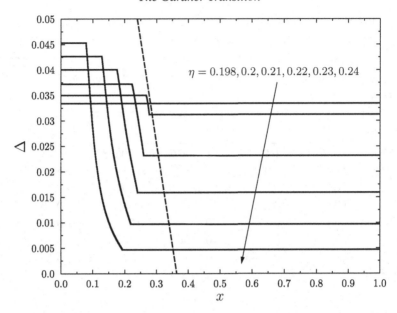

Figure 6.2 The function $\Delta(x)$ for initial density $\widehat{\varphi}_g = 7$ and for several values of $\eta = \log(\widehat{\varphi}/\widehat{\varphi}_g)$. The perturbative prediction for $\Delta(x)$ at the instability, $\Delta(x) \sim \Delta + \dot{\Delta}(x^*)(x - x^*)$, with parameters Δ, x^* and $\dot{\Delta}(x^*)$ given in Table 6.1, is reported as a dashed line. Once replica symmetry is broken, upon compression, the difference between the two plateaus $\Delta(0)$ and $\Delta(1)$ increases, and the shape of the continuous part becomes increasingly non-linear. The function of $\Delta(x)$ at larger values of η, close to jamming, is reported in Figure 9.6.

jamming.[6] Instead, in the region of densities $\widehat{\varphi}_d \leq \widehat{\varphi}_g \lesssim \widehat{\varphi}_g^\dagger$, the slope is positive, and the Gardner transition is a continuous transition towards a 1RSB phase. This transition is followed by a second continuous transition from 1RSB to a fullRSB solution, which then persists up to jamming.[7] The properties of the fullRSB phase that emerges beyond the Gardner transition are further discussed in Chapter 9.

6.3.2 Soft Spheres

The soft sphere potential $\bar{v}(h) = \varepsilon e^{-4h}$, which is the infinite-dimensional limit of a pure inverse power-law potential, $v(r) = \varepsilon(\ell/r)^{4d}$, was studied at the RS level in [315], as discussed in Section 4.4.2. In the following, we fix $\varepsilon = 1$ for simplicity. Because for this potential the unique control parameter is $\widehat{\Gamma} = \widehat{\varphi} T^{1/4}$

---

[6] The numerical solution of the fullRSB equations indicates that for low $\widehat{\varphi}_g \lesssim 6$, at some intermediate density, there might be a phase transition from a phase with continuous $\Delta(x)$ to another phase in which $\Delta(x)$ develops a discontinuity. This transition has been discussed in [108] for a spin glass model but has not been investigated systematically for hard spheres in $d \to \infty$.

[7] This second transition has not yet been studied in detail, and RSB results for $\widehat{\varphi}_d \leq \widehat{\varphi}_g \lesssim \widehat{\varphi}_g^\dagger$ are thus not reported in Figure 6.1. The solution is expected to display a discontinuity and a continuous part, as discussed in [108].

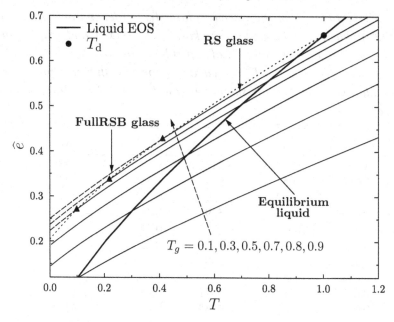

Figure 6.3 Following soft-sphere glasses with potential $\bar{v}(h) = \varepsilon e^{-4h}$, under heating and cooling, within the full replica symmetric breaking ansatz [315, 316]. The reduced energy $\widehat{e} = e/d$ is plotted versus temperature T, both expressed in units of the interaction strength ε, at fixed density $\widehat{\varphi} = 4.304$. As in Figure 4.5, the liquid equation of state is plotted as a full line, and the dynamical transition $T_d = 1$ is marked by a full circle. The equations of state of glasses prepared at $T_g < T_d$ are reported as full lines in the region where the replica symmetric ansatz is stable. Upon cooling, higher-energy glasses undergo a Gardner transition at a temperature $T_G(T_g)$, marked by a triangle. The 'Gardner line' obtained by connecting the Gardner transitions of different glasses is plotted as a dotted line. Only data for $T_g < T_g^{\dagger}$ are reported, for which glasses are described by a fullRSB ansatz below the Gardner transition (dashed lines). Lower-energy glasses, by contrast, display no Gardner transition. Note that once RSB is taken into account, no unphysical spinodal is present upon cooling[8] (unlike in the RS case of Figure 4.5), and all glasses can thus be followed down to $T = 0$.

(see Section 4.4.2), one can fix the density to $\widehat{\varphi} = 4.304$, corresponding to $T_d = 1$, and study the phase diagram as function of temperature [315]. The results including replica symmetry breaking effects[8] are reported in Figure 6.3 and Table 6.2. Additional results on the Gardner transition in soft harmonic spheres can be found in [58, 59, 316].

---

[8] Most of these results are taken from [315, 316], but as in the case of hard spheres, the numerical solution of the fullRSB equations has been improved, and new results have been produced for this book. Despite these improvements, for $(T_g = 0.9, T \lesssim 0.14)$ and $(T_g = 0.8, T \lesssim 0.03)$, the code is unstable, and convergence could not be reached. As for hard spheres, we believe that this is just a numerical instability of the code, and in Figure 6.3, the equations of state have thus been extended to zero temperature by a quadratic fit. By contrast, for $T_g = 0.7$, the code converges down to zero temperature.

Table 6.2 *Gardner transition temperature T_G, order parameters Δ and Δ_r, breaking point x^*, slope at the breaking point $\dot{\Delta}(x^*)$ and dynamical critical exponents γ and a for several T_g of soft spheres at constant density $\widehat{\varphi} = 4.304$. The first line corresponds to $T_g = T_d$ and is reported with higher precision because the calculation is numerically simpler. Below $T_g = 0.58$, no Gardner transition is observed.*

T_g	T_G	Δ_r	Δ	x^*	$\dot{\Delta}(x^*)$	$\gamma = 1/a$
1	1	1.31228	1.31228	0.66128	13.0198	2.92034
0.98	0.730	1.05	0.924	0.554	22.7	2.637
0.95	0.570	0.912	0.722	0.488	41.7	2.508
0.90	0.414	0.767	0.529	0.407	−124	2.382
0.85	0.306	0.667	0.399	0.344	−33.2	2.300
0.80	0.216	0.586	0.292	0.283	−23.7	2.231
0.75	0.150	0.518	0.209	0.225	−18.5	2.173
0.70	0.098	0.459	0.141	0.168	−16.5	2.123
0.65	0.0585	0.406	0.0871	0.112	−16.7	2.078
0.60	0.018	0.359	0.0282	0.0417	−33.8	2.027
0.58	0	0.338	0	0	∞	2

As for hard spheres, we find that the RS solution for states prepared at $T_g < T_d$ and heated to $T > T_g$ remains stable up to the spinodal point where the glass melts into the liquid. The results of Section 4.4.2 are therefore unchanged upon heating. Upon cooling, the RS solution for the lowest-energy states (corresponding to the lowest T_g) remains stable down to $T = 0$. These states, therefore, do not display any Gardner transition (see Figure 6.3), and also in this case, the results of Section 4.4.2 remain unchanged. A Gardner transition emerges continuously from $T = 0$ when $T_g \gtrsim 0.58$, and for $T_g \in [0.58, 1]$, the states display a Gardner transition at a finite $T_G(T_g)$ upon cooling. The values are reported in Table 6.2. In the vicinity of $T_g = 0.58$, when $T_G \rightarrow 0$, one can show analytically that $\Delta \rightarrow 0, x^* \sim 1.59\Delta \rightarrow 0$ and $\dot{\Delta}(x^*) \sim -0.76/\Delta \rightarrow -\infty$. Hence, the breaking point tends to zero, and the slope tends to negative infinity when $T_G = 0$. For $T_g > 0.58$, x^* always remains between 0 and 1, signalling a continuous transition, as for hard spheres. The slope $\dot{\Delta}(x^*)$ changes sign around $T_g^\dagger \approx 0.92$. For $T_g^\dagger < T_g < T_d$, the slope is positive, indicating a continuous transition towards a 1RSB phase.[9]

---

[9] It is currently not known whether, during the cooling process, this 1RSB phase undergoes another transition to fullRSB at a lower temperature. As for hard spheres, RSB results for $T_g^\dagger < T_g < T_d$ are thus not reported in Figure 6.3. Furthermore, as in the case of hard spheres, the numerical solution of the fullRSB equation suggests the existence of a phase in which $\Delta(x)$ has a discontinuity, on top of a continuous part [108], for relatively high T_g and low T.

6.4 Critical Properties of the Gardner Transition

To conclude the discussion of the Gardner transition, we discuss here (shortly and without proof) some of its critical properties. The Gardner transition is a continuous phase transition that takes place within one of the glass basins that appeared discontinuously at the dynamical transition.

In order to better illustrate this idea, the organisation of phase space in the Gardner phase is sketched in Figure 6.4. As discussed in Chapter 4, glass states emerge discontinuously at the dynamical transition, where diffusion stops. In phase space, this means that glass basins form somewhat before the dynamical arrest but remain connected by relatively narrow phase space bottlenecks that must be dynamically explored in order for the liquid to relax. A plateau in the mean square displacement is correspondingly observed at intermediate times (Figure 3.3). At the dynamical transition, however, these bottlenecks close; hence, the relaxation time becomes infinite, and the long-time limit of the MSD discontinuously jumps from infinity to a finite value. Equilibrium configurations at (φ_g, T_g) in the dynamically arrested phase thus select one of the stable glass states that are uniformly scattered

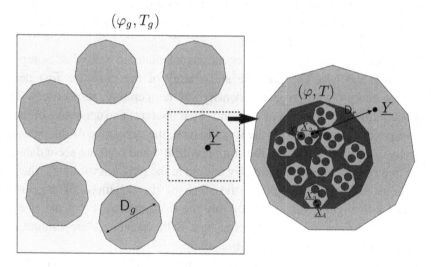

Figure 6.4 A schematic picture of the organisation of phase space in the Gardner phase. (Left) As in Figure 4.2, a glass state is selected by the configuration $\underline{Y}$ extracted in equilibrium at (φ_g, T_g), with associated mean square displacement D_g. Glass states are uniformly scattered in phase space. The connections between them close discontinuously at the dynamical transition. (Right) Once followed at a state point (φ, T) within the Gardner phase, the state selected by $\underline{Y}$ at (φ_g, T_g) becomes a metabasin, hierarchically organised in sub-states, as discussed in Chapter 5. The structure of sub-states emerges continuously at the Gardner transition. The replicas $\underline{X}_2, \ldots \underline{X}_{s+1}$ explore this structure, giving rise to a distribution of mean square displacements D_{ab}. The displacement between $\underline{Y}$ and any of the $\underline{X}_a$, however, remains D_r.

across phase space. The Gardner transition, by contrast, is a continuous transition that happens independently within each glass basin, when the basin is followed adiabatically from (φ_g, T_g) to (φ, T). It is this glass basin that becomes critical. Its bottom becomes flat and then fractures into many sub-basins, similarly to what happens to the paramagnetic state at the transition to ferromagnetism (Figure 1.4).

In the following, we denote by ϵ the distance from the Gardner transition point along a state following path, with $\epsilon > 0$ in the RS phase and $\epsilon < 0$ in the RSB phase. For example, in the hard-sphere case, we have $\epsilon = \widehat{\varphi}_G(\widehat{\varphi}_g) - \widehat{\varphi}$ for a given $\widehat{\varphi}_g$, while in soft spheres, we have $\epsilon = T - T_G(T_g)$ for a given T_g. Note that for generic interaction potentials, both density and temperature are control parameters, and one can move away from the Gardner transition point in both directions within a given state [316].

6.4.1 Thermodynamic Properties

The thermodynamic properties of the Gardner transition are quite unusual. In the Ehrenfest classification, the transition is of third order – i.e., all the derivatives of the free energy (with respect to density, temperature, shear strain, ...) up to second order included are continuous at the Gardner line, while higher order derivatives are discontinuous but not divergent.

As an example, consider the equations of state of the glasses reported in Figures 6.1 and 6.3. The pressure and energy of the glass are first derivatives of its free energy, with respect to density and temperature, respectively. They are always continuous functions of the control parameters. The derivative of the energy with respect to temperature is the specific heat, the derivative of the pressure with respect to density is related to the compressibility, and both are second derivatives of the free energy. While they display a jump at the glass transition T_g, they remain continuous at the Gardner point; the slope of the equations of state is indeed continuous in Figures 6.1 and 6.3. Only the third derivatives of the free energy, such as the derivative of the specific heat with respect to temperature, show a discontinuous jump at the Gardner transition.

A possible order parameter for this transition, as shown in Figure 6.2, is the difference $\delta\Delta = \Delta(0) - \Delta(1)$, which grows linearly, $\delta\Delta \propto |\epsilon|$, on the RSB side of the transition.

6.4.2 Dynamical Properties

The Gardner transition displays a dynamical critical behaviour similar to a standard second-order phase transition [288]. This dynamical criticality is also similar to that of the dynamical transition discussed in Section 3.4, with the important difference that, in this case, the system is confined into a glass state, and the dynamical mean square displacement $\Delta(t)$ has a plateau at long times, $\lim_{t \to \infty} \Delta(t) = \Delta$; hence, dynamics is arrested, and caging is permanent (Figure 4.1). At the Gardner transition, one has for large t

$$\Delta(t) - \Delta \propto -t^{-a}, \tag{6.26}$$

where the exponent a is expressed in terms of the breaking point by a relation similar to Eq. (3.85):

$$x^* = \frac{\Gamma^2(1-a)}{\Gamma(1-2a)}. \tag{6.27}$$

Upon approaching the transition from the RS phase, one has at large times

$$\Delta(t) - \Delta = -\epsilon f(t/\tau), \qquad \tau \propto \epsilon^{-\gamma}, \qquad \gamma = 1/a, \tag{6.28}$$

which reduces to Eq. (6.26) at the Gardner transition when $\epsilon \to 0$, under the assumption that $f(x \to \infty) \sim x^{-a}$. Eq. (6.28) indicates that the relaxation to the long time (restricted equilibrium) limit happens on a time scale τ that diverges at the Gardner transition. The critical exponent a is reported in Table 6.1 for hard spheres and in Table 6.2 for soft spheres.

Beyond the Gardner transition, in the RSB phase, the dynamics becomes extremely slow. Relaxation to the restricted equilibrium limit then takes an infinite time. If the system is prepared in the RS phase and suddenly cooled or compressed into the RSB phase, a stationary state is never observed, and the system ages forever. The properties of this ageing dynamics are particularly complex and will not be reported here. Yet, even though the system is unable to equilibrate within the glass basin in the RSB phase, the restricted equilibrium RSB calculation provides a good approximation of its physical properties.

6.5 Wrap-Up

6.5.1 Summary

In this chapter:

- We have applied the general RSB framework discussed in Chapter 5 to the calculation of the glass free energy in the Franz–Parisi framework. We have derived RSB expressions for the glass free energy; the corresponding variational equations; and the most important observables such as energy, pressure and radial distribution function (Section 6.1).
- We have obtained stability criteria for the RS solution, as well as the perturbative expression of the RSB solution around the RS instability. Using these criteria, we

have described a recipe for studying RSB effects in the Franz–Parisi framework (Section 6.2).

- We have applied this recipe to two representative potentials, hard spheres and soft spheres, thus obtaining phase diagrams that include the Gardner transition and RSB equations of state for the glass (Section 6.3).
- We have described the most important static and dynamical critical properties of the Gardner transition and provided values for the dynamical critical exponents (Section 6.4).

6.5.2 Further Reading

We provide here a list of references that can be consulted to further explore the subjects discussed in this chapter, selected according to the criteria discussed in Section 1.6.2.

The idea that glass basins are not simple minima but complex metabasins containing a hierarchical structure of sub-basins has a long history in the literature on the potential energy landscape of glasses. Some early papers and introductory reviews are

- Goldstein, *Viscous liquids and the glass transition: A potential energy barrier picture* [166]
- Stillinger and Weber, *Dynamics of structural transitions in liquids* [335]
- Debenedetti and Stillinger, *Supercooled liquids and the glass transition* [119]
- Wales, *Energy landscapes: Applications to clusters, biomolecules and glasses* [354]
- Heuer, *Exploring the potential energy landscape of glass-forming systems: From inherent structures via metabasins to macroscopic transport* [177]

Since the existence of a Gardner transition has been proposed in hard spheres in the limit $d \to \infty$, signatures of this transition in finite dimensional glasses have been searched for in many numerical and experimental studies. Because the field is moving rapidly, we only provide here a complete list of papers appeared until the completion of this book:

- Hicks, Wheatley, Godfrey, et al. *Gardner transition in physical dimensions* [178]
- Berthier, Charbonneau, Jin et al., *Growing timescales and lengthscales characterizing vibrations of amorphous solids* [44]
- Seguin and Dauchot, *Experimental evidence of the Gardner phase in a granular glass* [324]
- Jin and Yoshino, *Exploring the complex free-energy landscape of the simplest glass by rheology* [192]

- Seoane and Zamponi, *Spin-glass-like aging in colloidal and granular glasses* [326]
- Liao and Berthier, *Hierarchical landscape of hard disk glasses* [230]
- Jin, Urbani, Zamponi et al., *A stability-reversibility map unifies elasticity, plasticity, yielding and jamming in hard sphere glasses* [193]
- Scalliet, Berthier and Zamponi, *Absence of marginal stability in a structural glass* [315]
- Seoane, Reid, de Pablo et al., *Low-temperature anomalies of a vapor deposited glass* [327]
- Geirhos, Lunkenheimer and Loidl, *Johari-Goldstein relaxation far below* T_g: *Experimental evidence for the Gardner transition in structural glasses?* [162]
- Charbonneau, Corwin, Fu et al., *Glassy, Gardner-like phenomenology in minimally polydisperse crystalline systems* [94]

The first paper, which reports a study of a one-dimensional system of hard spheres, identified sub-basins associated to localised defects but pointed out that these defects could misleadingly give rise to Gardner-like physics in certain observables. The next six papers considered hard-sphere glasses in $d = 2$ and $d = 3$, both numerically and experimentally. These studies observed several anomalies potentially associated with a Gardner transition but could not establish whether these effects were associated to a true phase transition or to a sharp crossover. The defects associated with sub-basins are, however, certainly extended over large regions in space and are, therefore, of collective origin. The next three papers considered soft-potential models of structural glasses. Only localised defects have been identified, and no Gardner transition is detected numerically. Experimental studies remain inconclusive. The last paper reports the existence of Gardner-like anomalies in complex crystals of hard particles.

6.6 Appendix: Numerical Resolution of the RSB Equations

The fullRSB Eqs. (6.9), (6.11) and (6.13) constitute a set of closed equations for $f(x,h)$, $P(x,h)$, $\Delta(x)$ and Δ_r, which must be solved numerically. One possibility is to follow the procedure outlined in Section 6.1.2 and simply discretise the differential equations Eqs. (6.9) and (6.11) on a grid in (x,h).

Another possibility is to write explicitly the kRSB equations for finite k and then increase k until the results converge to the desired precision. For completeness, we report here the kRSB equations. These can be obtained from Eqs. (6.9), (6.11) and (6.13) by inserting a piecewise constant form of $\Delta(x)$, following the convention of Figure 5.4, with $s = m_{-1} = 0$. Also, we recall that while in Figure 5.4 the function $q(x)$ increases monotonically, the function $\Delta(x)$ decreases monotonically.

The procedure to solve the kRSB equations is as follows. Keeping the breaking points $\{m_i\}$ fixed (recalling that $m_{-1} = 0$ and $m_k = 1$), one starts by guessing the values of $\{\Delta_i\}$ and Δ_r. Then, the following steps are iterated until convergence.

1. For a piecewise constant $\Delta(x)$, introducing the function $q(\Delta, \beta; h)$ defined in Eq. (4.74), Eq. (6.9) takes the form:

$$f(m_k, h) = \log q(\Delta_k, \beta; h - \eta + \Delta_r - \Delta_k/2), \tag{6.29}$$

$$f(m_i, h) = \frac{1}{m_i} \log \gamma_{\Delta_i - \Delta_{i+1}} \star e^{m_i f(m_{i+1}, h)}, \qquad i = k - 1, \ldots, 0.$$

Similarly, applying the identity in Eq. (4.89), Eq. (6.11) takes the form:

$$P(m_0, h) = \widehat{\varphi}_g e^{h + \Delta_r - \Delta_0/2} q(2\Delta_r - \Delta_0, \beta_g; h + \Delta_r - \Delta_0/2),$$

$$P(m_i, h) = e^{m_{i-1} f(m_i, h)} \gamma_{\Delta_{i-1} - \Delta_i} \star [P(m_{i-1}, h) e^{-m_{i-1} f(m_{i-1}, h)}], \tag{6.30}$$

$$i = 1, \ldots, k.$$

Solve these two equations in the order in which they are written to obtain $f(m_i, h)$ and $P(m_i, h)$.

2. Compute

$$K_i = \frac{1}{2} \int dh\, P(m_i, h) f'(m_i, h)^2, \qquad i = 0, \ldots, k, \tag{6.31}$$

and from it, compute a new estimate of $\lambda(x)$ from a discrete version of Eq. (6.14):

$$\frac{1}{\lambda_0} = -\frac{1}{2} \int dh\, P(m_0, h)[f''(m_0, h) + f'(m_0, h)],$$

$$\frac{1}{\lambda_i} = \frac{1}{\lambda_{i-1}} + m_{i-1}(K_i - K_{i-1}), \qquad i = 1, \ldots, k. \tag{6.32}$$

3. Obtain a new estimate of $\Delta(x)$ and Δ_r using the relation

$$\Delta_k = \lambda_k,$$

$$\Delta_i = \Delta_{i+1} - \frac{1}{m_i}(\lambda_{i+1} - \lambda_i), \qquad i = k - 1, \ldots, 0, \tag{6.33}$$

$$\Delta_r = \frac{1}{2}(\Delta_0 + K_0 \lambda_0^2).$$

Once convergence is reached, the observables can be computed using the results of Section 6.1.3.

7

Counting Glass States
The Complexity

In Chapters 4 and 6, we studied the properties of individual glass states. In this chapter, we introduce the 'complexity', which counts the number of distinct glass states that can be found in a given sample. This function is similar to the standard entropy, which counts the number of typical microscopic configurations that contribute to a macroscopic state. We first introduce the equilibrium complexity, which is simply the difference between the liquid entropy and the internal entropy of typical glass states. The complexity vanishes at the Kauzmann transition, which sets a limit on the existence of the liquid state. At the Kauzmann point, a phase transition towards an ideal glass phase takes place. Next, we describe how to compute the out-of-equilibrium complexity and the properties of the ideal glass through the Monasson method. We provide the expression of the Monasson free energy at the replica symmetric level, and, as an example, we give results for the phase diagram of hard spheres. Finally, we briefly discuss spontaneous replica symmetry breaking effects in this context.

7.1 Equilibrium Complexity and the Kauzmann Temperature

Consider a simple liquid with potential $\bar{v}(h)$ of the class discussed in Section 2.3.2, in the limit $d \to \infty$. Its dynamically arrested phase, which was investigated in Chapter 4, is found in a regime with finite temperature $T = 1/\beta$ and scaled packing fraction $\widehat{\varphi} = 2^d \varphi/d$. As discussed in Chapter 3, when $T < T_d(\widehat{\varphi})$, with $T_d(\widehat{\varphi})$ given by Eq. (3.66), the liquid dynamics starting from an equilibrium configuration is fully arrested, and the system is confined to a glass state. In this section, we introduce the 'equilibrium complexity', which counts the number of such glass states.

7.1.1 Equilibrium Liquid and Glass Free Energy

Before proceeding, it is useful to recall some of our previous results. Recall that $\widehat{\varphi} = 2^d \varphi/d = \rho \ell^d \Omega_d/d^2$, where ρ is the number density, Ω_d is the solid angle and

ℓ is the typical interaction scale of the potential. At leading order in $d \to \infty$, with fixed $\widehat{\varphi}$, we then have

$$\rho \ell^d = \frac{\widehat{\varphi} d^2}{\Omega_d} \quad \Rightarrow \quad 1 - \log(\rho \ell^d) \sim \log \Omega_d \sim \frac{d}{2} \log\left(\frac{2\pi e}{d}\right), \qquad (7.1)$$

with corrections growing slower than d. In Chapter 2, it was shown that the liquid free energy per particle is given by Eq. (2.76) – which, using Eq. (7.1), becomes[1]

$$-\beta f_{\text{liq}}(\widehat{\varphi}, T) \sim \frac{d}{2} \log\left(\frac{2\pi e}{d}\right) + \frac{d\widehat{\varphi}}{2} \int_{-\infty}^{\infty} dh \, e^h [e^{-\beta \bar{v}(h)} - 1]. \qquad (7.2)$$

In Chapter 4, we computed the free energy of a typical glass state prepared in equilibrium at $(\widehat{\varphi}_g, T_g)$ and followed to a different state point $(\widehat{\varphi}, T)$ by using the replica method within the Franz–Parisi construction. It was further shown in Chapter 6 that when $(\widehat{\varphi}_g, T_g) = (\widehat{\varphi}, T)$, the replica symmetric calculation is correct. Specialising the replica symmetric result of Section 4.3.3 given in Eq. (4.75) to the case $(\widehat{\varphi}_g, T_g) = (\widehat{\varphi}, T)$, with $\Delta_r = \Delta$, we obtain for the typical glass free energy[2]

$$-\beta f_g(\widehat{\varphi}, T) = \frac{d}{2} \log\left(\frac{\pi e \Delta}{d^2}\right) + \frac{d}{2} + \frac{d\widehat{\varphi}}{2} \int_{-\infty}^{\infty} dh \, e^h \, q(\Delta, \beta; h) \log q(\Delta, \beta; h),$$
$$(7.3)$$

where Δ is the solution of Eq. (4.80). It was further shown in Chapter 4 (see Figures 4.4 and 4.5) that the equilibrium liquid pressure and energy coincide with those of the glass,[3] as given by Eqs. (4.77) and (4.79), respectively, computed at $(\widehat{\varphi}_g, T_g) = (\widehat{\varphi}, T)$. Note that, as discussed in Chapter 4, in equilibrium, the liquid and glass phases are not distinct, coexisting phases. The glass phase corresponds to the exploration of a restricted portion of phase space, in which the system is trapped for an infinite time in the limit $N \to \infty$ and $d \to \infty$. The liquid phase corresponds to the collection of all glasses, which is the total accessible phase space at this state point and is explored on time scales that diverge exponentially in N or d.

7.1.2 Equilibrium Complexity

Because the energies of the liquid and glass phases are identical in equilibrium, their difference in free energy, $f = e - Ts$, must originate from an entropy

---

[1] In this chapter, we omit the irrelevant kinetic free energy term $dT \log(\Lambda/\ell)$ from the ideal gas contribution. Because the most important terms of the free energy are proportional to d, we also neglect all terms that diverge slower than d.

[2] In this chapter, we drop the suffixes 'g' from $(\widehat{\varphi}_g, T_g)$ and 'r' from Δ_r to lighten the notation. Because we do not follow glasses at different state points, this does not create any ambiguity.

[3] Note that the glass pressure and energy are not obtained by differentiating Eq. (7.3). One should instead differentiate Eq. (4.75) with respect to $\widehat{\varphi}$ or T, respectively, at fixed $(\widehat{\varphi}_g, T_g)$ and compute the result in $(\widehat{\varphi}_g, T_g) = (\widehat{\varphi}, T)$.

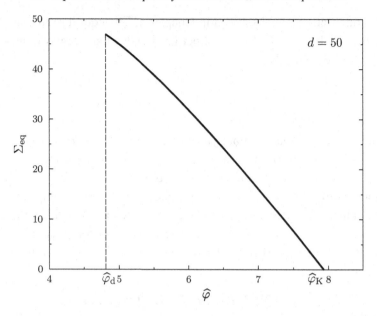

Figure 7.1 Equilibrium complexity for hard spheres given by Eq. (7.4), as a function of packing fraction $\widehat{\varphi}$. Because different terms in the complexity have different scaling with d, the curve is plotted for $d = 50$, neglecting all terms that diverge slower than d. The complexity jumps to a non-zero value at the dynamical transition $\widehat{\varphi}_d$ and then decreases continuously until it vanishes at the Kauzmann transition $\widehat{\varphi}_K$.

difference[4] [80, 119, 120, 201]. We thus define an equilibrium 'configurational entropy', or 'complexity', as

$$\Sigma_{eq}(\widehat{\varphi}, T) = -\beta[f_{liq}(\widehat{\varphi}, T) - f_g(\widehat{\varphi}, T)] = s_{liq}(\widehat{\varphi}, T) - s_g(\widehat{\varphi}, T)$$

$$= \frac{d}{2} \log d - \frac{d}{2} - \frac{d}{2} \log \left(\frac{\Delta}{2} \right) \qquad (7.4)$$

$$+ \frac{d\widehat{\varphi}}{2} \int_{-\infty}^{\infty} dh \, e^h [e^{-\beta \bar{v}(h)} - 1 - q(\Delta, \beta; h) \log q(\Delta, \beta; h)].$$

If $(\widehat{\varphi}, T)$ are kept finite (so that Δ is also finite) and $d \to \infty$, the first term dominates. Hence, $\Sigma_{eq}(\widehat{\varphi}, T)$ is always positive and diverges as $\frac{d}{2} \log d$ in this regime. In other words, the liquid entropy,

$$s_{liq}(\widehat{\varphi}, T) = s_g(\widehat{\varphi}, T) + \Sigma_{eq}(\widehat{\varphi}, T), \qquad (7.5)$$

is the sum of the entropy of a typical glass and of a positive contribution, $\Sigma_{eq}(\widehat{\varphi}, T)$, that captures the multiplicity of possible glasses selected by distinct equilibrium

---

[4] This can also be checked by explicitly computing the liquid and glass entropies.

liquid configurations, as discussed in Chapter 4. The equilibrium complexity thus counts the number $\mathcal{N}_{eq}(\widehat{\varphi}, T)$ of distinct glass states that compose the equilibrium liquid, or, more precisely,

$$\Sigma_{eq}(\widehat{\varphi}, T) = \lim_{N \to \infty} \frac{1}{N} \log \mathcal{N}_{eq}(\widehat{\varphi}, T). \tag{7.6}$$

Note that when the liquid is not dynamically arrested, for $T > T_d(\widehat{\varphi})$, glass states do not exist in equilibrium because there is no solution for the parameter Δ. In this case, the complexity is not defined by Eq. (7.4) but vanishes because the liquid is the only equilibrium state. At $T_d(\widehat{\varphi})$, the complexity thus jumps from zero to a positive value. Upon decreasing temperature or increasing density, the complexity decreases, as illustrated in Figure 7.1. The density (or temperature) at which the complexity vanishes is called the Kauzmann density (or temperature).

7.1.3 The Kauzmann Transition

The extrapolation of the complexity to densities above or temperatures below the Kauzmann point is formally negative. As noted already by Kauzmann [201], this result is physically inconsistent. According to Eq. (7.6), the complexity is the logarithm of the number of glass states, which is certainly $\mathcal{N}_{eq}(\widehat{\varphi}, T) \geq 1$, and hence, $\Sigma_{eq}(\widehat{\varphi}, T) \geq 0$. A formally negative complexity implies that the number of glass states is exponentially small in N – i.e., there are no glass states at all. Because in this region the liquid state is a superposition of glass states, the liquid phase thus ceases to exist at the Kauzmann point. A phase transition to another phase must then happen. Before investigating this transition in more detail (Section 7.2), it is illustrative to compute the Kauzmann point for some model potentials.

Because of the stronger divergence of the $\frac{d}{2} \log d$ term, the complexity remains positive for any finite $\widehat{\varphi}$, as $d \to \infty$. The density $\widehat{\varphi}$ should thus itself diverge when $d \to \infty$ for the complexity to vanish. According to Eq. (4.80), Δ decreases with increasing $\widehat{\varphi}$. When $\Delta \to 0$, one can show that

$$q(\Delta, \beta; h) \sim e^{-\beta \bar{v}(h)}[1 + \Delta q_1(\beta; h) + O(\Delta^2)]. \tag{7.7}$$

Plugging this expansion in Eq. (4.80), we conclude that $\widehat{\varphi}^{-1} = \Delta A(\beta) + O(\Delta^2)$, where $A(\beta)$ is some constant. We want to show that $\widehat{\varphi} \propto \log d$ and $\Delta \propto 1/\log d$. Under this assumption, plugging the leading order $\Delta = A(\beta)/\widehat{\varphi}$ and Eq. (7.7) in Eq. (7.4), we obtain at leading order

$$\Sigma_{eq}(\widehat{\varphi}, T) \approx \frac{d}{2} \log d + \frac{d\widehat{\varphi}}{2} \int_{-\infty}^{\infty} dh \, e^h [e^{-\beta \bar{v}(h)}(1 + \beta \bar{v}(h)) - 1]. \tag{7.8}$$

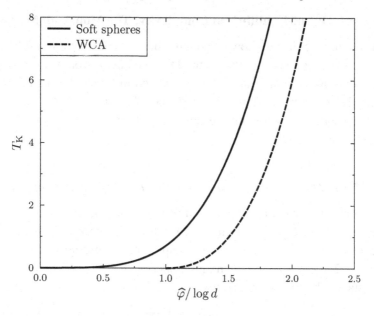

Figure 7.2 The Kauzmann temperature T_K plotted as a function of scaled density $\widehat{\varphi}/\log d$ for the soft-sphere potential $\bar{v}(h) = \varepsilon e^{-4h}$ and the WCA potential $\bar{v}(h) = \varepsilon(1 + e^{-4h} - 2e^{-2h})\theta(-h)$. In both cases, the unit of energy is set to $\varepsilon = 1$. For the soft-sphere potential, $T_K = 0.713\,(\widehat{\varphi}/\log d)^4$ because T and $\widehat{\varphi}$ are not independent. For the WCA potential, the Kauzmann temperature vanishes at $\widehat{\varphi}/\log d = 1$, which corresponds to the hard-sphere Kauzmann density.

From this result, it is straightforward to obtain an expression for the Kauzmann density by the condition $\Sigma_{\mathrm{eq}}(\widehat{\varphi}, T) = 0$,

$$\widehat{\varphi}_K = \frac{\log d}{g_K(\beta)}, \qquad g_K(\beta) = -\int_{-\infty}^{\infty} dh\, e^h [e^{-\beta \bar{v}(h)}(1 + \beta \bar{v}(h)) - 1]. \qquad (7.9)$$

Note that for the hard-sphere potential, $g_K(\beta) = 1$, and, thus, $\widehat{\varphi}_K = \log d$. The Kauzmann transition line in the $(\widehat{\varphi}, T)$ plane for two representative potentials is given in Figure 7.2.

It is important to stress that the Kauzmann transition is not always present in mean field models. Consider, for instance, the class of finite range potentials that vanish outside their interaction distance ℓ; hence, $\bar{v}(h) = 0$ for $h > 0$. These potentials reduce to the hard-sphere potential when $T \to 0$, and in that case, the complexity is positive for $\widehat{\varphi}_d = 4.8067\ldots < \widehat{\varphi} < \widehat{\varphi}_K = \log d$. Therefore, for all these potentials the Kauzmann transition line must vanish at $\widehat{\varphi} = \log d$, and if one studies them as a function of T for $4.8067\ldots < \widehat{\varphi} < \log d$, the complexity remains strictly positive down to zero temperature.

7.2 Out-of-Equilibrium Complexity: The Monasson Method

In this section, we discuss how to obtain a better understanding of the complexity function and describe a concrete method to compute it, from the decomposition of the partition function into glass states discussed in Chapters 4 and 5. We consider not only the equilibrium states that dominate the Gibbs–Boltzmann measure but also other states that dominate the out-of-equilibrium behaviour, using a method due to Monasson [260].

7.2.1 The Liquid Partition Function as a Sum over Glass States

We begin by discussing again the equilibrium liquid at a state point (φ, T). Each typical equilibrium liquid configuration $\underline{Y}$ falls into a glass state that can be defined by the Franz–Parisi construction of Chapter 4. According to the discussion of Section 7.1.2, there is an exponential number of such states. The partition function of each state, labelled by an index α, can then be written as

$$Z_\alpha = e^{-\beta N f_\alpha}. \tag{7.10}$$

As was shown in Chapter 4, distinct states are essentially disjoint. The probability that a configuration $\underline{Y}$ cannot be unambiguously attributed to a single state is exponentially small in N. Each configuration $\underline{Y}$ thus defines a state, with a typical free energy, $f_\alpha = f_g(\varphi, T)$, as computed in Chapter 4.

At a given state point (φ, T), there are, however, also atypical configurations $\underline{Y}$ that belong to atypical states, which have $f_\alpha \neq f_g(\varphi, T)$. By analogy with the typical glass states, we assume that the total number of glass states with free energy f, for a given state point (φ, T), is exponentially large in N in an interval $[f_{min}(\varphi, T), f_{th}(\varphi, T)]$:

$$\mathcal{N}(f; \varphi, T) = \sum_\alpha \delta(f - f_\alpha) \sim e^{N\Sigma(f; \varphi, T)}, \qquad f \in [f_{min}, f_{th}]. \tag{7.11}$$

Note that the interval $[f_{min}, f_{th}]$ is defined by the condition that the complexity $\Sigma(f; \varphi, T) \geq 0$. If $\Sigma(f; \varphi, T) < 0$, then the number of states is, on average, exponentially small in N, which means that, with probability 1 for $N \to \infty$, there are no states. An illustration of the complexity $\Sigma(f; \varphi, T)$ for hard spheres is given in Figure 7.5. In most cases, the complexity increases with increasing f. States with high energy, low entropy (small size) or both are more numerous than states with low energy and high entropy.

Under these assumptions, we can write the total contribution of glass states to the equilibrium (configurational) partition function as

$$Z_{eq} = e^{-\beta N f_{eq}(\varphi, T)} \sim \sum_{\alpha} e^{-\beta N f_{\alpha}} = \int df \sum_{\alpha} \delta(f - f_{\alpha}) e^{-\beta N f}$$

$$= \int df \mathcal{N}(f; \varphi, T) e^{-\beta N f} = \int_{f_{min}(\varphi, T)}^{f_{th}(\varphi, T)} df\, e^{N[\Sigma(f; \varphi, T) - \beta f]}. \tag{7.12}$$

For $N \to \infty$, evaluating the last integral by the maximum of the integrand following Laplace's method gives

$$-\beta f_{eq}(\varphi, T) = \max_{f \in [f_{min}(\varphi, T), f_{th}(\varphi, T)]} \{\Sigma(f; \varphi, T) - \beta f\}$$

$$= \Sigma(f_g(\varphi, T); \varphi, T) - \beta f_g(\varphi, T), \tag{7.13}$$

where $f_g \in [f_{min}, f_{th}]$ is such that $f - T\Sigma(f; \varphi, T)$ is minimal – i.e., it is the solution of

$$\frac{d\Sigma}{df} = \frac{1}{T}, \tag{7.14}$$

provided that it belongs to the interval $[f_{min}, f_{th}]$. Eq. (7.13) shows that the equilibrium value of the free energy f_g depends on a trade-off between lowering the individual free energy f_g of the glass, which comes at the price of having fewer states (lower complexity), and having more states that contribute (higher complexity) at the price of increasing their individual free energy. This trade-off is very similar to what happens in a standard thermodynamic equilibrium, except that the energy is here replaced by the free energy of individual states. Note that while Eq. (7.13) looks like a standard Legendre transform of the glass free energy with parameter β, the complexity $\Sigma(f; \varphi, T)$ itself depends explicitly on temperature, which complicates the inversion of the Legendre transform. The inversion would be greatly simplified if the Legendre parameter could be varied independently of the temperature that appears in the complexity. This is precisely the aim of the Monasson construction, as we discuss in Section 7.2.2.

For consistency, the maximum in Eq. (7.13) must be identified with the typical value of the free energy of glass states $f_g(\varphi, T)$ computed via the Franz–Parisi method, given by Eq. (7.3). Therefore, the equilibrium complexity that we computed explicitly in $d \to \infty$ in Section 7.1.2 should be identified with

$$\Sigma_{eq}(\varphi, T) = \Sigma(f_g(\varphi, T); \varphi, T). \tag{7.15}$$

We have seen, however, that $\Sigma_{eq}(\varphi, T)$ is only well defined for $\varphi_d < \varphi < \varphi_K$, for hard spheres and more generally for $T_d(\varphi) > T > T_K(\varphi)$. In the rest of this section, we will show that, generally speaking, in Eq. (7.13), one can encounter three distinct situations, depending on the specific state point (φ, T) [65, 79, 217, 250, 253].

- At high temperatures or low densities – or, more precisely, for $T > T_d(\varphi)$ – either there are no states at all and thus $\Sigma(f; \varphi, T)$ is always formally negative, or there is a regime within which $\Sigma(f; \varphi, T) > 0$ and the solution of the maximum condition in Eq. (7.13) is formally given by $f = f_{th}$. In the former case, f_{eq} is not defined, while in the latter case, it is usually found that $f_{eq} = f_{th} - T\Sigma(f_{th}; \varphi, T) > f_{liq}$, where f_{liq} is the liquid free energy computed in Chapter 2 and given by Eq. (7.2) in $d \to \infty$. Therefore, even if glass states exist, they are subdominant in the equilibrium measure, which is given by a single pure state – i.e., the liquid.

- At intermediate temperatures or densities – or, more precisely, for $T_d(\varphi) \geq T \geq T_K(\varphi)$ – a value $f_g \in [f_{min}, f_{th}]$ can be found, which is a solution of Eq. (7.14) such that

$$f_{liq}(\varphi, T) = f_{eq}(\varphi, T) = f_g - T\Sigma(f_g; \varphi, T). \tag{7.16}$$

In other words, the total free energy of all glass states coincides with the analytic continuation of the free energy of the liquid state below T_d. We already know from the study of the dynamics in Chapter 3 that in this regime the liquid dynamics is fully arrested. The interpretation is therefore that the liquid state has become a superposition of an exponential number of glass states of higher individual free energy density f_g. The Gibbs–Boltzmann measure is split on this exponential number of contributions. Yet no equilibrium phase transition happens at T_d because Eq. (7.16) guarantees that the free energy is analytic upon crossing T_d. Note that Eq. (7.16) is non-trivial because $f_{liq}(\varphi, T)$ is the result of a simple liquid virial calculation, while $f_{eq}(\varphi, T)$ follows from a complicated computation of the complexity of glass states. It will be justified a posteriori in Section 7.2.3.

- At low temperatures or high densities – or, more precisely, for $T < T_K(\varphi)$ – the partition function is dominated by the lowest free energy states, $f_g = f_{min}$, with $\Sigma(f_{min}) = 0$. In this case, the total glass free energy is $f_{eq}(T) = f_{min} - T\Sigma(f_{min}) = f_{min}$. At T_K, a phase transition takes place. The free energy and its first derivatives are continuous, but the second derivative of f_{eq} with respect to T (the specific heat) has a jump, as will be shown in Section 7.2.3.

Note that such an 'entropy crisis' scenario [65, 79, 80], where the number of glass states vanishes at a critical temperature T_K, is also realised in a class of completely solvable spin glass models, the simplest of which being the random energy model (REM) [125].

7.2.2 Computing the Complexity Curve

In Section 7.2.1, we introduced the complexity function $\Sigma(f; \varphi, T)$, which gives the entropy of glass states of free energy f at a state point (φ, T), and we have shown that this function encodes a lot of information about the glass phase. For instance, the typical free energy of glass states, the equilibrium complexity and both the dynamical and Kauzmann temperatures can be derived from the complexity. However, we did not yet provide a method to compute this quantity.

In order to do so, Monasson suggested [260] to consider m identical replicas of the original system, which (1) are constrained to be in the same glass state and (2) are uncorrelated within this glass state. We will discuss in Section 7.3 how to implement these two requirements in practice. For now, assuming that they can be implemented, we observe that the free energy of m copies inside a single glass state is just m times f_α, because these copies independently sample the state. Then, at low enough temperatures, the partition function of the replicated system is the sum over all states of the contribution of each state,

$$Z_m \sim \sum_\alpha e^{-\beta N m f_\alpha} = \int_{f_{min}}^{f_{th}} df\, e^{N[\Sigma(f; \varphi, T) - \beta m f]} \sim e^{N[\Sigma(f^*; \varphi, T) - \beta m f^*]}, \qquad (7.17)$$

where now $f^*(m; \varphi, T)$ is such that $mf - T\Sigma(f; \varphi, T)$ is minimal. It thus satisfies the equation

$$\frac{d\Sigma}{df} = \frac{m}{T}. \qquad (7.18)$$

The introduction of the m identical replicas thus adds a weight m to the term $-\beta f$ in Eq. (7.17). This weight can then be tuned, in order to extract the complexity function from a Legendre transform of the replicated free energy. Defining

$$-\beta \Phi(m; \varphi, T) = \frac{1}{N} \log Z_m = \max_{f \in [f_{min}(\varphi, T), f_{th}(\varphi, T)]} \{\Sigma(f; \varphi, T) - \beta m f\}$$

$$= \Sigma(f^*(m; \varphi, T); \varphi, T) - \beta m f^*(m; \varphi, T), \qquad (7.19)$$

one indeed has [253, 260]

$$f^*(m; \varphi, T) = \frac{\partial \Phi(m; \varphi, T)}{\partial m},$$

$$\Sigma(m; \varphi, T) = \Sigma(f^*(m; \varphi, T); \varphi, T) = m^2 \frac{\partial [m^{-1} \beta \Phi(m; \varphi, T)]}{\partial m}. \qquad (7.20)$$

The function $\Sigma(f; \varphi, T)$, for a given state point (φ, T), can then be reconstructed from the parametric plot of $\Sigma(m; \varphi, T)$ versus $f^*(m; \varphi, T)$ varying m.

Before proceeding, let us recapitulate the two main assumptions of the Monasson method:

1. Phase space can be partitioned in glass states labelled by α, with free energies f_α.
2. m replicas can be confined within the same glass state in such a way that their free energy is the Legendre transform of the complexity function, as expressed by Eqs. (7.19) and (7.20).

The method thus gives the complexity $\Sigma(f; \varphi, T)$ at any given state point, provided we are able to compute the free energy of m copies of the original system, constrained to be in the same glass state, and to perform the analytical continuation to real m, in such a way that we can take the derivatives in Eq. (7.20). This method was developed and tested in the context of spin glasses, in which a direct comparison with other methods to compute the complexity is possible, e.g., via the TAP equations [79, 250]. Such a comparison is unfortunately not possible in the context of particle systems, because there is no straightforward way of individually defining the atypical glass states and counting them.[5] The Monasson method is therefore the only method thus far available to compute the complexity in particle systems.

7.2.3 Properties of the Replicated Free Energy

To conclude the presentation of the Monasson method, we now discuss a few important properties of the replicated free energy $\Phi(m; \varphi, T)$ defined in Eq. (7.19).

The Equilibrium Line

The equilibrium line corresponds to $m = 1$, on which the partition function Z_m defined in Eq. (7.17) reduces to the equilibrium partition function in Eq. (7.12). We can therefore identify

$$f^*(1; \varphi, T) = f_g(\varphi, T),$$
$$\Sigma(1; \varphi, T) = \Sigma_{eq}(\varphi, T),$$
(7.21)

and, therefore,

$$\Phi(1; \varphi, T) = f_g(\varphi, T) - T\Sigma_{eq}(\varphi, T).$$
(7.22)

On the other hand, because in the Monasson construction $\Phi(m; \varphi, T)$ is the free energy of a system of m constrained replicas, for $m = 1$, it reduces to the free energy of a single replica, which is then equal to that of the equilibrium liquid,

$$\Phi(1; \varphi, T) = f_{liq}(\varphi, T).$$
(7.23)

Combining Eqs. (7.22) and (7.23) thus provides a proof of Eq. (7.16), which implies that the free energy is analytic at the dynamical transition. Note that the Monasson construction makes sense only if the complexity $\Sigma(m; \varphi, T)$ is positive. A formally negative complexity indicates a wrong choice of maximum in Eq. (7.19); the correct one must lie in the region $[f_{min}(\varphi, T), f_{th}(\varphi, T)]$, in which the complexity is positive. In particular, for $m = 1$, this implies that Eq. (7.22) can only hold when $\Sigma_{eq}(\varphi, T) \geq 0$.

---

[5] In principle, one should use density functional theory [206, 329] to do so, but this approach is technically difficult to implement. An exact treatment in $d \to \infty$ has therefore not yet been found.

The Kauzmann Transition

In order to lighten the notation, we do not consider here the dependence of the complexity on both temperature and density but focus instead on temperature. In the following, 'lower temperature' can, however, always be replaced by 'higher density'.

The complexity decreases with decreasing f. Usually,[6] upon lowering temperature, the value of $f_g(T)$ decreases towards the minimum $f_{min}(T)$. One can then expand the complexity around $f_{min}(T)$ as

$$\Sigma(f; T) = \sigma_1(T)(f - f_{min}(T)) + \frac{1}{2}\sigma_2(T)(f - f_{min}(T))^2 + \cdots \qquad (7.24)$$

Because $f_g(T)$ is the solution of Eq. (7.14), it coincides with $f_{min}(T)$ at a temperature T_K, such that

$$\left.\frac{d\Sigma}{df}\right|_{f=f_{min}(T_K)} = \sigma_1(T_K) = \frac{1}{T_K}. \qquad (7.25)$$

In equilibrium ($m = 1$) the total free energy of the system is then given by

$$f_{eq}(T) = \begin{cases} f_g(T) - T\Sigma(f_g(T); T) = f_{liq}(T), & T > T_K, \\ f_{min}(T), & T < T_K. \end{cases} \qquad (7.26)$$

A perturbative calculation for $T \gtrsim T_K$, using Eq. (7.24), under the assumption that $\sigma_1(T)$ and $\sigma_2(T)$ are analytic functions in the vicinity of the Kauzmann transition defined by $T_K\sigma_1(T_K) = 1$, shows that

$$f_g(T) - f_{min}(T) = \frac{1 - T\sigma_1(T)}{T\sigma_2(T)} + O((T - T_K)^2),$$

$$f_{eq}(T) - f_{min}(T) = \frac{(1 - T\sigma_1(T))^2}{2T\sigma_2(T)} + O((T - T_K)^3). \qquad (7.27)$$

Hence, the total free energy is given by $f_{min}(T)$ for $T < T_K$, while for $T > T_K$, it is given by $f_{min}(T) + O((T - T_K)^2)$. The entropy $s_{eq} = -df_{eq}/dT$ is therefore continuous at T_K, but the specific heat $c_V = T ds_{eq}/dT$ has a jump:

$$\Delta c_V = c_V(T_K^+) - c_V(T_K^-) = -\frac{T_K^2\sigma_1'(T_K)^2}{\sigma_2(T_K)}. \qquad (7.28)$$

Note that because usually $\sigma_2(T_K) < 0$, the specific heat drops upon lowering the temperature across T_K.

---

[6] With notable exceptions. For example, in hard spheres with an attractive short range interaction, an 'inverse freezing' transition can take place [321].

The Ideal Glass Phase

The equilibrium Kauzmann transition can be extended into a whole line $m_K(T)$ or $T_K(m)$ in the (m, T) plane, defined by the condition of vanishing complexity:

$$\Sigma(m_K(T); T) = 0 \qquad \Leftrightarrow \qquad \sigma_1(T) = \frac{m_K(T)}{T}, \qquad (7.29)$$

where the second condition is derived from Eq. (7.18). Because the temperature dependence of $\sigma_1(T)$ is usually mild, the Kauzmann transition shifts to lower temperatures (higher densities) for $m < 1$ and to higher temperatures (lower densities) for $m > 1$. See Figure 7.3 for an illustration.

For $m < m_K(T)$ – or equivalently $T > T_K(m)$ – the system is in the 'high-temperature' phase, where $f^* > f_{min}$ and $\Sigma > 0$. By contrast, for $m > m_K(T)$ or $T < T_K(m)$, it is in the 'low-temperature' phase, where $f^* = f_{min}$ and $\Sigma = 0$. Note that, because a phase transition happens on the line $m_K(T)$, calculations based on a high-temperature (low-density) expansion such as those of Chapters 2 and 4 only give access to the free energy $\Phi(m; T)$ in the high temperature region $m \leq m_K(T)$ [250, 253, 260]. In the low-temperature phase, because one has $\Phi(m; T) = m f_{min}(T)$, the free energy can instead be reconstructed from the value of $f_{min}(T)$, obtained using the relation

$$f_{min}(T) = \frac{\Phi(m_K(T); T)}{m_K(T)} = f^*(m_K(T); T). \qquad (7.30)$$

This equation is evaluated on the transition line, hence one can use the high-temperature expression of $\Phi(m; T)$. We will better illustrate this point with a concrete example in Section 7.3. The low-temperature phase at $T < T_K$, where the complexity is identically zero and the free energy is given by Eq. (7.30), is called the 'ideal glass' or 'condensed' phase.

7.3 The Monasson Construction in Infinite Dimensions

In this section, we discuss the practical implementation of the Monasson construction in $d \to \infty$, and we provide results for the free energy and for thermodynamic observables. To compute the free energy of m identical replicas of the original system, we use the methods of Chapter 4 – in particular, those of Section 4.2.4. Because these results are based on the truncation of the low-density virial expansion, they only hold in the equilibrium liquid phase of the system of m replicas for $T \geq T_K(m)$, where the complexity is positive, as discussed in Section 7.2.3.

7.3.1 Free Energy

Eq. (4.56) provides the free energy of a replicated liquid in terms of the matrix $\hat{\Delta}$ of mean square displacements between replicas. This result can be used to compute the complexity by straightforwardly renaming the number of replicas $n \to m$ and using again Eq. (7.1) to keep only the leading order in large d of the ideal gas term. One then has

$$-\beta \Phi(m; \widehat{\varphi}, T, \hat{\Delta}) = \frac{d}{2} \log\left(\frac{2\pi e}{d}\right) + \frac{d(m-1)}{2} \log\left(\frac{2\pi e}{d^2}\right)$$

$$+ \frac{d}{2} \log[2 \det(-\hat{\Delta}/2)(-\underline{1}^{\mathrm{T}}\hat{\Delta}^{-1}\underline{1})] - \frac{d\widehat{\varphi}}{2} \mathcal{F}(\hat{\Delta}), \qquad (7.31)$$

$$\mathcal{F}(\hat{\Delta}) = -\int_{-\infty}^{\infty} dh \, e^h \left\{ e^{-\frac{1}{2}\sum_{a,b=1}^{m} \Delta_{ab} \frac{\partial^2}{\partial h_a \partial h_b}} \left[e^{-\sum_{a=1}^{m} \beta \bar{v}(h_a)} - 1 \right] \right\}_{h_a=h},$$

where $\underline{1} = \{1, \ldots, 1\}$. Only terms proportional to $d \log(d)$ and to d have been kept; subleading terms have been discarded. As discussed in Section 7.2.2, the replicas must satisfy two requirements:

1. They must be in the same glass state; hence, Δ_{ab} should remain finite for all pairs ab.
2. They should be independently equilibrated within that glass state; hence, Δ_{ab} should be given by the unconstrained mean square displacement of the glass state.

Therefore, $\hat{\Delta}$ should be set to a local extremum[7] of the free energy in Eq. (7.31), with finite matrix elements [250, 253, 260, 292], similarly to the Franz–Parisi construction discussed in Chapter 4. Because m should be continued to real values, we need to make an ansatz for the matrix $\hat{\Delta}$ that allows for such an analytic continuation. As for the Franz–Parisi construction, we assume that the correct ansatz has the hierarchical structure introduced in Chapter 5.

7.3.2 Replica Symmetric Ansatz

We now specialise the calculation to the replica symmetric ansatz. Because all the replicas are equivalent, and $\Delta_{aa} = 0$, we have $\Delta_{ab} = \Delta(1 - \delta_{ab})$. The eigenvectors of $\hat{\Delta}$ are the vector $\underline{1}$, with eigenvalue $\lambda_1 = (m-1)\Delta$, and any vector orthogonal to $\underline{1}$, with $(m-1)$ degenerate eigenvalues $\lambda_2 = -\Delta$. Therefore, we have

$$\det\left(-\frac{\hat{\Delta}}{2}\right) = \frac{(1-m)\Delta}{2} \left(\frac{\Delta}{2}\right)^{m-1}, \qquad -\underline{1}^{\mathrm{T}}\hat{\Delta}^{-1}\underline{1} = \frac{m}{(1-m)\Delta}. \qquad (7.32)$$

Applying the same steps as in Sections 4.3.2 and 4.3.3 to the calculation of $\mathcal{F}(\hat{\Delta})$, we obtain

---

[7] A local minimum for $m > 1$, which is analytically continued to a local maximum for $m < 1$.

$$\mathcal{F}(\hat{\Delta}) = -\int_{-\infty}^{\infty} dh \, e^h \left\{ e^{-\frac{\Delta}{2}\frac{d^2}{dh^2}} g_{RS}(\Delta, \beta; h)^m - 1 \right\}$$

$$= -\int_{-\infty}^{\infty} dh \, e^h \left\{ q(\Delta, \beta; h)^m - 1 \right\}, \tag{7.33}$$

where the function $g_{RS}(\Delta, \beta; h)$ is defined in Eq. (4.70) and $q(\Delta, \beta; h)$ is defined in Eq. (4.74). Plugging these results in Eq. (7.31), we obtain the replica symmetric expression of the Monasson replicated free energy,

$$-\beta\Phi(m; \widehat{\varphi}, T, \Delta) = \frac{d}{2} \log\left(\frac{2\pi e}{d}\right) + \frac{d(m-1)}{2} \log\left(\frac{\pi e \Delta}{d^2}\right)$$

$$+ \frac{d}{2} \log m + \frac{d\widehat{\varphi}}{2} \int_{-\infty}^{\infty} dh \, e^h \left\{ q(\Delta, \beta; h)^m - 1 \right\}. \tag{7.34}$$

The equation for Δ is obtained by imposing a vanishing first derivative of $\Phi(m; \widehat{\varphi}, T, \Delta)$ with respect to Δ. The result is very similar to Eq. (4.80):

$$\frac{1}{\widehat{\varphi}} = \mathcal{F}_m(\Delta; \beta) = -\frac{\Delta m}{m-1} \int_{-\infty}^{\infty} dh \, e^h \, q(\Delta, \beta; h)^{m-1} \frac{\partial q(\Delta, \beta; h)}{\partial \Delta}. \tag{7.35}$$

Taking derivatives with respect to m according to Eq. (7.20) and keeping in mind that the derivative with respect to Δ should not be taken because Δ is set variationally, we obtain

$$-\beta f^*(m; \widehat{\varphi}, T) = \frac{d}{2} \log\left(\frac{\pi e \Delta}{d^2}\right) + \frac{d}{2m}$$

$$+ \frac{d\widehat{\varphi}}{2} \int_{-\infty}^{\infty} dh \, e^h \, q(\Delta, \beta; h)^m \log q(\Delta, \beta; h),$$

$$\Sigma(m; \widehat{\varphi}, T) = \frac{d}{2} \log d - \frac{d}{2} - \frac{d}{2} \log\left(\frac{\Delta}{2m}\right) \tag{7.36}$$

$$+ \frac{d\widehat{\varphi}}{2} \int_{-\infty}^{\infty} dh \, e^h [q(\Delta, \beta; h)^m - 1 - mq(\Delta, \beta; h)^m \log q(\Delta, \beta; h)].$$

One can now check explicitly that when $m \to 1$, which corresponds to equilibrium sampling of the glass states, the following properties hold:

- The replicated free energy Eq. (7.34) reduces to the liquid free energy Eq. (7.2) – i.e., $\Phi(1; \widehat{\varphi}, T) = f_{liq}(\widehat{\varphi}, T)$.
- The equation for Δ, Eq. (7.35), reduces to the equilibrium equation for Δ of the Franz–Parisi construction, Eq. (4.80), which itself coincides with the equation for the long time limit of the mean square displacement obtained from the equilibrium dynamical equations in the arrested phase in Section 3.3.2.

- The glass free energy in Eq. (7.36) coincides with that obtained from the Franz–Parisi construction given by Eq. (7.3) – i.e., $f^*(1; \widehat{\varphi}, T) = f_g(\widehat{\varphi}, T)$.
- The complexity in Eq. (7.36) coincides with the equilibrium complexity given in Eq. (7.4) – i.e., $\Sigma(1; \widehat{\varphi}, T) = \Sigma_{eq}(\widehat{\varphi}, T)$.

Some of these results have already been discussed in the general case in Section 7.2.3. They show that the Monasson method is fully consistent with the Franz–Parisi method and with equilibrium dynamics. In particular, in equilibrium, a finite solution for Δ of Eq. (7.35) appears at the dynamical transition density $\widehat{\varphi}_d(\beta)$. For lower densities or higher temperatures, the only solution is $\Delta = \infty$, which implies that replicas cannot be confined in a same glass state – i.e., there are no such states. For higher densities or lower temperatures, the Monasson method predicts the existence of glass states with free energy $f^*(1; \widehat{\varphi}, T) = f_g(\widehat{\varphi}, T)$, consistently with the Franz–Parisi method, and complexity $\Sigma(1; \widehat{\varphi}, T) = \Sigma_{eq}(\widehat{\varphi}, T)$. Furthermore, because

$$\Phi(1; \widehat{\varphi}, T) = f^*(1; \widehat{\varphi}, T) - T\Sigma(1; \widehat{\varphi}, T) = f_{liq}(\widehat{\varphi}, T), \qquad (7.37)$$

the total free energy is predicted by the Monasson method to be analytic at the dynamical transition $\widehat{\varphi}_d(\beta)$, as discussed in Section 7.2.3.

The results in Eqs. (7.34) and (7.36) hold only when the complexity $\Sigma(m; \widehat{\varphi}, T)$ is positive. In the equilibrium case of $m = 1$, as discussed in Section 7.1.3, the complexity remains positive up to $\widehat{\varphi}_K \sim \log d$. The case of hard spheres for $d = 50$, for which $\widehat{\varphi}_K \sim 8$, is illustrated in Figure 7.1.

7.3.3 Thermodynamic Observables

For fixed m and $(\widehat{\varphi}, T)$, the partition function is dominated by a set of glass states that have an identical free energy $f^*(m; \widehat{\varphi}, T)$ as well as identical values of all intensive observables – e.g., internal entropy, pressure or energy. Therefore, to each state point $(m; \widehat{\varphi}, T)$, one can equivalently associate a state point using other thermodynamic variables. The pressure, internal entropy and energy cannot be computed by taking derivatives of $f^*(m; \widehat{\varphi}, T)$, because this free energy corresponds to different glass states at each state point. In order to bypass this difficulty, we express these observables in terms of correlation functions, as discussed in Section 2.1.2. For instance, the radial distribution function $g(r | m; \widehat{\varphi}, T)$ of the glass states selected at the state point $(m; \widehat{\varphi}, T)$ gives the average potential energy per particle, according to Eq. (2.28),

$$
\begin{aligned}
e(m; \widehat{\varphi}, T) &= \frac{\rho \Omega_d}{2} \int_0^\infty dr\, r^{d-1}\, v(r) g(r | m; \widehat{\varphi}, T) \\
&= \frac{d\widehat{\varphi}}{2} \int_{-\infty}^\infty dh\, e^h\, \bar{v}(h)\, \bar{g}(h | m; \widehat{\varphi}, T),
\end{aligned} \qquad (7.38)
$$

where in the second equality we changed variables to $h = d(r/\ell - 1)$ and introduced $\bar{g}(h|m;\widehat{\varphi},T) = g(r|m;\widehat{\varphi},T)$. The pressure can be similarly written through the virial theorem [175, section 2.5],

$$
\begin{aligned}
p &= 1 - \frac{\beta \rho V_d}{2} \int_0^\infty dr\, r^d\, v'(r) g(r|m;\widehat{\varphi},T) \\
&= 1 - \frac{\beta d\widehat{\varphi}}{2} \int_{-\infty}^\infty dh\, e^h\, \bar{v}'(h) \bar{g}(h|m;\widehat{\varphi},T),
\end{aligned}
\tag{7.39}
$$

which, for hard spheres, gives [175, section 2.5]

$$
p = 1 + \frac{2^d \varphi}{2} g(\ell^+|m;\widehat{\varphi}) = 1 + \frac{d\widehat{\varphi}}{2} \bar{g}(0^+|m;\widehat{\varphi}).
\tag{7.40}
$$

The problem is then reduced to that of computing the radial distribution function. In the liquid phase, Eq. (2.29) expresses $g(\mathbf{r})$ as a functional derivative of the liquid free energy with respect to the pair potential. This result can be generalised to a system of m replicas. The radial distribution function of any of the replica can be written as the functional derivative of the replicated free energy with respect to the pair potential of that replica. This amounts to replacing, either in Eq. (7.31) or in its replica symmetric version (7.34), the pair potential $\bar{v}(h_a)$ by a replica-dependent potential $\bar{v}_a(h_a)$, taking the derivative with respect to one of these potentials, and then setting $\bar{v}_a(h_a) = \bar{v}(h_a)$. Because all replicas are equivalent, we can also use a functional analog of Eq. (4.71) to express the average of the partial derivatives with respect to $\bar{v}_a(h_a)$ in terms of the total derivative with respect to $\bar{v}(h)$. Changing variables from r to h, Eq. (2.29) generalises to

$$
\frac{\partial \Phi(m;\widehat{\varphi},T)}{\partial \bar{v}(h)} = m \frac{d\widehat{\varphi}}{2} e^h\, \bar{g}(h|m;\widehat{\varphi},T).
\tag{7.41}
$$

Using the replica symmetric expression (7.34) for $\Phi(m;\widehat{\varphi},T)$, we then obtain

$$
\bar{g}(h|m;\widehat{\varphi},T) = e^{-\beta \bar{v}(h)} \int_{-\infty}^\infty dz\, e^{z-h}\, q(\Delta,\beta;z)^{m-1} \gamma_\Delta(z - h + \Delta/2).
\tag{7.42}
$$

Plugging this result in Eqs. (7.38) and (7.39) provides the energy and pressure of the glass states as a function of $(m;\widehat{\varphi},T)$.

7.4 The Phase Diagram of Hard Spheres

We now turn to the study of the phase diagram, specialising to the case of hard spheres, for which the function $q(\Delta;h)$ does not depend on β and is given in Eq. (4.83).

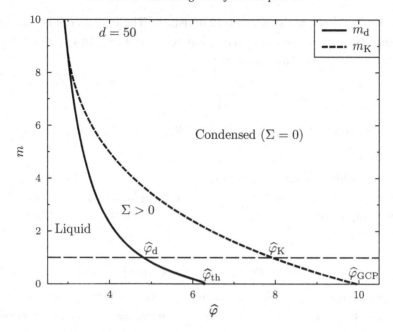

Figure 7.3 Replica symmetric phase diagram of hard spheres, in the plane $(m, \widehat{\varphi})$, for $d = 50$. The long-dashed line is the equilibrium line $m = 1$. The solid line is the dynamical line $m_{\mathrm{d}}(\widehat{\varphi})$, above which a finite solution for Δ exists and replicas can be constrained to the same state. The short-dashed line is the Kauzmann line $m_{\mathrm{K}}(\widehat{\varphi})$ above which the m replicas are stuck into the lowest free energy state. Because on this scale the Kauzmann line moves to infinite m for $d \to \infty$, it is plotted here for $d = 50$, but the other lines are independent of d.

7.4.1 Phase Diagram in the $(m, \widehat{\varphi})$ Plane

To compute the complexity as given by Eq. (7.36), one should first solve Eq. (7.35) to obtain Δ. A solution exists only for $m > m_{\mathrm{d}}(\widehat{\varphi})$, as given in Figure 7.3. In the region where a solution for Δ exists, the complexity $\Sigma(m; \widehat{\varphi})$ can be computed numerically, and one can obtain the Kauzmann line defined by $\Sigma(m_{\mathrm{K}}(\widehat{\varphi}); \widehat{\varphi}) = 0$. This line is also given in Figure 7.3 for $d = 50$. The two lines $m_{\mathrm{K}}(\widehat{\varphi})$ and $m_{\mathrm{d}}(\widehat{\varphi})$ intersect, defining three distinct regions in Figure 7.3:

- A region where no solution for Δ exists and, hence, replicas cannot be kept in a same state. Here the replicated system is liquid, and the complexity is not defined.
- A region where $\Sigma > 0$, and glass states thus exist. In this region, the low-density results are correct, and one can use Eqs. (7.36) to compute parametrically the curve $\Sigma(\mathrm{f}; \widehat{\varphi})$. Examples for some $\widehat{\varphi}$ are given in Figure 7.5.
- A region where, formally, $\Sigma < 0$ if one uses Eq. (7.34) but, in reality, the system is condensed in the lowest free energy state, with $\Sigma = 0$. In this regime, Eqs. (7.34) and (7.36) are invalid because the low-density expansion

has broken down at the Kauzmann transition. The free energy is instead given by $\Phi(m;\widehat{\varphi}) = m\,f_{\min}(\widehat{\varphi})$. One can compute $f_{\min}(\widehat{\varphi})$ by using the continuity of the free energy on the Kauzmann line:

$$f_{\min}(\widehat{\varphi}) = \left[\frac{\Phi(m;\widehat{\varphi})}{m}\right]_{\forall m \geq m_K(\widehat{\varphi})} = \frac{\Phi(m_K(\widehat{\varphi});\widehat{\varphi})}{m_K(\widehat{\varphi})}, \qquad (7.43)$$

as discussed in Section 7.2.3. On the Kauzmann line, one can use Eqs. (7.34) and (7.36).

On the equilibrium line, which corresponds to $m = 1$ in Figure 7.3, these three regions correspond to the regime $\widehat{\varphi} < \widehat{\varphi}_d$ (liquid phase), $\widehat{\varphi}_d < \widehat{\varphi} < \widehat{\varphi}_K$ (dynamically arrested liquid phase) and $\widehat{\varphi} > \widehat{\varphi}_K$ (thermodynamically stable ideal glass phase), respectively.

7.4.2 Phase Diagram in the $(\widehat{p},\widehat{\varphi})$ Plane

The parameter m can be used to bias the Gibbs–Boltzmann measure towards atypical states. Its physical meaning is, however, not fully transparent. One can interpret the quantity T/m as an 'effective temperature', conjugated to the internal free energy of the states, exactly as temperature is conjugated to energy [65, 109, 110]. The study of the off-equilibrium dynamics has indeed shown that in some regimes, this effective temperature controls the equilibration of the slowest degrees of freedom in the system [109]. Alternatively, one can convert m into another more directly measurable physical quantity. As an example, we focus on reduced pressure. To each state point $(m,\widehat{\varphi})$, we can associate a scaled pressure $\widehat{p}(m,\widehat{\varphi}) = p(m,\widehat{\varphi})/d$. Within the replica symmetric ansatz, substituting Eq. (7.42) into Eq. (7.40), we obtain at leading order

$$\widehat{p}(m,\widehat{\varphi}) = \frac{\widehat{\varphi}}{2}\int_{-\infty}^{\infty} dh\, e^h\, q(\Delta,\beta;h)^{m-1}\gamma_\Delta(h+\Delta/2). \qquad (7.44)$$

This result holds, as for Eq. (7.34), in the region $m < m_K(\widehat{\varphi})$. Using Eq. (7.44), we can compute the pressure in each point $(m,\widehat{\varphi})$, and then replace m by $\widehat{p}$ as the state variable.

The resulting phase diagram in the plane $(\widehat{p},\widehat{\varphi})$ is given in Figure 7.4. Note that for $m = 1$ and $\widehat{\varphi} \leq \widehat{\varphi}_K$, we get $\widehat{p}(1,\widehat{\varphi}) = \widehat{\varphi}/2 = \widehat{p}_{\mathrm{liq}}(\widehat{\varphi})$, which coincides with the equilibrium liquid equation of state (EOS), obtained in Chapter 2. This confirms that $m = 1$ corresponds to the equilibrium case. The dynamical and Kauzmann lines have similar shapes in the $(\widehat{p},\widehat{\varphi})$ plane. In this case, they also delimit three regions, but two of them are inaccessible. More precisely:

- In the region $\Sigma = 0$ of the $(m,\widehat{\varphi})$ plane, the m replicas are independent, and Δ is formally infinite. In this situation, each replica is an independent liquid, and they

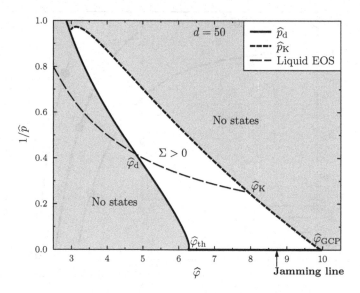

Figure 7.4 Replica symmetric phase diagram of hard spheres, in the plane $(1/\widehat{p}, \widehat{\varphi})$, for $d = 50$. The long-dashed line is the equilibrium line $\widehat{p}_{\text{liq}}(\widehat{\varphi}) = \widehat{\varphi}/2$. The full line is the dynamical line $\widehat{p}_{\text{d}}(\widehat{\varphi})$, above which a finite solution for Δ exists and replicas can be constrained to be in the same state. The short-dashed line is the Kauzmann line $\widehat{p}_{\text{K}}(\widehat{\varphi})$ corresponding to the lowest free energy states. Above this line there are no states. Because on this scale the Kauzmann line moves to infinite $1/\widehat{p}$ for $d \to \infty$, it is plotted here for $d = 50$, but the other lines are independent of d.

all have the same pressure $\widehat{p}_{\text{liq}}(\widehat{\varphi})$. Hence, in the $(\widehat{p}, \widehat{\varphi})$ plane, this whole region collapses onto the equilibrium liquid EOS, $\widehat{p}_{\text{liq}}(\widehat{\varphi}) = \widehat{\varphi}/2$.

- In the region $\Sigma > 0$, there are exponentially many glass states with given density and pressure $(\widehat{p}, \widehat{\varphi})$. The equilibrium states are still those with $\widehat{p} = \widehat{p}_{\text{liq}}$. The other state points with $\widehat{p} \neq \widehat{p}_{\text{liq}}$ are dominated by atypical states that can be visited if the system falls out of equilibrium.
- In the 'condensed' region of the $(m, \widehat{\varphi})$ plane, the system is stuck in the lowest free energy state. The pressure of this state gives the Kauzmann (ideal glass) line $\widehat{p}_{\text{K}}$. The whole negative complexity region of the $(m, \widehat{\varphi})$ plane thus collapses onto a single line in the $(\widehat{p}, \widehat{\varphi})$ plane.

In summary, in equilibrium the system follows the liquid line $\widehat{p}_{\text{liq}}(\widehat{\varphi})$ for $\widehat{\varphi} < \widehat{\varphi}_{\text{K}}$ and the Kauzmann (ideal glass) line $\widehat{p}_{\text{K}}(\widehat{\varphi})$ for $\widehat{\varphi} > \widehat{\varphi}_{\text{K}}$. In all other state points $(\widehat{p}, \widehat{\varphi})$ for which $\Sigma > 0$, there exists an exponential number of glass states that are never sampled in equilibrium but can be accessed if the system falls out of equilibrium. In the rest of the $(\widehat{p}, \widehat{\varphi})$ plane, no glass states are present.

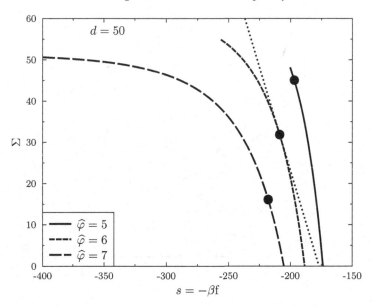

Figure 7.5 Examples of the complexity $\Sigma(f; \widehat{\varphi})$, plotted as a function of the entropy $-\beta f = s$ (because, for hard spheres, the energy is identically zero), for three representative values of $\widehat{\varphi}$. Because, according to Eq. (7.36), the complexity diverges when $d \to \infty$, it is plotted here for $d = 50$. For $\widehat{\varphi} = 7$, the entropy s can take values going from $s = -\infty$ (jammed states) to a maximal value where $\Sigma = 0$. For $\widehat{\varphi} = 5$ and $\widehat{\varphi} = 6$, the entropy can take values over a finite interval. The dotted line has slope -1, corresponding to the equilibrium condition given by Eq. (7.14). For all curves, the black dot denotes the equilibrium value of s and Σ.

7.4.3 The Edwards Ensemble for Hard Spheres

A particularly interesting region of the $(\widehat{p}, \widehat{\varphi})$ plane is the 'jamming line' corresponding to $\widehat{p} = \infty$ (or $1/\widehat{p} = 0$) in Figure 7.4. On this line, hard-sphere configurations contain a network of particles in direct contact – i.e., separated by a distance identical to their diameter. This network of mechanical hard core contacts is able to sustain the infinite external pressure. Such configurations are often called 'random close packings' [33, 34] or 'mechanically stable configurations' [6, 232].

From Eq. (7.44), it is possible to show that achieving infinite pressure requires both $\Delta = 0$ and $m = 0$. More precisely, one should take the joint limit $m \to 0$ and $\Delta \to 0$ with $\Delta/m = \vartheta$ [292]. In this 'jamming' limit, one can show from Eq. (7.44) that $\widehat{p} \to \infty$. The first condition, $\Delta = 0$, is natural. Because pressure is infinite and particles touch each other, no vibrations are possible. The second condition is less trivial but has an interesting interpretation. When $m = 0$, the modified partition function defined in Eq. (7.17) becomes a uniform sum over all glass states. At infinite pressure, these states coincide with the mechanically stable configurations. In other words, at infinite pressure, the Monasson ensemble reduces to the so-called "Edwards ensemble" [30], developed in the context of granular systems, which expresses the partition function as a uniform sum over all mechanically stable hard-sphere configurations.

The equation for ϑ in the jamming limit is obtained from Eq. (7.35) and reads

$$\frac{1}{\widehat{\varphi}} = \mathcal{F}_0(\vartheta) = \int_0^\infty dh \, e^{-h-\frac{h^2}{2\vartheta}} \frac{h^2}{2\vartheta}. \tag{7.45}$$

As for any finite m, this equation admits a solution only when $\widehat{\varphi}$ is large enough, $\widehat{\varphi} > \widehat{\varphi}_{\text{th}} = 1/[\max_\vartheta \mathcal{F}_0(\vartheta)] = 6.25812\ldots$, for $\vartheta = 0.60487\ldots$ For $\widehat{\varphi} > \widehat{\varphi}_{\text{th}}$, one can solve for ϑ and obtain the corresponding value of the complexity by taking the $m \to 0$, $\Delta = m\vartheta$ limit of Eq. (7.36):

$$\begin{aligned}
\Sigma_j^{\text{Ed}}(\widehat{\varphi}) &= \frac{d}{2}\log d - \frac{d}{2} - \frac{d}{2}\log\frac{\vartheta}{2} \\
&\quad + \frac{d\widehat{\varphi}}{2}\left[-1 + \int_0^\infty dh \, e^{-h-\frac{h^2}{2\vartheta}}\left(1+\frac{h^2}{2\vartheta}\right)\right].
\end{aligned} \tag{7.46}$$

This expression gives the complexity of mechanically stable configurations as a function of density along the jamming line – i.e., the 'Edwards complexity' [30]. The qualitative behaviour of the Edwards complexity is similar to that of the equilibrium complexity; see Figure 7.6. It jumps discontinuously from zero to a positive value at $\widehat{\varphi}_{\text{th}}$ and then decreases smoothly until it vanishes linearly at a density $\widehat{\varphi}_{\text{GCP}}$, called 'glass close packing' (GCP). This density represents the densest possible state that can be achieved by hard-sphere glasses. As for the Kauzmann transition (see Section 7.1.3), the GCP density scales as $\widehat{\varphi}_{\text{GCP}} \sim \log d$ at the leading order, while the difference $\widehat{\varphi}_{\text{GCP}} - \widehat{\varphi}_{\text{K}}$ remains finite when $d \to \infty$.

7.4.4 Atypical Glass States and Protocol Dependence

It is interesting to compare the phase diagram in the $(\widehat{p}, \widehat{\varphi})$ plane obtained via state following, Figure 6.1, with that obtained via the Monasson construction, Figure 7.4. In both cases, glass states exist in a region delimited by the jamming line at infinite pressure, by the Kauzmann line on the right side[8] and by a 'dynamical line' on the left side.[9] While the qualitative shape of the two phase diagrams is very similar, their physical interpretation is quite different. In the Monasson construction, for

---

[8] The equivalent of the Kauzmann line was not included in Figure 6.1 because the Kauzmann transition had not yet been introduced. In the state following construction, it corresponds to preparing an equilibrium state at $\widehat{\varphi}_{\text{K}}$ and following it in compression. Note that this line does not coincide with the Kauzmann line of Figure 7.4, which instead corresponds to the densest glass state for any fixed pressure. See [214] for details.

[9] In the state following construction, the dynamical line corresponds to following the states prepared at $\widehat{\varphi}_{\text{d}}$ in compression and to the envelope of the spinodal points of the different glass states in decompression.

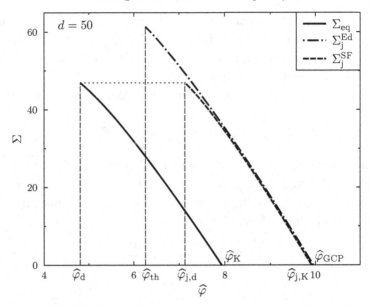

Figure 7.6 Equilibrium and jamming complexities for hard spheres, as a function of packing fraction $\widehat{\varphi}$. Because different terms in the complexity have different scaling with d, the curves are plotted for $d = 50$, neglecting all terms that diverge slower than d. The solid line $\Sigma_{eq}(\widehat{\varphi})$ is the equilibrium complexity ($m = 1$), as in Figure 7.1. The dot-dashed line $\Sigma_j^{Ed}(\widehat{\varphi})$ is the Edwards complexity of the jammed states ($m = 0$), given by Eq. (7.46) under the RS approximation. It jumps to a non-zero value at the threshold $\widehat{\varphi}_{th}$ and vanishes at the glass close packing point $\widehat{\varphi}_{GCP}$. Note, however, that this whole curve is unstable towards replica symmetry breaking. Finally, the dashed line $\Sigma_j^{SF}(\widehat{\varphi})$ is the equilibrium complexity,[12] plotted as a function of the jamming density, as defined in Eq. (7.47). It represents (neglecting RSB effects) the complexity of the jammed states obtained via state following.

each point $(\widehat{p}, \widehat{\varphi})$, one considers the 'typical' states for that point. These are the states that dominate the modified partition function defined in Eq. (7.17). In the state following construction, one instead prepares states that are typical of the equilibrium measure (i.e., at $m = 1$) and then follows them adiabatically in compression or decompression. Obviously, the two constructions coincide on the equilibrium line ($m = 1$). They thus provide the same predictions for the dynamical transition $\widehat{\varphi}_d$, the Kauzmann transition $\widehat{\varphi}_K$ and the equilibrium complexity. However, as soon as one leaves the equilibrium line, the states that were typical in equilibrium become atypical, and the Monasson construction is dominated by a distinct set of states. This leads to quantitative differences in the location of the lines that delimit the region of existence of glass states. In particular, the region obtained via state following must be strictly contained within that obtained by the Monasson construction, because the latter considers all possible states while the former considers only a subset of all possible states, namely those that can be obtained by adiabatically following an equilibrium state.

This is an example of a more general property of the $(\widehat{p}, \widehat{\varphi})$ phase diagram. Because of the proliferation of glass states with distinct properties, different out-of-equilibrium protocols are likely to visit different sets of states [65, 215, 292]. The state following construction corresponds to one specific protocol: prepare a state in equilibrium, and follow it adiabatically. The Monasson construction corresponds to another specific protocol: sample all glass states at given $(\widehat{p}, \widehat{\varphi})$ according to the modified measure. Many other protocols can be devised, and each would typically lead to distinct results as soon as one leaves the equilibrium line.[10] This specific example reflects a general fact: properties of out-of-equilibrium systems are strongly protocol dependent.

Another observable that depends sensitively on the protocol is the jamming complexity. The complexity of the packings obtained via state following can be defined by observing that the state following procedure maps each equilibrium density $\widehat{\varphi}_g$ to a corresponding jamming point $\widehat{\varphi}_j(\widehat{\varphi}_g)$ [86, 128, 292, 330]; see Figure 6.1. Therefore, under the assumption[11] that no states appear or disappear during the compression procedure from $\widehat{\varphi}_g$ to $\widehat{\varphi}_j(\widehat{\varphi}_g)$, one can define a jamming complexity of state following as

$$\Sigma_j^{\mathrm{SF}}(\widehat{\varphi}_j) = \Sigma_{\mathrm{eq}}[\widehat{\varphi}_g(\widehat{\varphi}_j)], \tag{7.47}$$

where $\widehat{\varphi}_g(\widehat{\varphi}_j)$ is the inverse function of $\widehat{\varphi}_j(\widehat{\varphi}_g)$, which maps each jamming density onto its corresponding equilibrium density. In other words, Eq. (7.47) indicates that in order to count the number of packings that can be constructed by state following at density $\widehat{\varphi}_j$, one can just count how many equilibrium states at density $\widehat{\varphi}_g(\widehat{\varphi}_j)$ exist, under the assumption that each of them would generate a single corresponding jammed state. It is interesting to observe that the Edwards complexity is distinct from the state following complexity of jammed packings, as shown[12] in Figure 7.6. While the two functions are numerically close in the region where they are both non-zero, they are clearly distinct. In addition, $\Sigma_j^{\mathrm{Ed}}(\widehat{\varphi})$ is non-zero for $\widehat{\varphi} \in [\widehat{\varphi}_{\mathrm{th}}, \widehat{\varphi}_{\mathrm{GCP}}]$, while $\Sigma_j^{\mathrm{SF}}(\widehat{\varphi})$ is non zero-over a smaller interval, $\widehat{\varphi} \in [\widehat{\varphi}_{j,d}, \widehat{\varphi}_{j,K}]$,

---

[10] For example, one can combine the Monasson and state following constructions to first prepare a typical state at a point $(\widehat{p}, \widehat{\varphi})$ out of the equilibrium line and then follow it adiabatically in compression or decompression.

[11] This assumption is correct only in the p-spin model [65], while it fails in all other known models, even in a RS phase. It is even more likely to be incorrect in presence of a RSB phase [230]. Yet it often provides a good approximation of the correct calculation.

[12] To construct the state following complexity of jammed packings in Figure 7.6, the function $\widehat{\varphi}_j(\widehat{\varphi}_g)$ has been obtained by fitting the fullRSB numerical data discussed in Section 6.3, which gives $\widehat{\varphi}_j(\widehat{\varphi}_g) \approx 3.324 + 0.730\widehat{\varphi}_g + 0.0127\widehat{\varphi}_g^2$. Note that these data are not very precise in the low $\widehat{\varphi}_g$ region, and especially for $\widehat{\varphi}_g \sim \widehat{\varphi}_d$, for the reasons discussed in Section 6.3.

where $\widehat{\varphi}_{j,d} = \widehat{\varphi}_j(\widehat{\varphi}_d)$ and $\widehat{\varphi}_{j,K} = \widehat{\varphi}_j(\widehat{\varphi}_K)$. This is another manifestation of the fact that the packings that can be reached by compressing an equilibrium state are only a subset of the full Edwards ensemble.

7.5 Replica Symmetry Breaking Instability in the Monasson Construction

We now discuss how the replica symmetry breaking (RSB) effects discussed in Chapter 5 modify the replica symmetric (RS) picture that has been discussed earlier. Unfortunately, the role of RSB in the Monasson construction has not been extensively studied. Because many problems remain open, in this section, we only briefly review the topic without entering into the full technical details.

7.5.1 Replica Symmetry Breaking Equations

For completeness, we provide here the basic equations needed to discuss RSB effects within the Monasson construction. The procedure to derive these equations is similar to that described in Chapter 6, so we do not repeat here the details of that derivation. Technically, the main difference with respect to Chapter 6 is that, in the Monasson construction, the number of replicas m is kept finite. The function $\Delta(x)$ is therefore defined over the interval $x \in [m, 1]$, while in state following, the number of replicas goes to zero and the function $\Delta(x)$ is defined in $x \in [0, 1]$.

Free Energy and Variational Equations

The expression for the free energy within the fullRSB ansatz can be derived starting from Eq. (7.31) and then inserting the hierarchical fullRSB ansatz for $\widehat{\Delta}$, as discussed in Chapter 5. In particular, the determinant of $\widehat{\Delta}$ is given by Eq. (5.29), and the term $\underline{1}^T \widehat{\Delta}^{-1} \underline{1}$ is given by Eq. (5.31), while the differential term can be treated as in Section 5.4.1. The details are very similar to Section 6.1. The result is then

$$-\beta \Phi(m; \widehat{\varphi}, T) = \frac{d}{2} \log \left(\frac{2\pi e}{d} \right) + \frac{d(m-1)}{2} \log \left(\frac{\pi e}{d^2} \right) + \frac{d}{2} \log m$$
$$- \frac{dm}{2} \int_m^1 \frac{dx}{x^2} \log \lambda(x) + \frac{d\widehat{\varphi}}{2} \int_{-\infty}^{\infty} dh \, e^h \left\{ e^{mf(m,h)} - 1 \right\}, \tag{7.48}$$

where

$$\lambda(x) = x \Delta(x) + \int_x^1 dy \, \Delta(y), \tag{7.49}$$

and

$$f(1, h) = \log \gamma_{\Delta(1)} \star e^{-\beta \bar{v}(h + \Delta(m)/2)},$$
$$\dot{f}(x, h) = \frac{1}{2} \dot{\Delta}(x) \left(f''(x, h) + x f'(x, h)^2 \right), \qquad m < x < 1. \tag{7.50}$$

The free energy should be optimised with respect to $\Delta(x)$, which can be done by introducing a Lagrange multiplier $P(x, h)$ conjugated to $f(x, h)$ as in Section 6.1.2. The resulting equations are

$$P(m, h) = \widehat{\varphi} e^{h + m f(m, h)}, \tag{7.51}$$

$$\dot{P}(x, h) = -\frac{1}{2} \dot{\Delta}(x) \left[P''(x, h) - 2x \left(P(x, h) f'(x, h) \right)' \right], \quad 0 < x < 1,$$

and

$$\frac{1}{\lambda(x)} = -\frac{1}{2} \int_{-\infty}^{\infty} dh\, P(x, h) [f''(x, h) + f'(x, h)]. \tag{7.52}$$

Eqs. (7.49), (7.50), (7.51) and (7.52) constitute a closed set that can be solved iteratively to obtain $\Delta(x)$. The solution should be inserted in Eq. (7.48) to obtain the Monasson free energy. The energy, pressure and other observables can be obtained along similar lines.

Instability of the RS Solution

One can follow the procedure outlined in Section 5.5 to study the linear instability of the RS solution and compute the RSB solution perturbatively around that instability. Note that, according to the definition in Eq. (4.74), the RS solution has

$$f(1, h) = \log q(\Delta, \beta; h), \qquad P(1, h) = \widehat{\varphi} e^h q(\Delta, \beta; h)^m. \tag{7.53}$$

From this, one obtains expressions for the replicon mode,

$$\lambda_R = 1 - \frac{\widehat{\varphi}}{2} \Delta^2 \int_{-\infty}^{\infty} dh\, e^h q(\Delta, \beta; h)^m f''(1, h)^2, \tag{7.54}$$

for the breaking point x^*,

$$x^* = \frac{\widehat{\varphi} \int_{-\infty}^{\infty} dh\, e^h q(\Delta, \beta; h)^m f'''(1, h)^2}{\frac{4}{\Delta^3} + 2\widehat{\varphi} \int_{-\infty}^{\infty} dh\, e^h q(\Delta, \beta; h)^m f''(1, h)^3}, \tag{7.55}$$

and for the slope $\dot{\Delta}(x^*)$ at the breaking point,

$$\dot{\Delta}(x^*) = \frac{\frac{4}{\Delta^3} + 2\widehat{\varphi} \int_{-\infty}^{\infty} dh\, e^h q(\Delta, \beta; h)^m f''(1, h)^3}{\frac{12(x^*)^2}{\Delta^4} - \widehat{\varphi} \int_{-\infty}^{\infty} dh\, e^h q(\Delta, \beta; h)^m A_{RS}(h)}, \tag{7.56}$$

$$A_{RS}(h) = f''''(1, h)^2 - 12x^* f''(1, h) f'''(1, h)^2 + 6(x^*)^2 f''(1, h)^4,$$

all evaluated on the RS solution. From these expressions, one can study RSB effects following the same recipe as in Section 6.2.2.

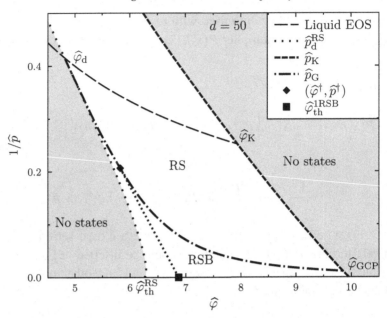

Figure 7.7 Full replica symmetry breaking phase diagram of hard spheres, in the plane $(1/\widehat{p}, \widehat{\varphi})$, for $d = 50$ [88, 221]. For better visualisation, this plot zooms (with respect to figure 7.4) on the region where fullRSB is observed. The long-dashed line is the equilibrium line $\widehat{p}_{\text{liq}}(\widehat{\varphi}) = \widehat{\varphi}/2$. The dotted line is the RS dynamical line $\widehat{p}_{\text{d}}(\widehat{\varphi})$. The portion of this line that connects $\widehat{\varphi}_{\text{d}}$ with $\widehat{\varphi}_{\text{th}}^{\text{RS}}$ is unstable because of RSB. The short-dashed line is the Kauzmann line $\widehat{p}_{\text{K}}(\widehat{\varphi})$. The dot-dashed line is the Gardner line $\widehat{p}_{\text{G}}(\widehat{\varphi})$, where the replicon eigenvalue vanishes. The black diamond marks the point $(\widehat{\varphi}^{\dagger}, \widehat{p}^{\dagger})$, above which a fullRSB solution can be constructed perturbatively around the Gardner line. The black square marks the location of the 1RSB threshold $\widehat{\varphi}_{\text{th}}^{\text{1RSB}}$.

7.5.2 Replica Symmetry Breaking Phase Diagram

In Figure 7.7, we report the phase diagram obtained within the fullRSB ansatz [88, 221]. The RS solution becomes unstable on a Gardner transition line $\widehat{p}_{\text{G}}(\widehat{\varphi})$, defined by $\lambda_R = 0$, which originates from the equilibrium dynamical transition, and extends all the way to the Kauzmann line, crossing it before GCP is reached. On this Gardner line, the breaking point x^* is always in $[0, 1]$, but the function $\Delta(x)$ is defined over $[m, 1]$. To obtain a consistent perturbative expansion around the Gardner line, one therefore has to require $x^* \in [m, 1]$. The values of x^* and m along the Gardner line are reported in Figure 7.8. The condition $x^* > m$ is only satisfied for $\widehat{\varphi} > \widehat{\varphi}^{\dagger} = 5.823$ (which corresponds to $m > m^{\dagger} = 0.4214$ and $\widehat{p} > \widehat{p}^{\dagger} = 4.838$) [88]. For $\widehat{\varphi} < \widehat{\varphi}^{\dagger}$, no RSB solution thus exist perturbatively around the Gardner line [303], and one concludes that the Gardner line is the limit of existence of the glass states. For $\widehat{\varphi} > \widehat{\varphi}^{\dagger}$, instead, a RSB solution can be constructed

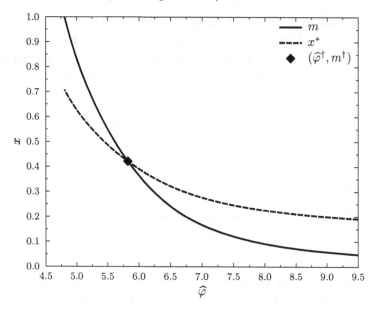

Figure 7.8   The parameter $m(\widehat{\varphi})$ and the breaking point $x^*(\widehat{\varphi})$ along the Gardner transition line of Figure 7.7. The function $\Delta(x)$ is defined, at any $\widehat{\varphi}$, for $x \in [m, 1]$; hence, it can be constructed perturbatively only when $x^* > m$. The point where $m(\widehat{\varphi}) = x^*(\widehat{\varphi})$ is the lower boundary where this condition is satisfied. The crossing gives $\widehat{\varphi}^\dagger = 5.823$, which corresponds to $m^\dagger = 0.4214$ and $\widehat{p}^\dagger = 4.838$.

perturbatively around the Gardner line. Because the slope $\dot{\Delta}(x^*) < 0$ in this region, the solution is described by a fullRSB ansatz in the whole region denoted by 'RSB' in Figure 7.7.

The crossing point between the Gardner transition line and the Kauzmann line defines an equilibrium Gardner transition [160, 171]. Above this density, the equilibrium glass phase is described by a fullRSB ansatz. Note that, as found in the state following framework, the whole jamming line ($1/\widehat{p} = 0$) falls into the fullRSB region. The consequences of the fullRSB structure of jamming will be discussed in Chapter 9. For now, we note that the fullRSB region in Figure 7.7 is delimited by the Gardner line on the top, by the Kauzmann line on the right and by the jamming line on the bottom. On the left, there must exist a 'dynamical fullRSB line', where the fullRSB solution becomes unstable and disappears, similarly to what happens to the RS solution on the dynamical RS line. Unfortunately, the dynamical fullRSB line has only been systematically studied for spin glass models [303], not for particle models. For hard spheres in $d \rightarrow \infty$, the only available calculation is an approximate one at the 1RSB level, on the jamming line $\widehat{p} = \infty$ [88]. This calculation provides a 1RSB approximation for the threshold density, $\widehat{\varphi}_{\mathrm{th}}^{\mathrm{1RSB}} = 6.870$ [88]. Joining the point $(\widehat{\varphi}^\dagger, 1/\widehat{p}^\dagger)$ with the point $(\widehat{\varphi}_{\mathrm{th}}^{\mathrm{1RSB}}, 0)$ by a straight line in Figure 7.7 provides a rough approximation of the dynamical fullRSB line.

It is conjectured [303] that in the phase diagram of Figure 7.7 no states exist to the left of the Gardner line for $1/\widehat{p} \geq 1/\widehat{p}^\dagger$ (because the RSB solution cannot be constructed perturbatively in this regime) and to the left of the dynamical fullRSB line (which, however, remains to be precisely located) for $1/\widehat{p} < 1/\widehat{p}^\dagger$.

7.5.3 On the Structure of Glass States

The interpretation of the Monasson construction when replica symmetry is broken is not straightforward. The method relies on the two assumptions formulated in Section 7.2.2:

1. Phase space can be partitioned in glass states.
2. m replicas can be confined within the same glass state.

When replica symmetry is broken, both assumptions must be revised [262, 263, 303]. While the partitioning in states is always possible [254], states are now organised in a hierarchical structure of sub-basins, as discussed in Section 5.3.3. In this situation, a state is fully identified by a series of indices $\{\alpha_1, \alpha_2, \ldots, \alpha_k\}$: a first index α_1 labelling the largest metabasins, a second index α_2 labelling the distinct sub-basins inside α_1 and so on, down to the last index α_k labelling individual states (see Figure 5.7). To each level of the hierarchy, one can associate a corresponding complexity: the complexity $\Sigma_k(f)$ counts the number of individual states with free energy f (labelled by α_k), the complexity $\Sigma_{k-1}(f)$ counts the number of sub-basins that group individual states and so on, up to the complexity $\Sigma_1(f)$ that counts the largest metabasins [262].

The second assumption must also be revised. In the simplest Monasson construction introduced in Section 7.2, we assumed that m replicas could be confined within a same state. However, in practical computations, it is hard to implement this constraint if replica symmetry is spontaneously broken. It is certainly straightforward to confine replicas into the same metabasin α_1 because escaping from such metabasin implies diffusion – hence, an infinite Δ. As long as $\Delta(x)$ remains finite for all x, the m replicas thus belong to a same metabasin. But if $\Delta(x)$ acquires a fullRSB structure, the replicas inside this metabasin then have a non-trivial distribution of mutual overlaps, and different glass states inside this metabasin get visited.

The correct interpretation of the Monasson construction in this case is that replicas are free to explore, independently in equilibrium, a given metabasin (labelled by α_1). Their free energy is therefore mf_{α_1}, where f_{α_1} is the equilibrium free energy of this metabasin. This free energy takes into account the contribution of the individual glass states within the metabasin and their complexity. Then, the partition function of the m replicas is given by Eq. (7.17), where the sum is over α_1, and the complexity $\Sigma_1(f)$ counts the number of metabasins. Hence, the Legendre transform with respect to m gives the complexity of the biggest metabasins when replica symmetry is spontaneously broken [262]. How to compute the complexities at lower levels in the hierarchy – i.e., $\Sigma_j(f)$ with $j = 2, \ldots, k$ – remains an open problem. In spin glass models, the calculation of $\Sigma_k(f)$ has been attempted either by counting directly the solution of the TAP equations [22, 81] or by using a more complicated RSB ansatz [287], but these methods are more involved than the Monasson method, and they have not yet been adapted to particle systems in the limit $d \to \infty$.

7.6 Wrap-Up

7.6.1 Summary

In this chapter, we have seen that

- A notion of equilibrium complexity $\Sigma_{eq}(\widehat{\varphi}, T)$ can be introduced as the difference between the liquid entropy and the internal entropy of typical glass states on the equilibrium line, in the dynamically arrested phase (Section 7.1). This quantity counts the number of typical glass states that contribute to the equilibrium partition function.
- For some models, the equilibrium complexity vanishes at high density or low temperatures, defining a Kauzmann transition line $T_K(\widehat{\varphi})$. For $T < T_K(\widehat{\varphi})$, the liquid state does not exist anymore, even as a dynamically arrested phase. A phase transition towards an ideal glass phase thus takes place at $T_K(\widehat{\varphi})$ (Section 7.1.3).
- The Monasson method (Section 7.2) makes use of m coupled replicas to compute the complexity $\Sigma(f; \widehat{\varphi}, T)$ of glass states of any free energy f at a given state point $(\widehat{\varphi}, T)$. The equilibrium case corresponds to a specific free energy $f = f_g(\widehat{\varphi}, T)$ dominating the equilibrium partition function.
- The lowest free energy states, $f_{min}(\widehat{\varphi}, T)$ such that $\Sigma(f; \widehat{\varphi}, T) = 0$, dominate the partition function beyond the Kauzmann transition. The ideal glass free energy, $f_{min}(\widehat{\varphi}, T)$, can also be computed from the Monasson method (Section 7.2.3).
- Explicit expressions of the Monasson free energy can be obtained in the limit $d \to \infty$ (Section 7.3).
- The Monasson phase diagram of hard spheres in $d \to \infty$ displays a region in which glass states exist. This region is delimited by the Kauzmann line, the dynamical line and the jamming line (Section 7.4). On the jamming line, where pressure diverges, glass states reduce to mechanically stable states, and the Monasson ensemble reduces to the Edwards ensemble.
- Spontaneous replica symmetry breaking is observed in the Monasson construction in part of the phase diagram. For hard spheres, it surrounds the jamming line. The interpretation of replica symmetry breaking in this context is not fully developed (Section 7.5).

7.6.2 Further Reading

We provide here a list of references that can be consulted to further explore the subjects discussed in this chapter, selected according to the criteria discussed in Section 1.6.2.

The existence of a Kauzmann transition and of an ideal glass phase in finite dimensions has been the subject of a long debate in the glass literature, and the equilibrium complexity has been measured in a wide variety of glass formers. While at the mean field level, dynamical arrest happens at $T_d > T_K$, in finite dimension, the dynamics is likely activated and thus very slow for $T < T_d$ (see Sections 3.5.2 and 4.5.2 for references). Equilibrating in the vicinity of T_K thus requires times growing exponentially in $1/|T - T_K|$, at variance with standard phase transitions for which equilibration times grow as power laws in the distance from the transition. As a consequence, state of the art methods can only measure the configurational entropy and the associated length scale quite far from T_K. Collections of experimentally available data can be found in

- Angell, *Entropy and fragility in supercooling liquids* [16]
- Richert and Angell, *Dynamics of glass-forming liquids. V. On the link between molecular dynamics and configurational entropy* [302]
- Capaccioli, Ruocco and Zamponi, *Dynamically correlated regions and configurational entropy in supercooled liquids* [75]

Note that in experiments, the complexity is usually estimated by the difference between the liquid and crystal entropies, under the assumption that the crystal and glass entropies are similar. This assumption has been questioned; see, e.g.,

- Johari, *A resolution for the enigma of a liquids configurational entropy-molecular kinetics relation* [194]
- Stillinger, Debenedetti and Truskett, *The Kauzmann paradox revisited* [336]

for a discussion. In numerical simulations of model systems, the glass entropy can be computed by several methods. The equilibrium complexity can then be directly estimated and compared with theoretical predictions. Early attempts are reported in

- Coluzzi, Mézard, Parisi et al., *Thermodynamics of binary mixture glasses* [102]
- Sciortino, Kob and Tartaglia, *Inherent structure entropy of supercooled liquids* [322]
- Sastry, *Evaluation of the configurational entropy of a model liquid from computer simulations* [313]
- Angelani and Foffi, *Configurational entropy of hard spheres* [13]

but were severely limited by the impossibility of equilibrating the liquid at temperatures below the dynamical arrest using standard molecular dynamics. More recently, thanks to the smart 'swap' algorithms already mentioned in Section 4.5.2, supercooled liquids have been equilibrated in computer simulations down to the laboratory glass transition and beyond (but still far from T_K). Thanks to these methods, the configurational entropy has been measured down to much lower temperatures, both in $d = 2$ and $d = 3$; see

- Berthier, Charbonneau, Ninarello et al., *Zero-temperature glass transition in two dimensions* [46]
- Berthier, Charbonneau, Coslovich et al., *Configurational entropy measurements in extremely supercooled liquids that break the glass ceiling* [45]

Extrapolation of the data in $d = 2$ suggests that $T_K = 0$, while in $d = 3$, it is consistent with a finite T_K. While the debate remains open and interesting, it is important to keep in mind that even at the mean field level, the existence of a Kauzmann transition is model dependent, and that, in any case, laboratory glasses are always trapped in free energy states much higher than f_{min}. The properties of the Kauzmann transition and of the ideal glass phase have, therefore, little practical relevance.

It has been theoretically established that the Kauzmann transition should necessarily be associated to a diverging thermodynamic length scale. This was later confirmed within the random first-order transition (RFOT) approach, by an analysis of the excitations around energy minima, by the study of Kac models, by a rigorous analysis, and by numerical simulations. Within the RFOT framework, this phenomenon is intimately connected with the non-convexity of the Franz–Parisi potential discussed in Section 4.5.2. Besides the references already mentioned in Section 4.5.2 on the non-convexity of the Franz–Parisi potential and nucleation in the RFOT approach, additional relevant references are

- Adam and Gibbs, *On the temperature dependence of cooperative relaxation properties in glass-forming liquids* [1]
- Bouchaud and Biroli, *On the Adam–Gibbs–Kirkpatrick–Thirumalai–Wolynes scenario for the viscosity increase in glasses* [63]
- Stillinger, *Supercooled liquids, glass transitions, and the Kauzmann paradox* [334]
- Montanari and Semerjian, *Rigorous inequalities between length and time scales in glassy systems* [261]
- Franz and Montanari, *Analytic determination of dynamical and mosaic length scales in a Kac glass model* [143]
- Biroli, Bouchaud, Cavagna et al., *Thermodynamic signature of growing amorphous order in glass-forming liquids* [57]
- Yaida, Berthier, Charbonneau et al., *Point-to-set lengths, local structure, and glassiness* [363]

Measuring the complexity out of equilibrium is much harder than in equilibrium because of the absence of a reference liquid state. Efforts have thus exclusively focused on the simplest case of jammed packings to test the validity of the Edwards ensemble. The literature on this problem is extremely large; see, e.g.,

- Mehta (ed.), *Granular matter: An interdisciplinary approach* [249]
- Barrat, Kurchan, Loreto et al., *Edwards measures: A thermodynamic construction for dense granular media and glasses* [29]
- Chakraborty, *Statistical ensemble approach to stress transmission in granular packings* [82]
- Bowles and Ashwin, *Edwards entropy and compactivity in a model of granular matter* [66]
- Asenjo, Paillusson and Frenkel, *Numerical calculation of granular entropy* [20]
- Martiniani, Schrenk, Stevenson et al., *Turning intractable counting into sampling: computing the configurational entropy of three-dimensional jammed packings* [247]

for introductory reviews and a sample of numerical results.

8

Packing Spheres in Large Dimensions

The sphere packing problem consists of finding the densest arrangement of equal-sized spheres in $\mathbb{R}^d$, i.e., the infinite d-dimensional Euclidean space. This geometrical problem is simply stated, which in part explains why it has attracted the attention of mathematicians since the ancient times. But it also has connections to other areas of mathematics, natural sciences and engineering [98, 103, 292, 344], which makes it far from being an abstract problem. In physics, this problem is also connected to the existence and stability of crystalline phases. Since Shannon's pioneering work, the large d limit of this problem is known to be connected to the practical problem of designing error-correcting codes in communication technology [103]. In this chapter, we review some of the known results and discuss how the results of the previous chapters can provide additional insight on this problem in the limit $d \to \infty$.

8.1 Statement of the Problem

8.1.1 Close Packing Density

The sphere close packing density of d-dimensional Euclidean space, denoted $\theta(d)$, is the supremum of the packing density over all equal-sized sphere packings $\mathcal{P}$ of $\mathbb{R}^d$. More precisely:[1]

- A sphere packing $\mathcal{P}$ of $\mathbb{R}^d$ is a countably infinite collection of points $\underline{X}(\mathcal{P})$ $= \{\mathbf{x}_i\}_{i \in \mathbb{Z}}$, such that $|\mathbf{x}_i - \mathbf{x}_j| \geq \ell$ for all pairs i, j of points. Here ℓ is the sphere diameter.
- Consider a continuous sequence of regions $\mathcal{V}(L) \subset \mathbb{R}^d$ parametrised by a characteristic linear size L and of volume $V(L)$. An example is a cubic volume

---

[1] The conventional notations for the sphere packing problem differ a lot in the mathematics and physics communities; here we mostly follow the physics conventions, with some exceptions, and we provide both notations when needed to avoid ambiguities.

$\mathcal{V}(L) = [-L/2, L/2]^d$ or a sphere of radius L centered at the origin. The packing density of $\mathcal{P}$ in this region is then

$$\varphi(L, \mathcal{P}) = \frac{V_c(L, \mathcal{P})}{V(L)}, \tag{8.1}$$

where $V_c(L, \mathcal{P})$ is the volume of $\mathcal{V}(L)$ covered by spheres of diameter ℓ centered at the points of $\mathcal{P}$ (note that spheres can cross the boundary of $\mathcal{V}$).

- The close packing density in d dimensions is defined as

$$\theta(d) = \sup_{\mathcal{P}} \limsup_{L \to \infty} \varphi(L, \mathcal{P}). \tag{8.2}$$

In order to solve the sphere packing problem in dimension d, one would like to know both the value of $\theta(d)$ and at least one packing $\mathcal{P}^*(d)$ that has a density asymptotically equal to $\theta(d)$ when $L \to \infty$.

8.1.2 Periodic versus Non-periodic Packings

Periodicity is an important property of packings. In fact, all known very dense packings are periodic; see Figure 8.1. The distinction between periodic and non-periodic packings is a recurrent theme in the literature. In physics, this corresponds to the distinction between amorphous (or 'disordered') and crystalline (or 'ordered') phases of matter. Disorder in this context is often intended as 'absence of order', and in this sense, any non-periodic packing can be considered as disordered. This grouping is fairly crude, but defining and quantifying disorder more precisely is particularly difficult [344]. This problem will not be discussed here, but instead, we list a few classes of packings that can be explicitly constructed, from the 'most ordered' to the 'most disordered' ones.

- **Bravais lattices** – The simplest class of periodic packings are 'Bravais lattices',[2] which represent crystals with a single–particle unit cell. They are defined by a set of d primitive vectors $a_1, \ldots, a_d$, with $a_i \in \mathbb{R}^d$ or, equivalently, by a $d \times d$ matrix $\hat{A} = \{a_1, \cdots, a_d\}$ having the primitive vectors as columns.[3] The points of the packings are obtained as

$$\mathbf{x}(n) = n_1 a_1 + \cdots + n_d a_d = \hat{A} n, \qquad n \in \mathbb{Z}^d, \tag{8.3}$$

where n is an arbitrary d-dimensional vector of integers. Because $\mathbf{x}(0) = \mathbf{0}$ is part of the packing, and by periodicity, the maximum allowed sphere diameter corresponds to the closest point to the origin, i.e.,

---

[2] In the mathematical literature, they are simply called 'lattice packings'.
[3] Many different matrices $\hat{A}$ can characterise the same lattice, but there is a well-defined region in the space of matrices that has a one-to-one correspondence with Bravais lattices.

$$\ell(\hat{A})^2 = \min_{n} |\hat{A}n|^2, \tag{8.4}$$

while the volume of a unit cell is $|\det \hat{A}|$. Because there is one sphere per unit cell, the packing fraction of a Bravais lattice is then simply

$$\varphi(\hat{A}) = \frac{V_d \ell(\hat{A})^d}{2^d |\det \hat{A}|}. \tag{8.5}$$

The calculation of $\ell(\hat{A})$ for a generic matrix $\hat{A}$, however, is an algorithmically hard problem when dimension increases [98]. Note that because of periodicity, the structure factor $S(\mathbf{q})$ of Bravais lattices exhibits Bragg peaks. These are delta peaks at values of $\mathbf{q}$ that belong to the reciprocal Bravais lattice, which is defined by $\mathbf{q} \cdot \mathbf{x}(n) = 2\pi p$, for some $p \in \mathbb{Z}$.

- **Non-Bravais lattices** – Lattices obtained by considering a unit cell containing k points in positions $\{b_1 = 0, b_2, \dots, b_k\}$, and repeating it periodically in space along a set of primitive vectors, are non-Bravais lattices. The lattice points then have the form

$$\mathbf{x}_\alpha(n) = b_\alpha + n_1 a_1 + \cdots + n_d a_d = b_\alpha + \hat{A}n, \tag{8.6}$$

with $n \in \mathbb{Z}^d$ and $\alpha = 1, \dots, k$. The unit cell vectors have to be chosen in such a way that $b_\alpha \neq \hat{A}n$ for all n. Non-Bravais lattices describe crystals with complex unit cells [21], and their packing fraction is easily computed by a generalisation of Eq. (8.5). In this case, periodicity also induces Bragg peaks in the structure factor, arranged in periodic positions in Fourier space.

- **Quasiperiodic packings** – Quasiperiodic packings, or 'quasicrystals', are non-periodic packings which still exhibit Bragg peaks in their structure factors [126, 190]. A simple way of constructing d-dimensional quasicrystals is to consider a lattice in some dimension $d' > d$ and project it on a plane of dimension d that is not one of the lattice planes. Despite the absence of periodicity, quasicrystals present patterns that repeat often, in such a way that the number of packings that can be created by a given rule is not exponential in the volume of the system [219].

- **Disordered packings** – Several proposals have been made to define disordered packings in terms of pattern repetitions [219, 309] or local order metrics [344], but these constructions are difficult to apply in practice. To bypass the problem of giving a general definition of disordered packings, one can instead define concretely a class of disordered packings by considering the Gibbs–Boltzmann measure of hard spheres, which is uniform over all possible packings of a given packing fraction φ. Whenever the system can be equilibrated in the fluid phase, typical equilibrium configurations of hard spheres are disordered sphere packings.

To obtain higher-density packings, one can compress the system adiabatically from the liquid phase, as discussed in Chapter 4, up to the jamming density. This particular class of disordered packings is discussed in more detail in the rest of this chapter.

An important remark is that the Gibbs–Boltzmann measure is usually defined by considering a cubic volume $\mathcal{V}(L) = [-L/2, L/2]^d \subset \mathbb{R}^d$ with periodic boundary conditions and then giving uniform probability to all configurations of N points in $\mathcal{V}(L)$ that satisfy the hard-sphere constraint. Therefore, all the configurations that belong to the Gibbs–Boltzmann measure can be considered, by periodic extension of the box $\mathcal{V}(L)$, as packings over the infinite Euclidean space $\mathbb{R}^d$. For finite L, one thus obtains non-Bravais lattices with N particles in a unit cell defined by the box $\mathcal{V}(L)$. Only in the limit $L \to \infty$, provided one is still in the liquid phase, does one obtain fully disordered packings over the infinite space. For the same reason, any disordered packing can be approximated up to a desired precision by a non-Bravais lattice with a large enough unit cell.

8.1.3 Constructive versus Non-constructive Approaches

As stated in Section 8.1.1, one would like to know the close packing density $\theta(d)$ and a packing that achieves it. Several strategies have been adopted to reach, at least partially, this goal:

- **Bounds –** One approach to the problem is to construct upper and lower bounds to $\theta(d)$, even if non-constructive, in the sense that they do not provide information about the best packing configuration, but only about the packing density. A review of available bounds will be given in Section 8.2. The strategies that have been followed to obtain them are diverse, spanning several fields of mathematics.
- **Deterministic packing construction –** Another approach is to directly exhibit a single (typically periodic) packing $\mathcal{P}$ of the infinite Euclidean space. The density $\varphi(\mathcal{P})$ is then a lower bound for $\theta(d)$. If $\varphi(\mathcal{P})$ happens to coincide with the best upper bound, then one has found the best packing. Sometimes packings can be constructed rather easily (e.g., in the case of Bravais lattices, it is enough to specify the primitive vectors), but the procedure to construct the packing can also be rather complex [268]. It is then also interesting to consider how many operations are needed to obtain the result.
- **Stochastic packing construction –** One can propose a stochastic procedure to construct an ensemble of packings. For example, one could sample from the equilibrium Gibbs–Boltzmann distribution at some constant density (if possible). In this case, one also encounters the algorithmic problem of the number of operations needed to achieve a proper sampling.

Examples of these approaches are given in the following sections.

8.2 Review of Rigorous Results

We present in this section a brief review of known rigorous results on the sphere packing problem. More detailed reviews can be found in [98, 103].

8.2.1 Best Known Packings

Currently, the value of $\theta(d)$ is rigorously known only for $d = 1, 2, 3, 8, 24$.

- The case $d = 1$ is trivial. The densest packing is a one-dimensional lattice, with particles in positions $x_n = \ell n, n \in \mathbb{Z}$, and $\theta(1) = 1$.
- The case $d = 2$ is elementary but not trivial. The densest packing is the hexagonal lattice and $\theta(2) = \pi\sqrt{3}/6 = 0.9069\ldots$ The first proof is attributed to Thue in 1892 [98, page 5].
- For $d = 3$, the close packing density $\theta(3) = \pi/(3\sqrt{2}) = 0.74048\ldots$ is achieved by any sequence of staggered hexagonal planes, including the face-centred cubic (FCC) and hexagonal close-packed (HCP) lattices. After centuries of efforts, a proof was finally achieved by Hales. The proof, which is very complex and heavily based on computer assistance, was first announced in 1998 and then went through a long review process to be finally published in 2005 [174]. It was subsequently verified at the level of formal logic [173].
- The cases $d = 8$ and $d = 24$ were proven in 2016, the former by Viazovska [353] ($d = 8$, the densest packing being a lattice called E_8) and the latter by Cohn, Kumar, Miller, Radchenko and Viazovska [101] ($d = 24$, the densest packing being the Leech lattice). The proof, which is much simpler than for $d = 3$, is based on deriving a series of upper bounds depending on auxiliary functions and then finding the (non-trivial) function such that the upper bound precisely coincides with the packing density of the candidate best lattice. A pedagogical introduction can be found in [99].

It is interesting to note that in all these cases, the best packing is a Bravais lattice – i.e., a crystal with a single-particle unit cell – but the optimal packing also depends sensitively on d. A list of the best known packings is available in [103], and more updated lists are also available online [271]. The density of the best known packings is reported as a function of d in Figure 8.1. The list in [271] and the figure clearly show that every dimension has its own geometric oddities. As a result, what is known about the densest packings in a certain dimension d is generally useless for other dimensions, even those very close to d. For instance, stacking d-dimensional good packings generally produces poor packings in $d + 1$.

Let us define a packing as 'saturated' when there is no room for adding an extra sphere. Clearly, saturated packings exist: they can be constructed by starting from a non-saturated packing, and adding spheres to fill the holes until the packing is saturated. An interesting observation is that a cubic lattice is not saturated for $d \geq 4$.

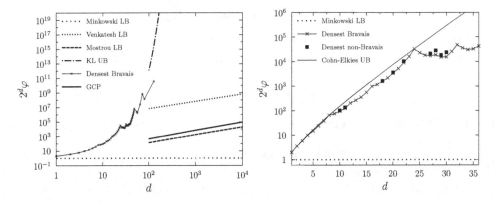

Figure 8.1 (Left) Scaled packing fraction $2^d \varphi$ of the best known Bravais lattice packings as a function of dimension d [271]. For comparison, the best upper bound (UB) by Kabatiansky and Levensthein [197]; the lower bounds (LB) by Minkowski, Venkatesh [352] and Mostrou [268]; and the glass close packing (GCP) density are also shown. Because these are only asymptotic results for $d \to \infty$, they are only shown for $d > 100$ for illustration. (Right) Scaled packing fraction $2^d \varphi$ of the best known packings as a function of d, zooming on the region of small d. Bravais lattices are shown as crosses joined by a full line; non-Bravais lattices are shown as full squares for a set of dimensions d where they are better than Bravais lattices [271]. Also shown is the best upper bound by Cohn and Elkies [100, table 3], which is saturated to very high precision in $d = 1, 2, 3, 8, 24$.

Indeed, the center of each cubic cell in the lattice is at distance $\ell\sqrt{d}/2$ from any point of the lattice, where ℓ is the lattice size. Therefore, a new sphere can be added there if $d \geq 4$. In sufficiently large d, no saturated Bravais lattice has been found, leading to the conjecture that no such lattice exists in large enough d [98]. If the conjecture is true, then dense periodic lattices in large d must have more than one particle per unit cell. For instance, the best known packing in $d = 10$ has 40 particles in the unit cell and is 8% denser than any known Bravais lattice. Non-Bravais lattices are also the densest known packings in several $d \geq 10$ (Figure 8.1). The best packings in asymptotically large d might thus have large unit cells or even be disordered [344].

8.2.2 Lower Bounds

When d grows, the unusual geometrical features of high-dimensional spaces become more pronounced. For instance, the volume of the unit diameter sphere vanishes exponentially with respect to the volume of the unit hypercube. In this asymptotic limit, only upper and lower bounds are known (see [98, 103, 344] for a more detailed review). We present here, in chronological order, a list of lower bounds.

- Minkowski proved in 1905 that the volume occupied by a saturated packing must satisfy $\varphi \geq 2^{-d}$; hence, because saturated packings certainly exist, we have $\theta(d) \geq 2^{-d}$. The proof is actually elementary. By definition, if we double the radius of each sphere, then space must be covered completely; otherwise, a new sphere would fit in the original packing. Doubling the radius multiplies volume by 2^d, and so the original packing must cover at least a 2^{-d} fraction of space.
- This lower bound was improved by a linear d factor by Rogers in 1947, who proved that $\theta(d) \geq d\, 2^{-d}$ [306, 307] (see also [283]).
- Lebowitz and Penrose in 1964 [226] proved that the virial series expansion of the entropy is convergent for $\varphi < 0.14467\ldots 2^{-d}$; hence, $\theta(d) \geq 0.14467\ldots 2^{-d}$. Additionally, the proof provides an exact expression for the excess equilibrium entropy per particle, as discussed in Chapter 2:

$$s^{\mathrm{ex}}(\varphi) = -\frac{2^d \varphi}{2}. \tag{8.7}$$

- Vance's 2011 bound is $\theta(d) \geq (6/e)d\, 2^{-d}$ (for d divisible by 4) [351].
- In 2012, Venkatesh dramatically stepped up this effort by proving that $\theta(d) \geq 65,963\, d\, 2^{-d}$ and that $\theta(d) \geq 0.5 \log(\log d)d\, 2^{-d}$ for infinitely many (but not all) dimensions [352].
- While most of these lower bounds are non-constructive, in 2017, Moustrou proved that, in infinitely many dimensions, packings with density $\varphi \geq 0.89 \log(\log d)\, d\, 2^{-d}$ can be constructed with $\exp(1.5\, d \log d)$ binary operations [268].
- The result of Lebowitz and Penrose was extended to higher densities in 2018 by Jenssen, Joos and Perkins [191]. They proved that at packing fraction $\varphi = \log(2/\sqrt{3})\, d\, 2^{-d}$, the entropy density of the Gibbs measure is $s^{\mathrm{ex}} \geq -2^d \varphi$, which also proves that there exist packings at this density. Note that this bound is compatible with the validity of Eq. (8.7). The missing factor 2 is likely due to technical details of the proof.

All the bounds mentioned have their own interest. Currently, the best lower bounds are those of Venkatesh and Moustrou, which give $\widehat{\varphi} \geq 0.89 \log(\log d)$ in infinitely many (but not all) dimensions, and $\widehat{\varphi} \geq 65,963$ in all other sufficiently large dimensions.

8.2.3 Upper Bounds

Proving upper bounds to $\theta(d)$ in the limit $d \to \infty$ is also quite challenging.

- In 1929, Blichfeldt proved that $\theta(d) \leq (d/2)2^{-d/2}$ [60]. This result was slightly improved by Rogers to $\theta(d) \leq (d/e)2^{-d/2}$ [306].
- In 1978, Kabatiansky and Levensthein proved that $\theta(d) \leq 2^{-0.5990\ldots d}$ [197].

To date, these are essentially the only two significant improvements that have been made to upper bounds in the asymptotic limit $d \to \infty$. However, in 2003, Cohn and Elkies [100] proved a very important related theorem. It can be stated roughly as follows. Let $f(\mathbf{r})$ be a function that is smooth enough and decays fast enough, and let $\widehat{f}(\mathbf{q}) = \int d\mathbf{r}\, e^{i\mathbf{q}\cdot\mathbf{r}} f(\mathbf{r})$ be its Fourier transform. If one can find a (non-identically zero) function such that

$$f(\mathbf{r}) \leq 0 \text{ for } |\mathbf{r}| \geq \ell, \qquad \widehat{f}(\mathbf{q}) \geq 0, \ \forall \mathbf{q}, \tag{8.8}$$

then for spheres of diameter ℓ one has[4]

$$\theta(d) \leq \varphi_{\mathrm{CE}}[f] = V_d \left(\frac{\ell}{2}\right)^d \frac{f(0)}{\widehat{f}(0)}, \qquad V_d = \frac{\pi^{d/2}}{\Gamma(d/2+1)}. \tag{8.9}$$

The idea of the proof is very simple, and it can be found in [98]. Note that the bound obtained from Eqs. (8.8) and (8.9) is invariant under multiplication of $f(\mathbf{x})$ by an arbitrary constant. The best upper bound from Eq. (8.9) can then be obtained by fixing, for example, a normalisation $\widehat{f}(0) = 1$ and then minimising $f(0)$ subject to the linear constraints in Eq. (8.8). This problem can be solved by efficient 'linear programming' algorithms. This theorem is at the basis of the proof of the optimal packing in $d = 8$ and $d = 24$ [99]. It has also provided many other interesting results in finite d, but thus far, for $d \to \infty$, it could only reproduce the asymptotic exponential scaling of the Kabatiansky–Levensthein 1978 bound, without improving it [98].

In a follow-up work, Cohn [97] made another interesting observation. Consider a function[5] $g(\mathbf{r})$ that satisfies, for a fixed ρ, the linear constraints

$$g(\mathbf{r}) = 0 \text{ for } |\mathbf{r}| < \ell, \qquad g(\mathbf{r}) \geq 0 \text{ for } |\mathbf{r}| \geq \ell,$$
$$S(\mathbf{q}) = 1 + \rho[\widehat{g}(\mathbf{q}) - (2\pi)^d \delta(\mathbf{q})] \geq 0, \ \forall \mathbf{q}. \tag{8.10}$$

Then, for any function satisfying the constraints in Eq. (8.8) one has

$$f(0) \geq \int d\mathbf{r}\, f(\mathbf{r})[\delta(\mathbf{r}) + \rho g(\mathbf{r})] = \int \frac{d\mathbf{q}}{(2\pi)^d} \widehat{f}(\mathbf{q})[1 + \rho \widehat{g}(\mathbf{q})] \geq \rho \widehat{f}(0), \tag{8.11}$$

and as a consequence,

$$\varphi_{\mathrm{CE}}[f] = V_d \left(\frac{\ell}{2}\right)^d \frac{f(0)}{\widehat{f}(0)} \geq V_d \left(\frac{\ell}{2}\right)^d \rho = \varphi. \tag{8.12}$$

---

[4] Note that according to Eq. (8.8), $\widehat{f}(\mathbf{q}) \geq 0$ and is not identically zero. It follows that $f(0) = \int \frac{d\mathbf{q}}{(2\pi)^d} \widehat{f}(\mathbf{q}) > 0$.
If $\widehat{f}(0) = 0$, then the bound is useless, as it gives $\theta(d) \leq \infty$.
[5] More precisely, a distribution. See [97] for details.

Note the following:

- The linear constraints in Eq. (8.10) are the minimal constraints one should impose on the radial distribution function of a system of hard spheres of diameter ℓ, number density ρ and packing fraction φ. The radial distribution function $g(\mathbf{r})$ should indeed vanish inside the hard core, be positive everywhere and have a positive structure factor $S(\mathbf{q})$ (see [175] and Chapter 2).
- Finding a solution to Eq. (8.10) at some density ρ does not imply that hard-sphere configurations exist at that density. Many other constraints have to be satisfied by pair and higher-order correlations [344].
- Finding a solution to Eq. (8.10) at some density ρ with associated packing fraction φ implies, according to Eq. (8.12), that the Cohn–Elkies method cannot provide a better upper bound than $\theta(d) \leq \varphi$. In this sense, the linear problem of finding the lowest $\varphi_{CE}[f]$ under the linear constraints in Eq. (8.8) and that of finding the highest ρ under the linear constraints in Eq. (8.10) are thus mathematically dual. Finding a solution to one implies an upper or lower bound on the solution of the other.[6]
- Parisi and Slanina [284] developed a procedure to obtain a solution to the linear problem in Eq. (8.12) up to a maximal value of packing fraction scaling asymptotically as $2^{-0.7786...d}$. This approach was simplified by Torquato and Stillinger [344], who proposed the simple test function $g(\mathbf{r}) = \theta(|\mathbf{r}|-A)+B\delta(|\mathbf{r}|-\ell)$, with adjustable parameters $A \geq \ell$ and $B \geq 0$, and found that this test function is a solution to Eq. (8.12) up to the same packing fraction. These works show that the Cohn-Elkies method cannot provide a better upper bound than $2^{-0.7786...d}$; hence, the gap with the best lower bound cannot be closed this way. Interestingly, the packing fraction $2^{-0.7786...d}$ is also where the resummation of ring diagrams in the virial expansion ceases to be valid, as discussed in Chapter 2.

In summary, while the best lower bound on $\theta(d)$ scales asymptotically as 2^{-d}, the best upper bound scales as $2^{-0.5990...d}$, and therefore, its ratio with the best lower bound grows exponentially in d (Figure 8.1). Moreover, there are currently no good ideas on how to improve the exponential scaling 2^{-d} of the lower bound, while the most effective method to prove upper bounds due to Cohn and Elkies surely cannot achieve anything better than $2^{-0.7786...d}$ (and it is still far from achieving it). This leaves a huge uncertainty in the asymptotic scaling of $\theta(d)$ for large d. In addition, it is by no means obvious that for large d there exist 'universal features' of optimal packings or that they should necessarily be periodic [98, 344].

---

[6] It is conjectured that there might be no gap between the best solutions of the two problems [97].

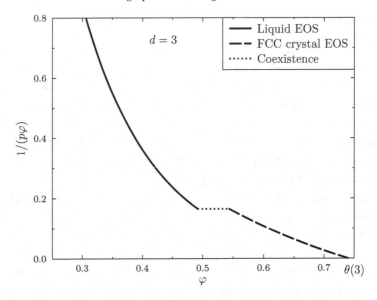

Figure 8.2   Equilibrium equation of state (EOS) of hard spheres in $d = 3$, showing the liquid and crystal phases and their coexistence. The inverse scaled pressure $\varphi p = \varphi \beta P / \rho = (\pi/6)\beta P \ell^3$ is shown as a function of the packing fraction φ. For the liquid, the Carnahan-Starling EOS $p = (1 + \varphi + \varphi^2 - \varphi^3)/(1 - \varphi)^3$ has been used [175]. For the crystal, the Speedy equation of state $p = 3/(1-z) - a(z-b)/(z-c)$ has been used, with $z = \varphi/\theta(3)$ and $a = 0.620735$, $b = 0.708194$, $c = 0.591663$ [25]. The coexistence pressure $\beta P_{co}\ell^3 = 11.5727$ is taken from [139].

8.3 Review of Non-rigorous Results

We now describe some of the results on packing problems in large dimension that have been found within the physics literature and currently lack a rigorous proof. We mostly focus on results obtained within statistical mechanics. More detailed reviews can be found in [30, 292, 344].

8.3.1 Crystals in High Dimensions

In physics, periodic packings correspond to crystals, while disordered packings correspond to liquids or glasses. It was first suggested by Kirkwood and Monroe in 1940 [209] and then shown numerically by Alder and Wainwright in 1957 [5, 175] that, upon increasing density or pressure, a hard-sphere liquid in $d = 3$ transforms into a crystal[7] via a first-order phase transition, as illustrated in Figure 8.2.

---

[7] Remarkably, the structure of the equilibrium hard-sphere crystal in $d = 3$ has been the subject of a debate because the free energy difference between the FCC and HCP crystals is extremely small [61, 358]. State-of-the-art numerical results suggest that the FCC lattice is slightly more stable [61, 211, 275]. Because the FCC and HCP lattices have the same close-packing density, however, which one of the two is more stable is irrelevant for the present discussion.

The crystal then becomes one of the close-packed structures in $d = 3$ (see Section 8.2.1) upon further compression to infinite pressure. This observation suggests that one can start by an ideal gas of hard spheres (i.e., throwing points uniformly at random) and then slowly compress the system by increasing the size of the spheres [235]. If the compression rate is slow enough, the system crystallises and eventually reaches the close packing density at $\varphi = \theta(3) = \pi/(3\sqrt{2})$. The same strategy would obviously work in $d = 1$, where there is no phase transition up to close packing, and in $d = 2$, where two phase transitions take place but crystallisation into a hexagonal lattice is nonetheless easily achieved [35]. Slow compression from the liquid is thus a simple way to obtain the densest packing in $d \leq 3$.

The equilibrium phase diagram has been studied in $d = 4, 5, 6$ as well [350]. The result is similar to that of $d = 3$: a first-order phase transition separates the low-density liquid phase from a high-density crystal phase whose lattice structure coincides with the one of the best packing. Crystallisation during a slow compression becomes, however, increasingly hard when dimension is increased. Already in $d = 4$, with current computational power, it is very hard to observe spontaneous crystallisation during a compression, and the situation is worse in $d = 5$ and $d = 6$, where no hint of crystallisation has been detected [330, 350]. This marked suppression of crystallisation is due to the growth of the surface tension between the two phases, which is related to the fact that the local geometry of the fluid becomes increasingly different from that of the crystal [350]. Instead of crystallising, upon compression, the system follows the metastable liquid equation of state and ultimately forms a glass, as it has been tested up to $d = 12$ [86, 330].

Conceptually, the fact that the barrier to crystallise increases with d implies that the liquid can be compressed at pressures above the coexistence pressure P_{co}, while remaining stable for very long times. The liquid at $P > P_{co}$ (see Figure 8.2) is usually deemed[8] 'supercooled' [120] and is metastable with respect to the crystal. For $d \geq 4$, the lifetime of the metastable supercooled liquid becomes extremely large and grows with d. The metastable liquid phase is therefore better and better defined upon increasing d. In the limit $d \to \infty$, as discussed in Chapter 1, the lifetime of the metastable state becomes infinite. In this limit, the liquid and the crystal are well defined and well separated minima of the free energy $F[\rho(\mathbf{x})]$ defined in Chapter 2. Within the theory, one can therefore very easily restrict the study to the liquid phase by assuming that the density profile is homogeneous and isotropic.

---

[8] Although when the control parameter is the density, 'overcompressed' would be a better nomenclature.

8.3.2 Algorithms to Construct Dense Packings

The difficulty to crystallise a system of hard spheres in $d \geq 4$ has an important algorithmic implication. Namely, it is impossible to construct good crystals by simple compression of low-density liquid configurations [330]. We now discuss a simple example to illustrate this effect.

Let us assume that the relaxation time of the equilibrium liquid is finite at the coexistence pressure, which is the case in all $d \leq 6$ where the problem has been studied [139, 350]. The density at which the liquid becomes metastable with respect to the crystal is then smaller than the dynamical glass transition density φ_d. Sampling equilibrium liquid configurations of N particles slightly below coexistence therefore requires a time of the order of Nd, the number of degrees of freedom in the system, because each degree of freedom must be updated a finite number of times (either by Monte Carlo or molecular dynamics) before equilibrium can be reached. If the system of N particles is confined in a box of linear size L and volume $V = L^d$ with periodic boundary conditions, the resulting packing can be periodically extended to the infinite Euclidean space, for which it would be a periodic packing of period L with N particles in the unit cell. For this procedure to be well defined, however, the linear size L of the system must be at least $L/\ell = 1 + O(1/d)$, in such a way that, for instance, a particle located in the origin can fit in the box without overlapping with itself. The periodic extension would otherwise not produce a packing.[9] In the following, to obtain a rough estimate of the scalings with dimension, we neglect all sub-exponential corrections in d. Recalling that the natural scale of packing fraction in the liquid phase is $\varphi \sim 2^{-d}$, and using Stirling's approximation for the volume $V_d \sim \exp[\frac{d}{2} \log(2\pi e/d)]$, the minimal number of particles must thus be of the order of

$$N \approx \frac{L^d \varphi}{V_d(\ell/2)^d} \approx \frac{2^d \times 2^{-d}}{V_d} \approx e^{\frac{d}{2} \log \frac{d}{2\pi e}} \approx d^{d/2}. \tag{8.13}$$

Note that this value of N diverges with d. Because finite size corrections vanish as $1/N$ in the liquid phase [86], in the large d limit, the resulting system is effectively in the thermodynamic limit despite the fact that its linear size L is finite.[10] The efficient sampling of liquid configurations then requires a time that scales proportionally to dN; hence,

$$\tau_{\text{liq}} \approx dN \approx d^{d/2} \tag{8.14}$$

---

[9] The stricter requirement that a particle cannot collide with itself due to the periodic boundary conditions would lead to $L/\ell = 2 + O(1/d)$, which does not change the leading order in Eq. (8.13).

[10] Note, however, that a d-dimensional cube of side $L = 2$ has 2^{d-1} diagonals of length $2\sqrt{d}$. Its linear size along the diagonals is thus very large when $d \to \infty$. Similarly, two random points in this cube have a mean distance $\propto \sqrt{d}$, and the distribution of distances is strongly concentrated around this mean.

at leading order in large d. One concludes that sampling configurations of a sufficiently large system to define a liquid phase requires a time scaling as $d^{d/2}$, which is why current studies are limited to $d \leq 12$ [86]. This scaling sets a 'natural' time scale for the problem of constructing large packings that is intrinsically hard to beat.

In addition, as discussed in Section 1.5.3, the crystallisation time is the exponential of a free energy barrier that scales proportionally either to N or to d, depending on the scaling of the size of the system with d. Because the relevant scaling here is $d \to \infty$ at fixed, small size L – in which case, the barrier is proportional to $N = L^d$ – one can estimate

$$\tau_{\mathrm{cryst}} \approx \tau_{\mathrm{liq}} e^N \approx e^{d^{d/2}}, \tag{8.15}$$

which is much worse than τ_{liq}. Although many other out-of-equilibrium compression algorithms have been proposed – see [30, 292, 344] for example – none can construct dense lattice packings in $d \geq 4$ much faster than slow compression.

An interesting alternative idea is to restrict the exploration of configurations to the space of Bravais lattices, which are fully specified by a $d \times d$ matrix. One can try to sample the space of these matrices looking for very dense lattices [12, 198, 199], but the number of possible lattices grows extremely fast with d. In addition, their sampling is particularly demanding because even obtaining the packing fraction of a given lattice is a computationally difficult problem. As a result, current studies only managed to reproduce the densest known packings for $d \leq 20$ and were unable to go beyond that limit. For a related statistical mechanics approach to lattice packings in large d, see [283].

8.4 Liquid, Glass and Packings in Infinite Dimensions

In this section, we show how the theory of liquids and glasses in $d \to \infty$ described in this book can provide some insight into the problem of packing spheres in $d \to \infty$.

8.4.1 Dynamical Glass Transition

In Chapter 3, it has been shown that the hard-sphere liquid dynamics has a finite relaxation time for $\widehat{\varphi} < \widehat{\varphi}_{\mathrm{d}} = 4.8067\ldots$ One can therefore start from an ideal gas configuration and compress the system [235] in order to achieve equilibration at any target density $\widehat{\varphi} < \widehat{\varphi}_{\mathrm{d}}$. As discussed in Section 8.3.2, this requires a number of particles N that scales as in Eq. (8.13) and a time scale (or inverse compression rate) that grows as Eq. (8.14). The solution of infinite dimensional liquids thus suggests a first interesting result:

R1 In $d \to \infty$, up to a packing fraction $\varphi = d2^{-d} \times 4.8067\ldots$, there exist packings of hard spheres that can be constructed in a time $\tau_{\text{liq}} \approx d^{d/2}$ by simple adiabatic compression of the ideal gas. Their excess entropy at packing fraction φ is given by $s^{\text{ex}} = -\frac{2^d \varphi}{2}$.

R1 is very similar to the rigorous result of Moustrou [268] discussed in Section 8.2.2. Moustrou proved that packings with density $\varphi \geq 0.89 \log(\log d) d\, 2^{-d}$ can be constructed with $\exp(1.5\, d \log d)$ binary operations – i.e., with the same scaling with d but with a prefactor of 1.5 instead of 0.5 in the leading exponential term. While Moustrou's result has a better asymptotic scaling of the density for $d \to \infty$, i.e., $\log(\log d)$ instead of constant, R1 is actually better as long as

$$0.89 \log(\log d) < 4.8 \qquad \Leftrightarrow \qquad d < e^{e^{4.8/0.89}} \approx 10^{95}. \qquad (8.16)$$

Because in practical applications one would never reach such high dimensions, R1 thus remains of comparable interest to Moustrou's result. Also, this procedure can construct an ensemble of packings with the liquid entropy at φ_{d}. The resulting entropy has the same scaling with d as the rigorous one of [191], with better prefactors.

The solution for the dynamical glass transition also suggests a way to slightly improve the prefactor in R1. Consider a liquid with interaction potential

$$\bar{v}(h) = \begin{cases} \infty & h < 0, \\ \bar{v}_+(h) & h \geq 0, \end{cases} \qquad (8.17)$$

where $\bar{v}_+(h)$ is a finite function that goes to zero for $h \to \infty$ fast enough that the hypotheses of Section 2.3.2 are satisfied. Any liquid configuration of the potential in Eq. (8.17) is also a valid hard sphere packing because particles satisfy the hard core constraint.[11] One can therefore choose an appropriate potential $\bar{v}_+(h)$ in such a way to push the dynamical glass transition to higher packing fractions. It was shown in [325] that it suffices to add a short range attractive potential

$$\beta \bar{v}_+(h) = \begin{cases} -1/\widehat{T} & 0 \leq h \leq \widehat{\sigma}, \\ 0 & \text{otherwise}, \end{cases} \qquad (8.18)$$

to push the dynamical transition up to $\widehat{\varphi}_{\text{d}} \approx 6.5$ (for $\widehat{\sigma} = 0.06$ and $\widehat{T} = 0.48$). Adding a sequence of attractive and repulsive steps and optimising the parameters to maximise $\widehat{\varphi}_{\text{d}}$ even gives $\widehat{\varphi}_{\text{d}} = 6.966\ldots$ [240]. Because the optimisation procedure of [240] converges upon increasing the number of steps, it is reasonable to conjecture that it achieves the maximum over all possible functions $\bar{v}_+(h)$ so that no

---

[11] Sampling from the liquid defined by Eq. (8.17), however, does not produce equilibrium hard-sphere configurations because the Gibbs–Boltzmann measure is biased by the potential $\bar{v}_+(h)$.

additional improvement can be obtained by this strategy. Sampling configurations of the liquid with optimised potential takes a time $\tau_{\text{liq}} \approx d^{d/2}$, as in R1, and the resulting liquid has finite entropy, thus leading to the improved result:

R1$_+$ In $d \to \infty$, up to a packing fraction $\varphi = d2^{-d} \times 6.966\ldots$, there exist packings of hard spheres that can be constructed in a time $\tau_{\text{liq}} \approx d^{d/2}$ by simple adiabatic compression of an ideal gas of hard spheres with an additional finite potential $\bar{v}_+(h)$. These packings have a finite excess entropy [240].

8.4.2 Adiabatic State Following

As discussed in Chapter 4, during a sufficiently slow compression, a liquid can be kept in equilibrium up to φ_{d}. More precisely, the time scale of the compression (the inverse of the compression rate) must be proportional to $\tau_{\text{liq}} \sim Nd$, as given in Eq. (8.14). In order to reach equilibrium beyond φ_{d}, a much smaller compression rate, scaling as $\exp(-N)$, must be employed, as we discuss in Section 8.4.3. A slow compression on a time scale proportional to τ_{liq} thus falls out of equilibrium at φ_{d}. Beyond that point, the pressure increases faster than that of the equilibrium liquid and diverges at a jamming point $\varphi_{\text{j}}(\varphi_{\text{d}})$. As discussed in Chapter 4, the out-of-equilibrium glass prepared at φ_{d} can be followed adiabatically through the state following construction in the limit $d \to \infty$. However, as shown in Chapter 6, this glass is always in the replica symmetry broken phase. It would then take a time $\exp(N)$ for the system to follow adiabatically the evolution of the state. If one is interested in a finite compression rate, the state following construction might then seem irrelevant. Yet experience with spin glass models shows that, for a continuous transition, the replica symmetry broken equation of state provides a good approximation of the actual non-equilibrium equation of state. A liquid of hard spheres compressed on a time scale given by Eq. (8.14) thus falls out of equilibrium around $\widehat{\varphi}_{\text{d}} = 4.8067$ and then remains close to the equation of state of the corresponding glass, which ends at a jamming point $\widehat{\varphi}_{\text{j}}(\widehat{\varphi}_{\text{d}})$. The precise jamming density is not known, but it is likely[12] that $\widehat{\varphi}_{\text{j}}(\widehat{\varphi}_{\text{d}}) > 7$. This analysis leads to a further improvement of the prefactor in R1 and R1$_+$:

R2 In $d \to \infty$, up to a packing fraction $\varphi \approx d2^{-d}\widehat{\varphi}_{\text{j}}(\widehat{\varphi}_{\text{d}})$, there exist packings of hard spheres that can be constructed in a time $\tau_{\text{liq}} \approx d^{d/2}$ by simple non-equilibrium compression of an ideal gas of hard spheres up to its jamming point.

---

[12] As already discussed in Section 7.4.4, the numerical results for $\widehat{\varphi}_{\text{j}}(\widehat{\varphi}_g)$ reported in Section 6.3 can be well approximated by $\widehat{\varphi}_{\text{j}}(\widehat{\varphi}_g) \approx 3.324 + 0.730\widehat{\varphi}_g + 0.0127\widehat{\varphi}_g^2$, which gives $\widehat{\varphi}_{\text{j}}(\widehat{\varphi}_{\text{d}}) \approx 7.1$.

Because the complexity of the packings obtained at $\widehat{\varphi}_j(\widehat{\varphi}_d)$ is positive, as discussed in Chapter 7, an exponential number in N of distinct packings can be constructed via this procedure.

It is important to stress that R2 is much harder to formalise than R1 and R1$_+$. Proving R1 'only' requires one to solve the equilibrium dynamics of the liquid, which has been done at the theoretical physics level. Its rigorous mathematical justification mostly requires proving that the truncation of the virial expansion is correct (which is, however, likely to be extremely difficult). By contrast, R2 is not even precise from the theoretical physics point of view, because the state following construction only provides an approximation to the non-equilibrium dynamics when replica symmetry is broken. A precise computation of the jamming point corresponding to slow compression would require the solution of the non-equilibrium dynamical equations in $d \to \infty$ derived in [3], which is very challenging already at the theoretical physics level. Transforming R2 in a rigorous mathematical statement would therefore be considerably harder. Note that R2 could perhaps be slightly improved by considering the non-equilibrium compression of a system with an additional potential $\bar{v}_+(h)$, as discussed in Section 8.4.1, but this possibility has not yet been explored.

8.4.3 Glass Close Packing

To conclude the discussion, we consider the behaviour of the liquid when it is compressed on a time scale (inverse compression rate) that scales as $\exp(N)$ – which means, according to the discussion of Section 8.3.2, a scaling with d of the form

$$\tau_{\text{act}} \approx e^N \approx e^{d^{d/2}}, \tag{8.19}$$

using again the minimal scaling of N with d discussed in Section 8.3.2. This time scale is slow enough that the system can jump over the barriers that separate the glass basins and thus also equilibrate in the dynamically arrested phase, $\widehat{\varphi} > \widehat{\varphi}_d$.

Because τ_{act} has the same scaling as the crystallisation time τ_{cryst} given by Eq. (8.15), it is possible that, at some point, the system could simply crystallise. Whether this would take place or not depends on the existence of a dense enough crystal phase and on the prefactors in the scaling of τ_{act} and τ_{cryst} [80]. If there is no crystal phase denser than the liquid, or if the prefactors are such that $\tau_{\text{cryst}} \gg \tau_{\text{act}}$ at all densities, then crystallisation would not be observed. Because we have no such information about crystals in large d, we leave this possibility as open and do not discuss it further.

Instead, we focus on what happens if the system stays in the amorphous phase. In this case, the system remains in equilibrium in the liquid phase at $\widehat{\varphi} > \widehat{\varphi}_d$

and samples the exponential number of glass basins in which the liquid is decomposed, as described in Chapter 7. Upon increasing density, the complexity of basins decreases up to the Kauzmann transition at $\widehat{\varphi}_K = \log d$, and, hence, $\varphi_K = d \log(d) \, 2^{-d}$. Beyond that density, the number of glass basins is no more exponential and the system transitions to an ideal glass phase. If one keeps compressing the system at rate $1/\tau_{\text{act}}$, equilibrium can be maintained in the ideal glass phase and the system then follows the equilibrium ideal glass pressure $\widehat{p}_K$, which can be computed using the Monasson method described in Chapter 7; see Figure 7.7. The equilibrium compression in the ideal glass terminates at the glass close packing (GCP) point, whose density $\widehat{\varphi}_{\text{GCP}} = \log d$ has the same scaling as the Kauzmann density at leading order. Because the number of ideal glass basins is sub-exponential in N, the number of distinct packings obtained at φ_{GCP} is also sub-exponential.

The solution of the infinite dimensional ideal glass of hard spheres discussed in Chapter 7 thus suggests the following result:

> R3 In $d \to \infty$, up to a packing fraction $\varphi_{\text{GCP}} = d \log(d) \, 2^{-d}$, there exist packings of hard spheres that can be constructed in a time $\tau_{\text{act}} \approx e^{d^{d/2}}$ by very slow adiabatic compression of an ideal gas of hard spheres up to its ideal glass phase. The number of distinct packings that can be constructed in this way is sub-exponential in N.

Note that this result provides quite a substantial improvement over the best rigorous lower bound, which predicts a scaling of $\widehat{\varphi} \sim 0.89 \log(\log d)$. R3 improves this bound to $\widehat{\varphi} \sim \log d$. Yet the time that would be needed to construct these packings has an extremely poor scaling with d, which makes a practical implementation impossible already in low d. R3 is therefore mostly an existence result. Interestingly, the leading exponential scaling of the packing fraction, 2^{-d}, cannot be improved by this approach. Improving it likely requires considering entirely different packings that have nothing to do with compressing ideal gas configurations. These packings are thus unlikely to be fully disordered [98, 344].

Finally, note that R3 cannot be improved by considering an additional potential $\bar{v}_+(h)$ as in Eq. (8.17). The expression of $\widehat{\varphi}_K$ for the potential in Eq. (8.17) is indeed deduced from Eq. (7.9):

$$\widehat{\varphi}_K = \frac{\log d}{g_K}, \qquad g_K = 1 + \int_0^\infty dh \, e^h [1 - e^{-\beta \bar{v}_+(h)}(1 + \beta \bar{v}_+(h))]. \qquad (8.20)$$

Because the integrand function $1 - e^{-x}(1 + x) \geq 0$ for all x, one has $g_K \geq 1$ (with $g_K = 1$ corresponding to the hard-sphere case $\bar{v}_+(h) = 0$), and the leading-order scaling of the Kauzmann density can only be decreased by adding an additional potential. One can also check that the expression for the Edwards complexity, discussed in Section 7.4.3, is independent of $\bar{v}_+(h)$. The presence of an additional finite potential thus cannot change the structure of jammed packings at infinite pressure, as expected. As a consequence, the expression of φ_{GCP} is independent of $\bar{v}_+(h)$. This result is physically intuitive. The glass close packing is the highest possible density of a well-defined class of amorphous packings, and as such, only depends on the geometric properties of the hard-core potential.

8.5 Wrap-Up

8.5.1 Summary

In this chapter, we have seen that

- The sphere packing problem of finding the highest packing fraction $\theta(d)$ of identical spheres in $\mathbb{R}^d$ is very difficult. Its solution is only known in a small set of low dimensions $d = 1, 2, 3, 8, 24$. The structure of the large d Euclidean space is such that it is also very hard to find regularities in the solution. As a consequence, no results for good packings are known in large dimensions (Section 8.2.1).
- The best lower bound on $\theta(d)$ scales as 2^{-d}, and decades of work were only able to improve it by polynomial or logarithmic factors in d (Section 8.2.2). The best upper bound instead scales as $2^{-0.5990d}$, which leaves an exponentially large range within which $\theta(d)$ can be located (Section 8.2.3). It can be proven that the best current techniques to obtain upper bounds (i.e., the linear programming method of Cohn and Elkies) cannot achieve anything better than $2^{-0.7786d}$.
- Spontaneous crystallisation upon compression from the liquid phase becomes increasingly difficult upon increasing d. Current computers cannot observe it for $d > 4$. It is therefore unknown whether a stable thermodynamic crystal phase exists for large enough d. For the same reason, the metastable liquid acquires a very long lifetime when d increases. It can thus be treated as a stable thermodynamic phase in computations (Section 8.3.1).
- Algorithmically, the sphere packing problem is complex. Constructing the liquid phase requires at least $N \sim d^{d/2}$ particles, and as a consequence, the natural equilibration time for the liquid is $\tau_{\text{liq}} \sim N \sim d^{d/2}$. The crystallisation time (provided a stable crystal exists) is likely to scale as $\tau_{\text{cryst}} \sim e^N \sim e^{d^{d/2}}$ in large d. Even sampling the restricted class of Bravais lattices in high d has a poor scaling with d. As a result, Bravais lattices can only be sampled up to $d \sim 20$. This approach thus provides no improvement over the best known packings (Section 8.3.2).
- The exact solution of the liquid dynamics in $d \to \infty$ discussed in this book leads to the conjecture that sphere packings exist and can be constructed in time $\tau_{\text{liq}} \sim d^{d/2}$ by slow equilibrium compression of the liquid, up to at least $\varphi = \widehat{\varphi}_{\text{d}} d \, 2^{-d}$. The natural value of $\widehat{\varphi}_{\text{d}} = 4.8067$ for hard spheres can be improved

up to $\widehat{\varphi}_d = 6.966$ by optimising the potential (Section 8.4.1) and up to a higher (but not precisely known) value, $\widehat{\varphi}_j(\widehat{\varphi}_d)$, by considering out of equilibrium compressions via the state following technique (Section 8.4.2).

- The exact calculation of the complexity and of the ideal glass phase thermodynamics leads to the conjecture that amorphous sphere packings exist up to $\varphi_{GCP} = d \log d \, 2^{-d}$. However, their construction requires a time $\tau_{act} \sim e^{d d/2}$, which is practically impossible to achieve even in low d. The density φ_{GCP} thus provides a limit of existence of packings that are structurally similar to a liquid. Beyond this density, if packings still exist, they likely have a very different structure (Section 8.4.3).

8.5.2 Further Reading

We provide here a list of references that can be consulted to further explore the subjects discussed in this chapter, selected according to the criteria discussed in Section 1.6.2.

The literature on the sphere packing problem is extremely vast and scattered across many disciplines. The reader interested in the problem can start from reviews of mathematical results in

- Rogers, *Packing and covering* [306]
- Conway and Sloane, *Sphere packings, lattices and groups* [103]
- Cohn, *Packing, coding, and ground states* [98]

while the reviews

- Torquato and Stillinger, *Jammed hard-particle packings: From Kepler to Bernal and beyond* [344]
- Baule, Morone, Herrmann et al., *Edwards statistical mechanics for jammed granular matter* [30]

attempt to connect results from physics to those from the mathematical literature.

A natural extension of the sphere packing problem is the problem of packing identical non-spherical particles and mixtures of different particles. The literature on this problem is also vast and partially covered in the previously cited reviews. Some additional results can be found in

- Yerazunis, Cornell and Wintner, *Dense random packing of binary mixtures of spheres* [365]
- Pinson, Zu, You et al., *Coordination number of binary mixtures of spheres* [295]
- Hopkins, Stillinger and Torquato, *Disordered strictly jammed binary sphere packings attain an anomalously large range of densities* [179]

- Donev, Cisse, Sachs et al., *Improving the density of jammed disordered packings using ellipsoids* [127]
- Torquato and Jiao, *Dense packings of the Platonic and Archimedean solids* [345]
- Haji-Akbari, Engel, Keys et al., *Disordered, quasicrystalline and crystalline phases of densely packed tetrahedra* [172]

All these problems can be, in principle, tackled by the formalism developed in this book, as in

- Biazzo, Caltagirone, Parisi et al., *Theory of amorphous packings of binary mixtures of hard spheres* [49]
- Ikeda, Miyazaki, Yoshino et al., *Decoupling phenomena and replica symmetry breaking of binary mixtures* [184]

9

The Jamming Transition

In Chapters 4 and 6, we have shown that a glass of hard spheres, prepared at equilibrium at scaled packing fraction $\widehat{\varphi}_g$, can be compressed up to a point $\widehat{\varphi}_j(\widehat{\varphi}_g)$, at which pressure diverges and spheres touch their neighbours, thus forming a rigid contact network. The values of $\widehat{\varphi}_j(\widehat{\varphi}_g)$ span an interval $[\widehat{\varphi}_{j,d}, \widehat{\varphi}_{j,K}]$ upon varying $\widehat{\varphi}_g \in [\widehat{\varphi}_d, \widehat{\varphi}_K]$. In Chapter 7, we have seen that stable amorphous packings can be found over an even larger interval, which goes from the threshold $\widehat{\varphi}_{th}$ up to the glass close packing density $\widehat{\varphi}_{GCP}$ (see Figure 7.6). In Chapter 8, we have seen that periodic and quasi-periodic packings might exist in the same density range as amorphous packings and even possibly at higher densities. In the limit $d \to \infty$, however, the crystallisation time diverges, and the amorphous branch of the phase diagram (liquids and glasses) is then disconnected from the crystalline branch. Amorphous packings can thus be studied without worrying about crystallisation or other structural transitions, as long as the amorphous state is locally stable.

The purpose of this chapter is to describe in more detail the properties of such jammed amorphous packings. We show that if the hard-sphere constraint is relaxed slightly, thus allowing spheres to overlap, jammed configurations separate mechanically floppy structures from mechanically rigid ones, and the associated jamming transition is critical. We discuss several protocols that are used to obtain jammed packings, focusing on the physical observables that characterise their properties. We show that the Gardner transition and full replica symmetry breaking induce a scaling behaviour close to the jamming transition and, thus, provide a set of associated universal critical exponents.

9.1 The Jamming Transition as a Satisfiability Threshold

The packing problem introduced in Chapter 8 consists in finding a configuration of N spheres of diameter ℓ with centers $\mathbf{x}_i \in \mathcal{V}$, where $\mathcal{V}$ is a region of the

d-dimensional Euclidean space with volume V, such that the spheres do not overlap.[1] Defining the (scaled) gap between two spheres, i and j, as

$$h_{ij} = d \left(\frac{|\mathbf{x}_i - \mathbf{x}_j|}{\ell} - 1 \right), \tag{9.1}$$

the packing problem consists in finding a configuration $\underline{X} = \{\mathbf{x}_i\}_{i=1,\dots,N}$, such that

$$h_{ij} \geq 0, \qquad \forall i \neq j, \tag{9.2}$$

which provides a set of $M = N(N-1)/2$ constraints[2] on the variables $\underline{X}$ (because $h_{ij} = h_{ji}$). The packing problem therefore consists in finding a configuration of N variables that simultaneously satisfies M constraints. Such problems are known as 'satisfiability' problems. See [19, 50, 251] for a historical introduction, many examples and a very complete discussion of the theory and algorithms for satisfiability. Interestingly, most of the satisfiability problems to which statistical mechanics tools have been applied involve discrete variables, but the packing problem involves continuous variables, which brings the criticality of the satisfiability transition [147, 151].

The simplest class of algorithms to solve satisfiability problems are 'local search algorithms'. These algorithms start from a random assignment of the variables. If this assignment satisfies all constraints, then the algorithm stops. Otherwise, each variable is updated according to its 'local' environment – i.e., the set of variables involved in the same constraints. While local search algorithms are extremely simple to implement, they may sometimes fail to find solutions even when such solutions do exist. In the rest of this section, we give some examples of local search algorithms for the packing problem and discuss their properties.

9.1.1 Gradient Descent

A classical strategy to solve satisfiability problems is to transform them in optimisation problems by introducing a 'cost function' associated with the unsatisfied constraints. In the sphere packing problem, the cost function is simply the interaction energy defined in Eq. (2.1) with a particular choice of pair potential, such that if and only if two spheres overlap, then the system pays a positive energy cost – i.e.,

$$V(\underline{X}) = \sum_{i<j} \bar{v}\left(h_{ij}\right), \qquad \text{with} \quad \begin{cases} \bar{v}(h) > 0 & \text{for } h < 0, \\ \bar{v}(h) = 0 & \text{for } h \geq 0. \end{cases} \tag{9.3}$$

---

[1] In Chapter 8, we discussed the problem when $\mathcal{V}$ coincides with the infinite space $\mathbb{R}^d$, but in this chapter, we also discuss finite size effects.
[2] Even if the total number of constraints is $M = N(N-1)/2$, in any finite d, the number of neighbours surrounding each sphere is finite. The effective number of relevant constraints is thus proportional to N.

Configurations $\underline{X}$ with $V(\underline{X}) = 0$ are solutions to the packing problem, while configurations with $V(\underline{X}) > 0$ contain at least one pair of overlapping spheres. Having introduced the cost function in Eq. (9.3), finding solutions to the packing problem amounts to minimising $V(\underline{X})$, hoping for a zero-energy outcome. A typical choice of cost function is the soft repulsive potential

$$\bar{v}(h) = \frac{\varepsilon}{\alpha}h^{\alpha}\theta(-h). \tag{9.4}$$

This potential is not only convenient for the numerical minimisation of Eq. (9.3) but also a good model of real materials (see Section 2.3.2).

The simplest initialisation for the system is a random Poisson point process [276, 277], which amounts to choosing $\mathbf{x}_i$ independently and uniformly at random over $\mathcal{V}$. Each point $\mathbf{x}_i$ is then associated to a sphere of diameter ℓ, which gives a packing fraction $\varphi = V_d(\ell/2)^d N/V$. In order to minimise the energy, the configuration is typically evolved according to a gradient descent dynamics

$$\zeta\frac{\mathrm{d}\mathbf{x}_i}{\mathrm{d}t} = -\frac{\partial V}{\partial \mathbf{x}_i} = -\sum_{j(\neq i)}\frac{\partial}{\partial \mathbf{x}_i}\bar{v}\left(h_{ij}\right). \tag{9.5}$$

Eq. (9.5) is the zero-noise limit of the overdamped ($m = 0$) Langevin dynamics of Eq. (3.4), where the parameter ζ defines the unit of time. This algorithm defines a local dynamics, because the net force acting on particle i depends only on neighbouring particles – i.e., particles that overlap with particle i. Note that $\bar{v}(h)$ should be differentiable at least once – i.e., $\alpha \geq 1$ in Eq. (9.4) – for Eq. (9.5) to be well defined. Note also that, although the dynamics in Eq. (9.5) is deterministic, the result of each energy minimisation is stochastic because it depends on the initial configuration, which is randomly generated.

Final configurations reached by the gradient descent dynamics for $t \rightarrow \infty$ have either positive or zero energy. The former are deemed 'overjammed', while the latter are 'unjammed', in reference to their athermal mechanical behaviour. For a given packing fraction φ, from N_s random initial conditions, one can measure the number $N_j \leq N_s$ of overjammed final configurations, and for $N_s \rightarrow \infty$, one has $N_j/N_s \rightarrow f_j(\varphi; N)$, which provides the fraction of overjammed configurations. A schematic plot of $f_j(\varphi; N)$ is presented[3] in Figure 9.1. For finite N, $f_j(\varphi; N)$ increases smoothly from zero to one upon increasing φ. Upon increasing N, the curve becomes steeper, and for $N \rightarrow \infty$, it converges to a step function. The condition $f_j(\varphi; N) = 1/2$ defines a finite-size jamming transition $\varphi_j(N)$, which

---

[3] Figure 9.1 is obtained by assuming $f_j(\varphi; N) = \Theta[(\varphi - \varphi_j(N))/(\sqrt{2}w(N))]$, with $\Theta(x)$ given in Eq. (4.83), $w(N) = 0.05N^{-1/2}$ and $\varphi_j(N) = 0.64 - 0.1N^{-1/2}$, for $N = 2^n$ and $n = 4, 5, \ldots, 11$. The approximate parameter values are taken from [24, 277].

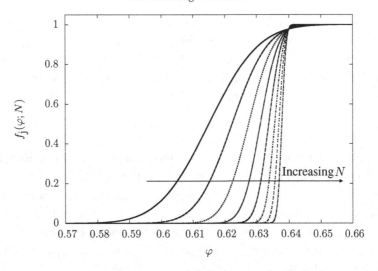

Figure 9.1 A schematic plot[3] of the fraction $f_j(\varphi; N)$ of overjammed configurations obtained after energy minimisation from a random initial configuration, as a function of packing fraction φ. The figure is roughly consistent with numerical data in $d = 3$ [24, 277]. Qualitatively similar results are obtained in $d = 2$ [347]. Increasing the system size N, the curves become steeper and approach a step function. The jamming transition φ_j^∞ can be defined as the limit for $N \to \infty$ of the φ such that $f_j(\varphi; N) = 1/2$. Above (below) φ_j^∞, energy minimisation from an infinite size random initial configuration leads to overjammed (unjammed) configurations.

converges in the thermodynamic limit to the thermodynamic jamming transition at packing fraction

$$\varphi_j^\infty = \lim_{N \to \infty} \varphi_j(N). \tag{9.6}$$

Note that the suffix in φ_j^∞ does not here refer to the thermodynamic limit taken in Eq. (9.6) but to the fact that initial configurations are prepared uniformly at random, which corresponds to an infinite temperature initial configuration, as further discussed in Section 9.1.2. Note also that φ_j^∞ then depends mostly on dimension d, but the details of the algorithm – e.g., the value of the exponent α in the potential and the choice of minimisation dynamics – can play a small role. Numerical simulations give $\varphi_j^\infty(d = 2) \simeq 0.84$ [347], $\varphi_j^\infty(d = 3) \simeq 0.64$ [276, 277], and $\varphi_j^\infty(d)$ has been measured up to $d = 13$ [87]. Correspondingly, the energy density $e(\varphi)$ of the final configurations, averaged over the initial configurations (here denoted by an overline),

$$e(\varphi) = \lim_{N \to \infty} \lim_{t \to \infty} \frac{1}{N} \overline{V(\underline{X}(t))}, \tag{9.7}$$

is such that $e(\varphi) = 0$ for $\varphi \leq \varphi_j^\infty$ and $e(\varphi) > 0$ for $\varphi > \varphi_j^\infty$.

9.1.2 Simulated Annealing

The pure gradient descent algorithm discussed in Section 9.1.1 may get trapped in local minima with finite energy even when plenty of zero-energy configurations exist. To find higher-density packings, we thus consider a generalisation of the gradient descent algorithm [9] similar to the simulated annealing protocol [202]. In this protocol, a white noise term at temperature $T(t)$ is added to Eq. (9.5), thus recovering the full (overdamped, with $m = 0$) Langevin dynamics given by Eq. (3.4), with a time-dependent temperature. The function $T(t)$ can then be optimised to improve the performance of the algorithm. For concreteness, we consider here the simplest possible choice.

1. The system is kept at finite temperature for a long enough time to reach equilibrium at state point (φ, T). The initial configuration for the subsequent dynamics is thus an equilibrium configuration at (φ, T).
2. From time $t = 0$, temperature decreases with annealing rate r, such that $T(t) = T(1 - rt)$, while keeping φ constant.
3. When $t = 1/r$, temperature reaches zero, and for $t \geq 1/r$, the dynamics follows a pure gradient descent as in Eq. (9.5).

Fluctuations due to thermal noise can then overcome local energy barriers and find configurations with a lower energy [9, 202].

We begin the analysis of this more general protocol by considering the particular case $r = \infty$, which corresponds to a gradient descent from an equilibrated configuration at (φ, T). In Section 9.1.1, we have considered the case of a uniformly random initial configuration, which corresponds to $T = \infty$. A sharp jamming transition at φ_j^∞ then separates unjammed and overjammed configurations for $N \to \infty$. The same happens at any finite temperature [347], thus defining a jamming transition line $\varphi_j(T)$, and its inverse $T_j(\varphi)$, in the (φ, T) phase diagram. The line $T_j(\varphi)$ could, in principle, be computed from the exact solution of the out-of-equilibrium dynamics obtained in [3], but such a study has not yet been attempted. The result for the line $T_j(\varphi)$ is thus only schematically illustrated[4] in Figure 9.2 for harmonic soft spheres in $d \to \infty$. Above the line, configurations

---

[4] The schematic line $T_j(\widehat{\varphi})$ in Figure 9.2 has been obtained by fitting the part of the line obtained via RS state following in [316] to a second-order polynomial,

$$1/T_j(\widehat{\varphi}) = A(\widehat{\varphi} - \widehat{\varphi}_j^\infty) + B(\widehat{\varphi} - \widehat{\varphi}_j^\infty)^2.$$

The resulting $\widehat{\varphi}_j^\infty \approx 5.8$ is consistent with a linear extrapolation of $2^d \varphi_j^\infty(d)/d$ to $d \to \infty$ in the soft harmonic sphere Mari-Kurchan model (unpublished data). The value of $2^d \varphi_j^\infty(d)/d$ for the regular soft harmonic sphere model [87, 267] can also be extrapolated to the same value when $d \to \infty$ but with much larger finite d corrections.

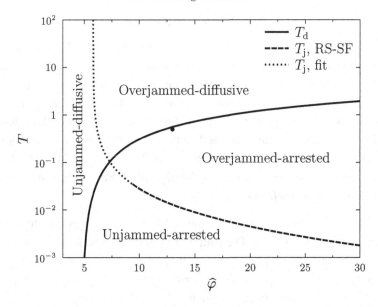

Figure 9.2 Phase diagram of harmonic soft spheres in $d \to \infty$. The dynamical transition line $T_d(\widehat{\varphi})$ (solid line) that separates the diffusive and dynamically arrested regimes is computed exactly by solving Eq. (3.66) [59, 316]. Energy minimisation from states above (below) the $T_j(\widehat{\varphi})$ line leads to overjammed (unjammed) configurations. The dashed line is a rough estimate of $T_j(\widehat{\varphi})$ from a RS state following calculation [316], which disappears before crossing $T_d(\widehat{\varphi})$ because of an unphysical spinodal (Section 6.3). The dotted line extrapolates by fitting these results[4]. An example of state following starting from the black dot is given in Figure 9.3.

reached by gradient descent from (φ, T) are overjammed, while below the line, they are unjammed.

The line $T_j(\widehat{\varphi})$ crosses the dynamical transition line $T_d(\widehat{\varphi})$, computed for soft harmonic spheres by solving Eq. (3.66) [59, 316]. Note that $T_d(\widehat{\varphi}) \to 0$ for $\widehat{\varphi} \to \widehat{\varphi}_d^{HS} = 4.8067\ldots$, where $\widehat{\varphi}_d^{HS}$ is the dynamical transition point of hard spheres, as discussed in Section 4.4.1. The intersection of the two lines defines four regions in Figure 9.2. In the unjammed-diffusive and overjammed-diffusive regions, configurations prepared in equilibrium at $(\widehat{\varphi}, T)$ have diffusive equilibrium dynamics and are, respectively, unjammed or overjammed upon gradient descent minimisation. In the unjammed-arrested and overjammed-arrested regions, by contrast, configurations prepared in equilibrium at $(\widehat{\varphi}, T)$ are located inside a glass state that traps the equilibrium dynamics (see Chapter 4). If the glass state is a simple harmonic energy basin with a single energy minimum,[5] then the annealing dynamics

---

[5] This is the case in the spherical p-spin glass model [65], but essentially all other models have a more complicated energy landscape.

starting from $(\widehat{\varphi}, T)$ should eventually reach the unique energy minimum at $T = 0$, independently of the annealing rate r. Under this assumption, in the arrested region of Figure 9.2, the line $T_{\rm j}(\widehat{\varphi})$ should be independent of the annealing rate and can thus be computed by considering a slow annealing $r \to 0$, which corresponds to following the glass from an initial state $(\widehat{\varphi}, T)$ to $(\widehat{\varphi}, 0)$ within the RS construction introduced in Chapter 4. This calculation has been performed in [316] and the result is reported in Figure 9.2. In the vicinity of the dynamical transition, however, the RS solution undergoes an unphysical spinodal instability (see Chapter 4) and disappears, impeding the calculation of $T_{\rm j}(\widehat{\varphi})$ within the RS ansatz.

It is important to stress that states prepared in the vicinity of the putative RS line $T_{\rm j}(\widehat{\varphi})$ actually undergo a Gardner transition upon cooling [316], beyond which the structure of energy basins is complex. As an example, the RS state following phase diagram of a glass prepared in equilibrium at a state point $(\widehat{\varphi}_g, T_g)$ located inside the overjammed-arrested region is reported in Figure 9.3. Other examples corresponding to different $(\widehat{\varphi}_g, T_g)$ can be found in [316]. This glass state can be adiabatically followed ($r \to 0$) upon cooling or (de)compression. The endpoint of a constant-density annealing is an overjammed configuration, with energy density $e_g(\widehat{\varphi}, 0 | \widehat{\varphi}_g, T_g) > 0$, given by Eq. (4.77). Once this zero-temperature overjammed

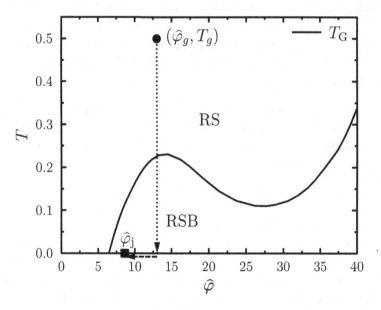

Figure 9.3 Adiabatic evolution of the glass prepared at the state point $(\widehat{\varphi}_g, T_g)$ marked by a black point in Figure 9.2, under cooling and (de)compression at $(\widehat{\varphi}, T)$ [316]. The energy at zero temperature is positive above $\widehat{\varphi}_{\rm j}(\widehat{\varphi}_g, T_g)$, and vanishes below it. Below the Gardner line $T_{\rm G}(\widehat{\varphi})$, the RS solution is unstable. The arrows mark an adiabatic cooling at constant density, which leads to an over-jammed final state, followed by decompression towards the jamming transition.

configuration is reached, decompressing the system leads to unjamming at a sharp transition density $\widehat{\varphi}_j(\widehat{\varphi}_g, T_g)$, which depends on the initial point $(\widehat{\varphi}_g, T_g)$. However, a Gardner transition surrounds the jamming point in Figure 9.3 and is generically found for all initial states [59, 316], indicating that the RS calculation is unstable around jamming. Crossing the Gardner transition makes the annealing dynamics more complex. In particular, the line $T_j(\widehat{\varphi})$ in Figure 9.2 then depends on the annealing rate even in the arrested region. A more detailed investigation of the RSB phase, however, remains to be performed. See [59, 316] for additional details.

In summary:

- In the diffusive regime of Figure 9.2, a sharp jamming transition line $T_j(\widehat{\varphi})$, originating from the infinite-temperature jamming transition $\widehat{\varphi}_j^\infty$, separates initial states $(\widehat{\varphi}, T)$ that lead to overjammed configurations from those that lead to unjammed configurations upon gradient descent energy minimisation ($r = \infty$). In this regime, the line $T_j(\widehat{\varphi})$ strongly depends on the annealing rate r. This line and its dependence on r should be computed via dynamical methods [3], which have not yet been attempted.
- In the dynamically arrested regime of Figure 9.2, under the assumption of a simple harmonic glass basin, the line $T_j(\widehat{\varphi})$ does not depend on r and can then be computed through the RS state following method. However, the RS solution is unstable in the vicinity of jamming because the energy landscape is then more complex. A weak r dependence of the line $T_j(\widehat{\varphi})$ is expected in the fullRSB regime, and its computation requires either dynamical methods or solving the fullRSB equations.

To conclude, we recall that while the dynamical transition is sharp in $d \to \infty$, in any finite dimension, glass states have a finite lifetime. The line $T_j(\widehat{\varphi})$ is then expected to always depend on r and exhibit only a crossover in the vicinity of the avoided dynamical transition.

9.1.3 Compression of Hard Spheres

The local search algorithms described in Sections 9.1.1 and 9.1.2 seek solutions of the packing problem by minimising a cost function at constant density. An alternative approach consists in compressing pure hard spheres – i.e., imposing that constraints be satisfied at all times. Recall that in the hard-sphere case, the energy is either zero (for valid configurations) or infinity (in presence of overlaps) and, therefore, only the space of zero-energy configurations is accessible. No energy barrier is then present, but at high density, 'entropic barriers' – i.e., narrow bottlenecks in phase space – can prevent the system from exploring the full phase space, trapping it into configurations with suboptimal density [215]. Thermal agitation then plays a similar role as in harmonic soft spheres. It provides the system with a mechanism to go through entropic bottlenecks and explore wider regions of phase space.

Hard-sphere configurations are easily constructed in the ideal gas limit $\varphi \to 0$. In finite-dimensional numerical simulations, the compression algorithm can be conveniently implemented by progressively increasing the sphere diameter $\ell(t)$ under periodic boundary conditions [235], which provides a homogeneous compression mechanism and avoids boundary effects that would be present if the box volume were reduced instead. The resulting packing fraction, $\varphi(t)$, and the reduced pressure, $p(t)$, then evolve along the compression, resulting in a line in the plane $(1/p, \varphi)$ of Figures 6.1 and 7.7.

A protocol similar to that of Section 9.1.2 consists in compressing adiabatically an ideal gas up to density φ, thus producing an equilibrium hard-sphere configuration at φ and, from there, compressing at a fast rate[6] [278, 279]. The compression then ends at a jamming density $\varphi_j(\varphi)$. The value of $\varphi_j(\varphi \to 0)$ is close, but not identical, to φ_j^∞ obtained by gradient descent from random configurations. A similar protocol consists in using a constant compression rate from $\varphi = 0$. In this case, the final jamming density $\varphi_j(r)$ increases continuously upon decreasing r [86, 128, 292, 330].

The analysis of this algorithm in $d \to \infty$ follows that of Section 9.1.2.

- In the simplest case, the compression starts from an equilibrium dynamically arrested configuration at $\widehat{\varphi}_g > \widehat{\varphi}_d$. Then, if the glass basin is simple, a replica symmetric state following calculation provides the endpoint $\widehat{\varphi}_j(\widehat{\varphi}_g)$ of the compression, as in Figure 4.4.
- As shown in Figure 6.1, a Gardner transition always happens before jamming is actually reached. A fullRSB state following calculation then provides the endpoint $\widehat{\varphi}_j(\widehat{\varphi}_g)$ of a slow enough compression, such that equilibration inside the glass basin can be achieved. This requires a rate that vanishes with increasing system size N. A compression with finite rate would instead fall out of equilibrium around the Gardner transition and end at a slightly lower jamming density.
- If the compression is initiated from an equilibrium liquid configuration at $\widehat{\varphi} < \widehat{\varphi}_d$, or if it is initiated at $\widehat{\varphi} = 0$ with finite compression rate, then the analysis of the algorithm requires solving the out-of-equilibrium dynamical equations [3]. It is not then obvious to compare the results with state following calculations [214, 216, 263].

9.1.4 Jamming as a Satisfiability Transition with Disorder

The analysis of local search algorithms shows that the jamming transition is characterised by the following properties.

---

[6] The compression rate should remain finite, because hard spheres otherwise get stuck in extremely sub-optimal configurations. In practice, a wide range of rates produce the same result [278].

- It takes place when a local search algorithm ceases to find solutions to the packing problem. It is therefore an algorithmic satisfiability transition. The jamming density, however, depends sensitively on the details of the algorithm (cooling or compression rate, choice of the potential, starting point, etc.) even in the thermodynamic limit.
- It separates unjammed configurations with zero energy from overjammed configurations with positive energy.
- Configurations at jamming are disordered because the initial configurations are disordered, and local search algorithms are typically unable to crystallise the system. This is precisely the case when $d \to \infty$ and approximately true[7] in finite d, as discussed in Chapter 8. The phase space regions that correspond to disordered and crystalline configurations are well separated in $d \to \infty$ so that one can focus on amorphous packings as if none other existed.
- We stress that, as discussed in Chapter 8, disordered packings exist up to $\widehat{\varphi}_{GCP} = \log d$, which is therefore the thermodynamic satisfiability transition. Beyond this point, no solutions exist. However, in general, local search algorithms jam at a finite density $\widehat{\varphi}_j \ll \log d$. There is, therefore, a large range of densities at which a large number of disordered packings exists, but no local search can find them. The existence of such an 'algorithmically hard' region is well known in random satisfiability problems [50, 370].

In summary, the jamming point can be seen as the satisfiability threshold of a constraint satisfaction problem that involves disorder and continuous variables. This analogy can be further exploited by constructing more general constraint satisfaction problems of the same kind, a research program that was started in [147]. Interested readers can find more details in [138, 150, 151].

9.2 Criticality of Jamming

It has been established that the jamming transition is critical, in the sense that several physical observables display a power-law singular behaviour close to the transition [93, 232, 234, 269, 277], and associated critical exponents satisfy scaling relations [167, 361]. In this section, we describe the critical properties of jamming when it is approached both from the overjammed (positive-energy) and the unjammed (zero-energy) sides.

9.2.1 Energy, Pressure and Coordination

Recall that physical observables associated to the jamming protocols described in Section 9.1 are fluctuating quantities because they depend on the initial starting

---

[7] In low dimensions, binary or polydisperse mixtures must sometimes be used to prevent crystallisation.

configuration. As discussed in Chapter 4, a double average is then needed, first over the thermal noise (if present) and then over the starting configuration, which can be either fully random or thermalised at a given state point. Unless otherwise specified, to simplify the notation, in the rest of this chapter, physical quantities are assumed to be averaged over both sources of fluctuations.

Energy, Pressure and Entropy

A first indication of the critical nature of jamming is obtained by looking at the average energy $e(\varphi)$ of the final configurations reached by jamming algorithms. By definition, this energy is zero for $\varphi \leq \varphi_j$ and positive for $\varphi > \varphi_j$; hence, the function $e(\varphi)$ must be singular at φ_j. In fact, for the soft harmonic potential in Eq. (9.4) and for $\varphi \to \varphi_j^+$, the energy scales as [277]

$$e(\varphi) \sim \left(\varphi - \varphi_j\right)^\alpha . \tag{9.8}$$

Another interesting observable is pressure. In zero-temperature overjammed configurations, particles are mechanically in contact and thus exert a force on the boundary of the confining volume. Pressure is thus finite and scales as

$$P(\varphi) = \rho^2 \frac{de}{d\rho} \propto \varphi^2 \frac{de}{d\varphi} \sim \left(\varphi - \varphi_j\right)^{\alpha-1} , \tag{9.9}$$

where the last equality holds for $\varphi \to \varphi_j^+$. In particular, for $\alpha = 2$, pressure is linear in the distance from jamming and, thus, provides a convenient control parameter in the overjammed phase. Unjammed configurations, by contrast, have zero energy and pressure. Because they are valid hard-sphere configurations, however, adding thermal noise leads to collisions and, as in hard spheres, makes pressure proportional to temperature [42]. The reduced pressure is then the appropriate observable, and, for $\varphi \to \varphi_j^-$ and $T \to 0$, one finds [128, 292]

$$p(\varphi) = \frac{\beta P(\varphi)}{\rho} \sim \left(\varphi_j - \varphi\right)^{-1}. \tag{9.10}$$

Eqs. (9.9) and (9.10) can be combined into a single scaling relation for the reduced pressure, which holds over the whole (φ, T) plane in the vicinity of jamming. For $\varphi \sim \varphi_j$ and $T \sim 0$ [43, 123, 182],

$$p(\varphi, T) = T^{-1/\alpha} \mathcal{P}\left[T^{-1/\alpha}(\varphi - \varphi_j)\right]. \tag{9.11}$$

In order to match Eqs. (9.9) and (9.10), the asymptotic behaviour of the scaling function must be

$$\mathcal{P}(x) \sim \begin{cases} |x|^{-1} & \text{for } x \to -\infty, \\ x^{\alpha-1} & \text{for } x \to \infty. \end{cases} \tag{9.12}$$

Similar scaling relations can be obtained for other observables, such as the energy [43, 123, 182].

Note that on the unjammed (hard-sphere) side of the transition, the reduced pressure coincides with the derivative of the entropy with respect to density [292]. Hence, for $T \to 0$ and $\varphi < \varphi_j$,

$$p(\varphi) = -\varphi \frac{\mathrm{d}s}{\mathrm{d}\varphi} \qquad \Rightarrow \qquad s(\varphi) \sim \log(\varphi_j - \varphi), \qquad \varphi \to \varphi_j^-. \tag{9.13}$$

Because the internal entropy is the logarithm of the phase space volume of the solutions to the packing problem, its divergence towards minus infinity indicates that the phase space volume shrinks continuously to zero [161]. This critical scaling of the entropy is a consequence of the continuous nature of the variables involved in the packing problem – i.e., the sphere positions. It is very different from what is observed around the satisfiability threshold of discrete constraint satisfaction problems [370]. These continuous variables are thus responsible for the appearance of critical scaling near jamming [147].

Coordination

The real-space structure of the configurations around jamming also displays an interesting criticality. Consider approaching the transition from the overjammed phase. Some spheres then overlap, and the energy density is positive. Having defined the gap variables as in Eq. (9.1), contacts correspond to negative gaps. The number z_i of spheres in contact with sphere i is then

$$z_i = \sum_{j(\neq i)} \theta\left(-h_{ij}\right). \tag{9.14}$$

Because each contact is shared by two spheres, the total number of contacts is

$$N_c = \frac{1}{2} \sum_{i=1}^{N} z_i. \tag{9.15}$$

If $z_i \geq d + 1$ and if the z_i neighbouring spheres are not all on the same side of any hyperplane going through $\mathbf{x}_i$ (a very unlikely event in random packings), then sphere i is blocked and belongs to the rigid contact network. Conversely, if $z_i \leq d$, the sphere is not blocked and can thus rattle inside the cage formed by its neighbours. It is then called a 'rattler' [128, 344]. The number N_r of rattlers depends on the packing fraction [279], the protocol used to create jammed packings and, most strongly, dimension [87]. Numerically, using gradient descent from random configurations, it is found that the number of rattlers at jamming scales roughly as $N_r/N \sim \exp(-d/d_r)$ with $d_r \approx 2$ and, thus, vanishes exponentially for $d \to \infty$ [87].

Being disconnected from the rigid contact network, rattlers are not relevant for determining the mechanical properties of jammed packings [227, 344] and are thus typically omitted from the analysis of packings. The contact network is then defined as the graph having as vertices the spheres with degree $z_i \geq d + 1$ and as links the contacts between two such spheres. The number of spheres belonging to the contact network is $N_{cn} = N - N_r$, and the average degree of the contact network is

$$z = \frac{1}{N_{cn}} \sum_{i=1}^{N_{cn}} z_i = \frac{2N_c}{N_{cn}}, \tag{9.16}$$

where N_c only counts the particles in the contact network – i.e., rattlers are assigned $z_i = 0$. Note that z is self-averaging over the ensemble of packings generated according to a given protocol (see Section 9.1) – i.e., it is equal to its average value with probability going to one in the thermodynamic limit $N_{cn} \to \infty$. At jamming, it is found that $z = 2d$ [128, 234, 277, 344], and in the overjammed phase, z is a function of the packing fraction. Upon approaching jamming, it scales as [277]

$$z - 2d \propto (\varphi - \varphi_j)^{\nu_z} \propto P^{\frac{\nu_z}{(\alpha-1)}}, \tag{9.17}$$

with a non-universal exponent ν_z that will be further discussed in section 9.2.3. In the unjammed phase, by contrast, the spheres do not touch. The gaps are all positive and $z = 0$. Hence, z jumps from zero to $2d$ at jamming and then increases as a power law in the overjammed phase.

9.2.2 Isostaticity

The property that $z = 2d$ at jamming is called 'isostaticity'. In this section, we explain the origin of this terminology and discuss some theoretical arguments that highlight the special role played by isostaticity in determining the properties of jamming.

Isostaticity, Hypostaticity and Pre-stress

We begin by reproducing an argument originally developed in [359, 360] to explain why sphere packings are isostatic. The argument generalises a simple counting argument due to Maxwell [248]. We focus for simplicity on a harmonic potential – i.e., we fix $\alpha = 2$ in Eq. (9.4) – but the argument can be easily extended to more general potentials [91, 122]. We denote $C = \{\langle ij \rangle : h_{ij} < 0\}$ the set of contacts, whose cardinality $|C| = N_c$ is the total number of negative gaps. Defining a contact index $c = \langle ij \rangle \in C$, the energy is then

$$V(\underline{X}) = \frac{\varepsilon}{2} \sum_{c \in C} h_c(\underline{X})^2. \tag{9.18}$$

The argument discussed in this section applies to a generic configuration $\underline{X}$ with N_{dof} degrees of freedom, and for arbitrary gap functions $h_c(\underline{X})$. The particular case of spheres is recovered when $N_{\text{dof}} = dN_{\text{cn}}$, and the gap is given in Eq. (9.1), but the argument also applies to particles of arbitrary shape (e.g., ellipsoids) [70, 318].

Let us consider an overjammed configuration $\underline{X}$ at $\varphi > \varphi_j$. By definition, such a configuration is a local minimum of the potential energy, hence $\partial V/\partial \underline{X} = 0$, with positive energy, $V(\underline{X}) > 0$. If the configuration is perturbed, $\underline{X} \to \underline{X} + \delta\underline{X}$, the energy variation is then controlled by the $N_{\text{dof}} \times N_{\text{dof}}$ Hessian matrix,

$$\mathcal{H} = \frac{\partial^2 V(\underline{X})}{\partial \underline{X}\, \partial \underline{X}}. \tag{9.19}$$

At order $|\delta\underline{X}|^2$, the energy variation can conveniently be written as

$$\delta V(\underline{X}) = V(\underline{X} + \delta\underline{X}) - V(\underline{X}) \approx \frac{1}{2}\delta\underline{X} \cdot \mathcal{H} \cdot \delta\underline{X} \tag{9.20}$$

$$= \underbrace{-\frac{\varepsilon}{2}\sum_{c\in C}|h_c|\,\delta\underline{X} \cdot \frac{\partial^2 h_c}{\partial \underline{X}\, \partial \underline{X}} \cdot \delta\underline{X}}_{\text{pre-stress}} + \underbrace{\frac{\varepsilon}{2}\sum_{c\in C}\left(\delta\underline{X} \cdot \frac{\partial h_c}{\partial \underline{X}}\right)^2}_{\text{harmonic}}.$$

Following [359, 360], the energy variation has been split in two contributions.

- The harmonic term is a semidefinite positive matrix, which is non-zero only if at least a contact c is such that $\delta\underline{X} \cdot \frac{\partial h_c}{\partial \underline{X}} \neq 0$. Each contact thus stabilises the direction parallel to its gradient, by providing a positive energy variation in that direction. Note that if there are N_c contacts, under the assumption that the vectors $\frac{\partial h_c}{\partial \underline{X}}$ are linearly independent, the harmonic term has max $[N_{\text{dof}} - N_c, 0]$ directions $\delta\underline{X}$ along which it vanishes – i.e., zero modes.
- The pre-stress term is a sum of terms proportional to $|h_c|$ over the negative gaps, which vanish at jamming. Note that for $\varphi \to \varphi_j^+$, energy and pressure scale as

$$e(\varphi) \propto \overline{\frac{1}{N_c}\sum_{c\in C}h_c^2}, \qquad P(\varphi) \propto \frac{de}{d\ell} \propto \overline{\frac{1}{N_c}\sum_{c\in C}|h_c|}. \tag{9.21}$$

In other words, energy is proportional to the average squared negative gap, while pressure is proportional to the average negative gap. Upon approaching jamming, the pre-stress term vanishes proportionally to pressure, while the harmonic term stays finite.

The pre-stress term determines the properties of the jamming transition [359, 360]. There are two possible situations.

1. If the matrix $\frac{\partial^2 h_c}{\partial \underline{X} \partial \underline{X}}$ is positive definite for each contact $c \in C$, then the pre-stress term is negative definite. By definition, the configuration $\underline{X}$ is an energy minimum in the overjammed phase, and the total Hessian matrix must then be positive definite. Suppose now that $N_c < N_{\text{dof}}$. The harmonic term then has $N_{\text{dof}} - N_c > 0$ zero modes, and the pre-stress term is generically negative on these modes. The total Hessian matrix is then negative along these modes, which contradicts the initial assumption that $\underline{X}$ is a local energy minimum. The hypothesis $N_c < N_{\text{dof}}$ must thus be rejected. One concludes that, if the pre-stress term is negative definite, one must necessarily have $N_c \geq N_{\text{dof}}$ in the overjammed phase. By continuity, this condition must also hold at jamming. Note that in this case, all the eigenvalues of the Hessian matrix remain finite at jamming. If the bound is saturated, $N_c = N_{\text{dof}}$, such a system is said to be isostatic.

2. If instead the matrix $\frac{\partial^2 h_c}{\partial \underline{X} \partial \underline{X}}$ is not positive definite, then the pre-stress term can stabilise some of the zero modes of the harmonic term. In this case, some eigenvalues of the Hessian matrix (those associated with the harmonic term) remain finite at jamming, while others vanish proportionally to P. The number of contacts is then unconstrained, and, in particular, it can be $N_c < N_{\text{dof}}$. Its precise value is system dependent; it is fixed by the properties of the pre-stress matrix. Such a system is said to be hypostatic.

The critical properties of jamming are very different in the isostatic and hypostatic case. Spheres belong to the first category, as we show next, but for more general potentials, jamming can be hypostatic. A notable example of the latter case are ellipsoids [70, 127, 129, 318]. In the following, we restrict our discussion to the isostatic case.

Application to Soft Harmonic Spheres

We here show that for soft harmonic spheres the pre-stress term is always negative [359, 360]. Suppose for simplicity that the spheres are within a d-dimensional cubic volume with periodic boundary conditions.[8] Recall that ∂i denotes the set of spheres in contact with i and rattlers are excluded from the analysis. The number of degrees of freedom is then $N_{\text{dof}} = d N_{\text{cn}}$. Each contact particle exerts a force on particle i,

$$\boldsymbol{F}_i = -\sum_{j \in \partial i} \frac{\partial \bar{v}(h_{ij})}{\partial \boldsymbol{x}_i} = \sum_{j \in \partial i} f_{ij} \boldsymbol{n}_{j \rightarrow i}, \qquad f_{ij} = -\frac{d}{\ell} \bar{v}'(h_{ij}) = \frac{d\varepsilon}{\ell} |h_{ij}|, \quad (9.22)$$

---

[8] The general case, in which walls and external forces are present, is discussed in [91].

where f_{ij} is the modulus of the force exchanged by spheres i and j, and

$$n_{j \to i} = \frac{\mathbf{x}_i - \mathbf{x}_j}{|\mathbf{x}_i - \mathbf{x}_j|}, \tag{9.23}$$

is a d-dimensional unit vector pointing from the center of sphere j towards the center of sphere i.

In a local minimum of the potential energy, forces vanish, $\mathbf{F}_i = \mathbf{0}, \forall i = 1, \ldots, N_{\mathrm{cn}}$. Writing explicitly the second-order variation of the energy, Eq. (9.20), in terms of particle displacements, one obtains [359, 360]

$$\delta V(\underline{X}) = -\frac{\varepsilon d}{2\ell} \sum_{\langle ij \rangle \in \mathcal{C}} |h_{ij}| \frac{|\delta \mathbf{r}_{ij}|^2 - (\delta \mathbf{r}_{ij} \cdot n_{j \to i})^2}{2 r_{ij}} + \frac{\varepsilon d^2}{2\ell^2} \sum_{\langle ij \rangle \in \mathcal{C}} (\delta \mathbf{r}_{ij} \cdot n_{j \to i})^2$$

$$= \underbrace{-\frac{\varepsilon d}{2\ell} \sum_{\langle ij \rangle \in \mathcal{C}} |h_{ij}| \frac{|\delta \mathbf{r}_{ij}^{\perp}|^2}{2 r_{ij}}}_{\text{pre-stress}} + \underbrace{\frac{\varepsilon d^2}{2\ell^2} \sum_{\langle ij \rangle \in \mathcal{C}} (\delta \mathbf{r}_{ij} \cdot n_{j \to i})^2}_{\text{harmonic}}, \tag{9.24}$$

where $\mathbf{r}_{ij} = \mathbf{x}_i - \mathbf{x}_j$, $r_{ij} = |\mathbf{r}_{ij}|$, $\delta \mathbf{r}_{ij} = \delta \mathbf{x}_i - \delta \mathbf{x}_j$ and $\delta \mathbf{r}_{ij}^{\perp}$ is the projection of $\delta \mathbf{r}_{ij}$ on the plane orthogonal to $n_{j \to i}$. For soft harmonic spheres, the pre-stress term is thus always negative, and, hence, jamming is always isostatic.

It is important to emphasise that, because the system is translationally invariant (it is confined in a cubic box with periodic boundary conditions), any uniform translation, $\delta \mathbf{x}_i = \mathbf{a}, \forall i$, is a zero mode of the Hessian. This is manifest in Eq. (9.24) because $\delta \mathbf{r}_{ij} = \mathbf{0}$, and then $\delta V(\underline{X}) = 0$ for such a mode. Once the d zero modes associated to translations are excluded, the effective number of degrees of freedom is reduced by d, and the Hessian is effectively a $d(N_{\mathrm{cn}} - 1) \times d(N_{\mathrm{cn}} - 1)$ matrix. The stability condition then becomes $N_c \geq d(N_{\mathrm{cn}} - 1)$. Using Eq. (9.16), in the thermodynamic limit $N_{\mathrm{cn}} \to \infty$, this condition corresponds to $z \geq 2d$, as numerically observed [277, 359, 360]. In the next section, we discuss more precisely the isostatic value of N_c for a finite system.

Determination of the Contact Forces at Isostatic Jamming

For a given contact network entirely specified by the contacts $\langle ij \rangle$ and the contact vectors $n_{j \to i}$, the force balance equations read

$$\sum_{j \in \partial i} f_{ij} n_{j \to i} = \mathbf{0}, \tag{9.25}$$

where $\mathbf{0}$ is the null vector. Eq. (9.25) can be thought of as a homogeneous linear system for the forces f_{ij}. Obviously, if the potential $\bar{v}(h_{ij})$ is known, the contact forces are then functions of the particle positions as in Eq. (9.22), but one can wonder if Eq. (9.25) alone suffices to determine these forces. It turns out that if jamming is isostatic, then the contact forces f_{ij} are entirely determined by the contact network, independently of the interaction potential.

The system in Eq. (9.25) contains dN_{cn} equations – i.e., one for each spatial coordinate of each particle in the contact network – and N_c unknowns – i.e., one force per contact. Because $f_{ij} = f_{ji}$, and $\mathbf{n}_{i \to j} = -\mathbf{n}_{j \to i}$, one has

$$\sum_{i=1}^{N_{cn}} \sum_{j \in \partial i} f_{ij} \mathbf{n}_{j \to i} = \mathbf{0},$$

(9.26)

because the terms ij and ji cancel. This global force balance follows from the invariance of the potential energy under a global translation of the system. One of the vectorial equations in Eq. (9.25), corresponding to d scalar equations, is thus linearly dependent on the others. In presence of additional symmetries – e.g., in crystals – there might be additional linear dependencies, but for disordered configurations, it is reasonable to assume that the remaining $d(N_{cn} - 1)$ equations are linearly independent. The linear system in Eq. (9.25) thus admits $\max[0, N_c - d(N_{cn} - 1)]$ linearly independent solutions. Note that for $N_c = d(N_{cn} - 1)$, which is the minimal number of contacts allowed by the stability condition on the Hessian, the force equation has no solution. Thus, this value must be excluded. The minimal value of N_c for a stable finite system is then [91, 227]

$$N_c = d(N_{cn} - 1) + 1 = N_{iso},$$

(9.27)

which defines the isostaticity condition in finite systems. This condition still corresponds to $z = 2d$ in the thermodynamic limit. At isostaticity, Eq. (9.25) has a unique solution for which the forces are not identically zero; the forces are therefore determined by this solution. If the system is hyperstatic, $N_c > N_{iso}$, then Eq. (9.25) has multiple solutions and the forces have to be determined from the interaction potential.

Note that Eq. (9.25) does not fix the overall normalisation of contact forces. If an isostatic jammed configuration is reached from above using a soft harmonic potential, then $f_{ij} = d|h_{ij}|/\ell$, and the average force is proportional to pressure P, which vanishes at jamming. By contrast, if jamming is reached from below by compressing hard spheres, then the contact forces can be defined as the average momentum transfer due to collisions between particles ij [128], and the average force diverges proportionally to the reduced pressure p at jamming. If forces are scaled by requiring that their average is unity,

$$\frac{1}{N_c} \sum_{\langle ij \rangle \in \mathcal{C}} f_{ij} = 1,$$

(9.28)

then one concludes that the forces f_{ij} associated with an isostatic jammed configuration are independent of the potential and the protocol that were used to prepare it. They are then given by the unique solution of Eq. (9.25) normalised as in Eq. (9.28) and depend only on the geometry of the contact network.

9.2.3 Critical Exponents of Isostatic Jamming

The jamming transition, of both isostatic and hypostatic nature, is characterised by a critical scaling of the energy, pressure and average coordination (Section 9.2.1). The associated critical exponents for pressure and energy are simple functions of the exponent α of the interaction potential. The exponent for coordination is $\nu_z = 1/2$ (isostatic) [277, 359, 360] or $\nu_z = 1$ (hypostatic) [70]. Isostatic jammed packings display additional universal critical exponents that are not simple rational numbers. We now describe these additional exponents.

Contact Force Distribution

The probability density of contact forces is

$$P(f) = \overline{\frac{1}{N_c} \sum_{\langle ij \rangle \in \mathcal{C}} \delta\left(f - f_{ij}\right)} \tag{9.29}$$

and, following Eq. (9.28), is normalised by

$$\int_0^\infty \mathrm{d}f\, P(f) = 1, \qquad \int_0^\infty \mathrm{d}f\, f\, P(f) = 1. \tag{9.30}$$

This distribution can be numerically measured either by constructing the contact network at jamming and then solving the linear system in Eq. (9.25) or by directly measuring the forces at $\varphi > \varphi_j$ and then taking the limit $\varphi \to \varphi_j^+$ (the former strategy is more numerically accurate) [91]. It turns out that at the jamming transition [91, 227]

$$P(f) \sim f^\theta, \qquad f \to 0^+, \tag{9.31}$$

which defines a critical exponent θ. This 'pseudo-gap' in the force distribution affects broadly the mechanical stability of packings at jamming, which is controlled by sphere contacts carrying an extremely small force [269, 361]. If the packing is perturbed, these small-force contacts are indeed likely destabilised.

Gap Distribution

The radial distribution function of a jammed packing can be computed from Eq. (2.30),

$$g(r) = \overline{\frac{1}{\rho N_{\text{cn}}} \sum_{i \neq j} \delta \left(\mathbf{r} - \mathbf{x}_i + \mathbf{x}_j \right)}. \tag{9.32}$$

Recall that rattlers are excluded from the analysis [87, 128], so the sums over $i \neq j$ in Eq. (9.32) only run over the spheres that belong to the contact network. The cumulative structure function,

$$Z(r) = \rho \Omega_d \int_0^r ds \, s^{d-1} g(s), \tag{9.33}$$

then gives the number of particle centers contained within a ball of radius r centered on a reference particle. At jamming, the gap constraint implies $Z(r) = 0$ for $r < \ell$. In the jamming limit, isostaticity also implies that

$$\lim_{r \to \ell^+} Z(r) = 2d . \tag{9.34}$$

Hence, at jamming, $Z(r)$ jumps from 0 to $2d$ in $r = \ell$, and $g(r)$ has a delta peak in $r = \ell$. It is also found that [87, 128, 361]

$$g(r) \sim (r - \ell)^{-\gamma}, \qquad Z(r) - 2d \sim (r - \ell)^{1-\gamma}, \qquad r \to \ell^+, \tag{9.35}$$

which defines a new critical exponent γ. Therefore, for $h = d(r/\ell - 1) \to 0^+$, the gap distribution $\overline{g}(h) = g[\ell(1 + h/d)]$ diverges as $\overline{g}(h) \sim h^{-\gamma}$. Isostatic jammed configurations thus display a large number of very small gaps. If the packing is perturbed, new contacts between almost touching particles are likely to form [269, 361].

Mean Square Displacement of the Hard-Sphere Glass

In a hard-sphere glass, the long time limit of the MSD, Δ, is finite and continuously decreases upon compression, as discussed in Section 4.4.1. In a replica symmetric glass phase, Δ has a unique value, while in a RSB phase, a function $\Delta(x)$ describes the different MSD plateaus corresponding to the exploration of different levels in the hierarchy of sub-basins (see Chapter 5). The smallest of them, $\Delta_M = \Delta(1)$, corresponds to the MSD plateau inside an individual glass state. In the jamming limit, each glass state becomes a jammed configuration; hence, Δ_M must vanish when the reduced pressure diverges, $p \to \infty$. It is found that [89, 93, 122, 269]

$$\Delta_M \sim p^{-\kappa}, \qquad p \to \infty. \tag{9.36}$$

In other words, upon approaching jamming, the cage size in which hard spheres are trapped shrinks to zero as a power law of reduced pressure, which defines a third critical exponent κ.

Critical Exponents in Finite Dimensions

The critical exponents γ and κ have been measured in numerical simulations and appear to be independent of dimension, within numerical accuracy [87, 89, 128, 330]. The exponent θ, however, depends on dimension. The underlying reason has been clarified in [91, 227]. In finite dimension, there is a finite probability that a sphere has $z_i = d + 1$ contacts, and there is also a finite probability that, among the $d + 1$ forces that stabilise such particle, d are almost coplanar. The extra force can then be anomalously small, thus allowing these particles to easily buckle in and out of this plane [91, 227]. This buckling mechanism produces an anomalous abundance of small forces. For simplicity, all particles having $z_i = d + 1$ are called 'bucklers', and the probability distribution of contact forces is then separated in the contribution coming from bucklers and from the rest of the system [91],

$$P_{\text{bucklers}}(f) \sim f^{\theta_l}, \qquad f \to 0^+,$$
$$P_{\text{non-bucklers}}(f) \sim f^{\theta_e}, \qquad f \to 0^+. \tag{9.37}$$

Denoting the fraction of bucklers as $f_b = N_b/N$, the total probability distribution of all forces is

$$P(f) = f_b P_{\text{bucklers}}(f) + (1 - f_b) P_{\text{non-bucklers}}(f). \tag{9.38}$$

Numerical simulations in finite dimensions found that $\theta_l < \theta_e$ [91, 227]. Hence, bucklers dominate and $P(f) \sim f^{\theta_l}$, as long as

$$f_b f^{\theta_l} \gg (1 - f_b) f^{\theta_e} \qquad \Leftrightarrow \qquad f \ll \left(\frac{f_b}{1 - f_b}\right)^{\frac{1}{\theta_e - \theta_l}}. \tag{9.39}$$

Because f_b is small, numerical simulations may observe a value of θ intermediate between θ_l and θ_e, unless small enough values of f can be probed. And because the smallest observable force decreases upon increasing the system size, large systems are needed to reach the asymptotic regime.

For large d, the density of bucklers goes to zero exponentially – i.e., $f_b \sim \exp(-d/d_b)$ with $d_b \sim 2.0$ – as does the density of rattlers [91]. In the $d \to \infty$ limit, the probability distribution of z_i in fact strongly concentrates around the average value, $z = 2d$, and the probability of observing any deviation from this value goes to zero exponentially [87]. The contribution of bucklers to the total force distribution hence vanishes, and

$$P(f) = P_{\text{non-bucklers}}(f) \sim f^{\theta_e}, \qquad f \to 0^+, \qquad d \to \infty. \tag{9.40}$$

In infinite dimensions, therefore, $\theta = \theta_e$. Numerical measurements further suggest, within current numerical precision, that both θ_l and θ_e are independent of dimension [91].

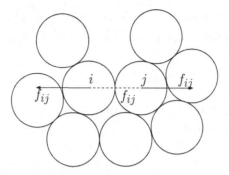

Figure 9.4 A schematic illustration of a dipolar force applied to two spheres i and j in a jammed packing.

9.2.4 Marginal Stability and Scaling Relations

For a system with negative definite pre-stress, such as soft harmonic spheres, isostaticity at jamming implies that the system is close to a mechanical instability. To analyse this instability, we reproduce here briefly an argument originally developed by Wyart [227, 361]. For simplicity, we assume[9] that bucklers are absent and $\theta = \theta_e$, which is correct in $d \to \infty$. The argument is based on

- Identifying the elementary excitations of the system
- Selecting those with smallest energy cost
- Perturbing the system along those excitations and looking for stabilisation mechanisms

This procedure is standard to assess the stability of many-body low-temperature systems in condensed matter. Similar approaches have been successfully applied to a variety of systems ranging from spin glasses to electron glasses and low-temperature particle glasses [269]. When applied to isostatic jammed packings, it provides a scaling relation between the critical exponents γ and θ.

Let us consider an isostatic jammed packing with $N_c = (N_{cn} - 1)d + 1 = N_{iso}$ total contacts. As discussed in Section 9.2.2, this is the minimal requirement for mechanical stability in isostatic systems, in which the pre-stress term is negative definite. Imagine applying a dipolar force on a pair of spheres in contact, as illustrated in Figure 9.4. It can be shown that if $N_c \geq N_{iso}$, then this dipolar force induces a well-defined response δx_i of the particles in the packing, which can be calculated in terms of the Hessian matrix [361] (see [91] for a detailed discussion). Wyart's argument assumes that these are the elementary excitations of a jammed packing [361].

---

[9] The argument can be extended to take into account bucklers, which then give rise to localised excitations [227].

We then study the stability of the lowest-energy dipolar excitations. Consider a system kept at constant pressure P. The work done to open a contact carrying a force f by a distance $s > 0$ is

$$P \, \delta V(s) \simeq f s - A N s^2, \qquad (9.41)$$

where $\delta V(s)$ is the variation of the volume V due to this perturbation, and the constant A is proportional to the average force $\langle f \rangle$ [361]. This work is minimal when the contact carries the smallest force. In a packing of size N, according to extreme value statistics,[10] if $P(f) \sim f^\theta$, then the smallest force scales as

$$f_{\min} \sim N^{-\frac{1}{1+\theta}}. \qquad (9.42)$$

With this choice of f, $\delta V(s)$ becomes negative for $s > s_{\min}$, with

$$s_{\min} \sim N^{-\frac{2+\theta}{1+\theta}}. \qquad (9.43)$$

The packing is then destabilised in favour of a denser configuration. However, in the process of opening this contact, a new contact could be formed elsewhere in the system. If this happens before s reaches $s_{\min}$, then the packing cannot be destabilised. Because a new contact is likely formed by closing a very small gap, we consider the scaling of the smallest gap between two spheres,

$$h_{\min} \sim N^{-\frac{1}{1-\gamma}}. \qquad (9.44)$$

If we assume that contacts close proportionally to s when the packing is perturbed, then the stability condition becomes [361]

$$h_{\min} \leq s_{\min} \qquad \Leftrightarrow \qquad \gamma \geq \frac{1}{2+\theta}. \qquad (9.45)$$

Numerical simulations show that the bound in Eq. (9.45) is saturated [91, 227], which provides a scaling relation $\gamma = 1/(2+\theta)$ between γ and θ, and suggests that jammed amorphous packings are indeed very close to a mechanical instability – i.e., they are marginally stable. Other scaling relations between the critical exponents can be obtained with similar arguments [122, 269]. In particular,

$$\kappa = 2 - \frac{2}{3+\theta}; \qquad (9.46)$$

hence, a single exponent – e.g., θ – remains undetermined.

---

[10] The probability to extract from $P(f)$ a force smaller than f is $\int_0^f \mathrm{d}f' P(f') \sim f^{1+\theta}$. If N forces are extracted independently, this probability is $\approx N f^{1+\theta}$, which is of order unity if Eq. (9.42) holds.

9.3 The Unjammed Phase: Hard Spheres

When hard spheres are compressed towards the jamming point, they undergo a Gardner transition. At this point, the glass basin continuously fractures into a large number of hierarchically organised sub-basins (Chapter 6). The fullRSB phase that appears beyond the Gardner transition is characterised by critical long-range correlations and marginal stability, in the sense of a diverging susceptibility (Section 5.5.3). The equation of state of these fullRSB hard-sphere glasses terminates at the jamming transition, which is also a critical point characterised by mechanical marginal stability (Section 9.2.4). The relation between these two notions of marginal stability is not a priori obvious. In this section, we show that at least in $d \to \infty$, they are, in fact, deeply related. The criticality of the jamming transition is indeed described by the fullRSB equations, which provide exact analytical results for the critical exponents. Isostaticity and the scaling relations in Eqs. (9.45) and (9.46) then emerge as direct consequences of the marginal stability of the fullRSB solution in the jamming limit.

9.3.1 *FullRSB Equations for Hard Spheres*

We begin by considering hard-sphere glasses that are adiabatically compressed towards jamming, as in Chapter 6. The fullRSB function $\Delta(x)$ is constant outside the interval defined by the two breaking points, x_m and x_M, as shown in Figure 6.2. The plateau value, Δ_M, for $x > x_M$ represents the long time limit of the MSD within an individual glass state, while the function $\Delta(x)$ for $x < x_M$ describes the MSD associated to the exploration of different states in the free energy landscape, as discussed in Chapter 5. The plateau value Δ_m for $x < x_m$ encodes the largest possible MSD within a same glass metabasin and, thus, captures the width of that metabasin. At jamming, the cage of individual glass states vanishes – i.e., $\Delta_M \to 0$.

Recall that, as described in Chapter 6, a slow compression (at finite rate) of a glass state selected by an equilibrium liquid configuration at $\widehat{\varphi}_g$ is unable to follow the glass adiabatically up to the Gardner transition and beyond because the equilibration time scale of the metabasin diverges at $\widehat{\varphi}_G(\widehat{\varphi}_g)$. Any finite-rate compression would then explore the glass metabasin out of equilibrium. The value of $\Delta_M(\widehat{\varphi})$ computed using the state following method is thus only an approximation of what would be obtained during a finite-rate compression. The critical exponents obtained for jamming nonetheless compare very well with those observed by the protocols described in Section 9.1, which justifies the approximation a posteriori.

In order to characterise the behaviour of the fullRSB state following solution in the limit $\Delta_M \to 0$, it is convenient to write the fullRSB equations in a different form. For $x \in [x_m, x_M]$, $\Delta(x)$ is monotonously decreasing and can be inverted. Its inverse function $x(\Delta)$ is defined over the interval $\Delta \in [\Delta_M, \Delta_m]$; see Section 5.3.2 and Figure 5.6. Expressing $\lambda(x)$ as a function of Δ, $\lambda(\Delta) = \lambda(x(\Delta))$, gives

$$\lambda(\Delta) = \Delta_M + \int_{\Delta_M}^{\Delta} d\Delta' x(\Delta'), \qquad x(\Delta) = \dot{\lambda}(\Delta), \tag{9.47}$$

where the dot here denotes a derivative with respect to Δ. Introducing[11]

$$f(\Delta, h) = f(x(\Delta), h + \eta - \Delta_r), \qquad P(\Delta, h) = P(x(\Delta), h + \eta - \Delta_r), \tag{9.48}$$

the set of fullRSB Eqs. (6.9), (6.11) and (6.13) derived in Chapter 6 can be rewritten as

$$
\begin{cases}
f(\Delta_M, h) &= \log \gamma_{\Delta_M} \star \theta(h), \\
\dot{f}(\Delta, h) &= \frac{1}{2} \left(f''(\Delta, h) + x(\Delta) f'(\Delta, h)^2 \right),
\end{cases}
$$
$$
\begin{cases}
P(\Delta_m, h) &= \widehat{\varphi}_g \, \gamma_{2\Delta_r - \Delta_m} \star e^{h + \eta - \Delta_r} \theta(h + \eta - \Delta_r), \\
\dot{P}(\Delta, h) &= -\frac{1}{2} \left[P''(\Delta, h) - 2x(\Delta) \left(P(\Delta, h) f'(\Delta, h) \right)' \right],
\end{cases}
\tag{9.49}
$$

where the differential equations are defined over the interval $\Delta \in [\Delta_M, \Delta_m]$, and

$$
\frac{1}{\lambda(\Delta_m)} = -\frac{1}{2} \int_{-\infty}^{\infty} dh \, P(\Delta_m, h)[f''(\Delta_m, h) + f'(\Delta_m, h)],
$$
$$
\frac{2\Delta_r - \Delta_m}{\lambda(\Delta_m)^2} - \int_{\Delta_m}^{\Delta} d\Delta' \frac{1}{\lambda(\Delta')^2} = \frac{1}{2} \int_{-\infty}^{\infty} dh \, P(\Delta, h) f'(\Delta, h)^2.
\tag{9.50}
$$

Restricting the analysis to hard spheres entails replacing $e^{-\beta \bar{v}(h)} = \theta(h)$ in the expressions for $f(\Delta_M, h)$ and $P(\Delta_m, h)$ in Eq. (9.49).

The jamming limit corresponds to $\Delta_M \to 0$, while Δ_r and the function $\Delta(x)$ for $x < x_M$ remain finite and positive. In the rest of this section, the scaling of Eqs. (9.49) in this limit is derived and compared to a direct numerical resolution of the fullRSB equations, following the procedure of Section 6.6 with a finite number k of RSB steps [298]. It has been checked [88, 298] that upon increasing k, the curves become smoother and converge towards the continuum fullRSB limit.

9.3.2 Jamming Scaling Ansatz for Hard Spheres

The fullRSB equations depend on $\widehat{\varphi}$ – or, equivalently, on $\eta = \log(\widehat{\varphi}/\widehat{\varphi}_g)$ – as a control parameter. The jamming density $\widehat{\varphi}_j$ or η_j is then the point at which $\Delta_M(\widehat{\varphi}) \to 0$.

---

[11] For later convenience, we have shifted the variable $h \to h + \eta - \Delta_r$. Note that this shift only explicitly enters in the form of the boundary functions $f(\Delta_M, h)$ and $P(\Delta_m, h)$.

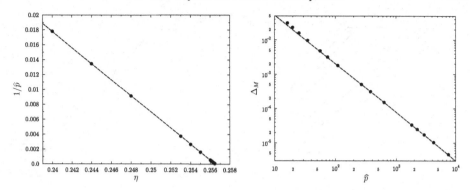

Figure 9.5 State following results for a hard-sphere glass prepared at $\widehat{\varphi}_g = 7$ obtained by numerical integration of the kRSB equations with $k = 99$. (Left) Inverse reduced pressure $1/\widehat{p} = d/p$ as a function of $\eta = \log(\widehat{\varphi}/\widehat{\varphi}_g)$. Circles are numerical results; the dashed line is a fit to $1/\widehat{p} = A(\eta_j - \eta)$ with $\eta_j = 0.2565$ and $A = 1.0767$. (Right) Mean square displacement plateau Δ_M as a function of $\widehat{p}$. Circles are numerical results, the dashed line is a fit to $\Delta_M = \Delta_j/\widehat{p}^\kappa$ with the exponent $\kappa = 1.41574$ fixed to the analytically predicted value (Section 9.3.3), and $\Delta_j = 1.3766$.

Recall that all the results also depend on $\widehat{\varphi}_g$, as discussed in Chapter 6, but for the rest of this chapter, we will not indicate this dependence explicitly. The reduced pressure can be computed from Eq. (6.17), and, as shown in Figure 9.5, its inverse vanishes linearly when $\eta \to \eta_j^-$, consistently with numerical simulations in finite dimensions [86, 128, 330]. It is then convenient to use $1/\widehat{p} = d/p$ as a control parameter because it encodes the linear distance from jamming. Note, however, that the scaling $1/\widehat{p} \propto \eta_j - \eta$ cannot be proven analytically from the scaling analysis. It has to be inferred from the numerical solution of the fullRSB equations (Figure 9.5).

For the scaling analysis, it is convenient to introduce

$$y(\Delta) = x(\Delta)\widehat{p},$$
$$\widehat{\lambda}(\Delta) = \lambda(\Delta)\widehat{p},$$
$$\widehat{f}(\Delta, h) = f(\Delta, h)/\widehat{p}, \qquad (9.51)$$
$$m(\Delta, h) = \widehat{\lambda}(\Delta)\widehat{f}'(\Delta, h) = \lambda(\Delta)f'(\Delta, h).$$

Numerical inspection of the fullRSB solution shows that these functions develop a scaling regime close to jamming [88, 151, 298]. In Figure 9.5, Δ_M shows a power-law behaviour as a function of $\widehat{p}$, with exponent κ as in Eq. (9.36). If the breaking point x_M remains finite, then $y_M = x_M\widehat{p}$ diverges linearly, while $\Delta_M \sim \widehat{p}^{-\kappa} \sim y_M^{-\kappa}$. This observation suggests a power-law scaling of $y(\Delta)$,

$$y(\Delta) \sim y_j \Delta^{-1/\kappa}, \qquad \Delta \to 0^+, \qquad (9.52)$$

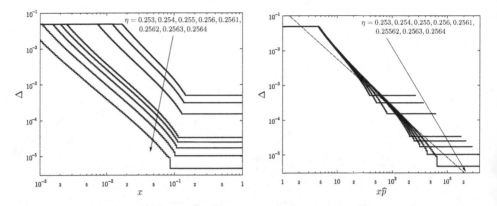

Figure 9.6 State following results for a hard-sphere glass prepared at $\widehat{\varphi}_g = 7$ obtained by numerically integrating the kRSB equations with $k = 99$. The small discontinuities disappear in the limit $k \to \infty$ [88]. (Left) The function $\Delta(x)$ for several values of $\eta = \log\left(\widehat{\varphi}/\widehat{\varphi}_g\right)$. See Figure 6.2 for lower values of η. (Right) Same results, but as a function of $y = x\widehat{p}$. In this case the collapse of $\Delta(y)$ to a master curve for finite y is clearly observed, and the cutoff on y diverges proportionally to $\widehat{p}$. The dashed line gives the asymptotic behaviour of $\Delta(y) \sim y^{-\kappa}$ at large y.

with some constant y_j, as confirmed numerically in Figure 9.6. In the jamming limit, $f(\Delta, h)$ and $P(\Delta, h)$ also develop a scaling regime for $\Delta \to 0$. The following scaling is then conjectured [88, 151, 298]:

$$
P(\Delta, h) \sim
\begin{cases}
\Delta^{\frac{1-\kappa}{\kappa}} p_-\left(h\Delta^{\frac{1-\kappa}{\kappa}}\right) & \text{for } h \sim -\Delta^{\frac{\kappa-1}{\kappa}}, \\
\Delta^{-\frac{a}{\kappa}} p_0\left(h\Delta^{-\frac{1}{2}}\right) & \text{for } |h| \sim \Delta^{\frac{1}{2}}, \\
p_+(h) & \text{for } h \gg \Delta^{\frac{1}{2}},
\end{cases}
\tag{9.53}
$$

$$
m(\Delta, h) = -\Delta^{\frac{1}{2}} \mathcal{M}\left(h\Delta^{-\frac{1}{2}}\right).
$$

The scaling functions $p_-(t)$, $p_0(t)$, $p_+(t)$ and $\mathcal{M}(t)$ are yet to be determined, and a is a new critical exponent. Rather than deriving this scaling ansatz, we show here that it provides a consistent scaling description of the fullRSB solution and agrees with numerical results. More technical details on the derivation can be found in [88, 151].

The asymptotic behaviour of $f(\Delta_M, h)$ for $\Delta_M \to 0$ is

$$
f(\Delta_M, h) = \log \gamma_{\Delta_M} \star \theta(h) \sim -\frac{h^2}{2\Delta_M}\theta(-h), \qquad \Delta_M \to 0, \tag{9.54}
$$

and then, recalling that $\lambda(\Delta_M) = \Delta_M$,

$$
m(\Delta_M, h) = \lambda(\Delta_M) f'(\Delta_M, h) \sim -h\theta(-h), \qquad \Delta_M \to 0. \tag{9.55}
$$

The differential equation for $f(\Delta, h)$ implies that a similar equation holds for $m(\Delta, h)$,

$$\dot{m}(\Delta, h) = \frac{1}{2}m''(\Delta, h) + \frac{y(\Delta)}{\hat{\lambda}(\Delta)}m(\Delta, h)\left[1 + m'(\Delta, h)\right]. \qquad (9.56)$$

This equation admits a solution that scales asymptotically as $m(\Delta, h) \sim -h\theta(-h)$ for $h \to \pm\infty$ for any Δ. Consistency with Eq. (9.53) therefore requires the boundary conditions for the scaling function $\mathcal{M}(t)$ to be

$$\mathcal{M}(t \to \infty) = 0, \qquad \mathcal{M}(t \to -\infty) \sim t. \qquad (9.57)$$

The scaling functions $p_-(t)$ and $p_+(t)$ that appear in Eq. (9.53) for $P(\Delta, h)$ can be understood as follows. For $h \to \infty$, because $m(\Delta, h \to \infty) \to 0$, the differential equation for $P(\Delta, h)$ reduces to the heat equation $\dot{P}(\Delta, h) = -\frac{1}{2}P''(\Delta, h)$. The initial condition $P(\Delta_m, h)$ is a regular, finite function for $h > 0$. Then, $P(\Delta, h)$ is also expected to be finite for large enough $h > 0$, which justifies $p_+(h)$. The behaviour for $h \to -\infty$ is slightly more complicated. In this limit,

$$P(\Delta, h) \propto \sqrt{A(\Delta)} \exp\left[-A(\Delta)h^2\right], \qquad (9.58)$$

which is compatible with the initial condition $P(\Delta_m, h)$. Plugging this Gaussian ansatz within the differential equation for $P(\Delta, h)$, in the limit $h \to -\infty$, one obtains a differential equation for $A(\Delta)$,

$$\dot{A}(\Delta) = 2A(\Delta)^2 - 2A(\Delta)\frac{y(\Delta)}{\hat{\lambda}(\Delta)}. \qquad (9.59)$$

For $\Delta \to 0$, using $\Delta_M \sim \Delta_j \hat{p}^{-\kappa}$ and $y(\Delta) \sim y_j \Delta^{-1/\kappa}$, one has

$$\hat{\lambda}(\Delta) = \hat{p}\Delta_M + \int_{\Delta_M}^{\Delta} d\Delta' y(\Delta')$$

$$\sim \left(\Delta_j - \frac{y_j \kappa \Delta_j^{1-1/\kappa}}{\kappa - 1}\right)\hat{p}^{1-\kappa} + \frac{y_j \kappa}{\kappa - 1}\Delta^{1-1/\kappa}. \qquad (9.60)$$

In the regime where $\Delta \gg \Delta_M \sim \hat{p}^{-\kappa}$, the second term in Eq. (9.60) dominates, and, thus,

$$\frac{y(\Delta)}{\hat{\lambda}(\Delta)} \sim \frac{\kappa - 1}{\kappa}\frac{1}{\Delta}, \qquad \Delta \to 0 \text{ and } \Delta \gg \Delta_M. \qquad (9.61)$$

This scaling behaviour is confirmed by the numerical solution of the fullRSB equations, as shown in Figure 9.7. Substituting Eq. (9.61) into Eq. (9.59), the term $2A(\Delta)^2$ is negligible provided $\kappa < 2$ (which will be verified a posteriori), leading to

$$A(\Delta) \propto \Delta^{2(1-\kappa)/\kappa}, \qquad \Delta \to 0 \text{ and } \kappa < 2. \qquad (9.62)$$

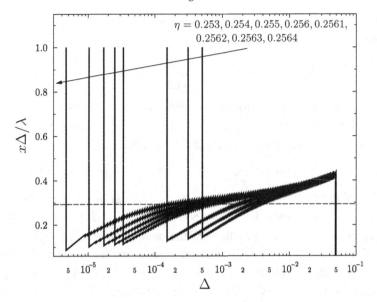

Figure 9.7 The ratio $\Delta x(\Delta)/\lambda(\Delta) = \Delta y(\Delta)/\widehat{\lambda}(\Delta)$ for a hard-sphere glass prepared at $\widehat{\varphi}_g = 7$, obtained numerically integrating the kRSB equations with $k = 99$. The dashed horizontal line corresponds to $(\kappa - 1)/\kappa$, and the convergence to Eq. (9.61) in the region $\Delta_M \ll \Delta \ll \Delta_m$ is visible.

Plugging this result into Eq. (9.58) implies that for $h \to -\infty$ and $\Delta \to 0$, $P(\Delta, h) \sim \Delta^{(1-\kappa)/\kappa} p_-(h\Delta^{(1-\kappa)/\kappa})$, with $p_-(t \to -\infty) \propto e^{-t^2}$, which justifies the scaling form of $P(\Delta, h)$ for $h \to -\infty$ in Eq. (9.53).

The intermediate regime $|h| \sim \sqrt{\Delta}$ is needed in Eq. (9.53) to match the scaling behaviour at positive and negative h. The scaling variable in this matching regime is the same as in the scaling of $m(\Delta, h)$; it is the only possible choice that leads to a non-trivial scaling equation for $p_0(t)$. Matching $p_+(t \to 0^+)$ and $p_0(t \to \infty)$ then gives

$$p_+(t \to 0^+) \sim t^{-\gamma}, \quad p_0(t \to \infty) \sim t^{-\gamma} \quad \Rightarrow \quad \gamma = \frac{2a}{\kappa}, \qquad (9.63)$$

while matching $p_-(t \to 0^-)$ and $p_0(t \to -\infty)$ gives

$$p_-(t \to 0^-) \sim |t|^\theta, \quad p_0(t \to -\infty) \sim |t|^\theta \quad \Rightarrow \quad \theta = \frac{1 - \kappa + a}{\kappa/2 - 1}. \quad (9.64)$$

The scaling in Eq. (9.53) is thus compatible with the asymptotic behaviour of all the functions in the scaling regime, but the exponents a and κ as well as the scaling functions remain undetermined at this stage.

9.3.3 Analytical Computation of the Jamming Exponents

Plugging the scaling ansatz of Eq. (9.53) in Eq. (9.56), using Eq. (9.61), one obtains an equation for $\mathcal{M}(t)$, with boundary conditions given in Eq. (9.57),

$$\begin{cases} \mathcal{M}(t) - t\mathcal{M}'(t) = \mathcal{M}''(t) + 2\frac{1-\kappa}{\kappa}\mathcal{M}(t)\left[1 - \mathcal{M}'(t)\right], \\ \mathcal{M}(t \to \infty) = 0, \qquad \mathcal{M}(t \to -\infty) \sim t. \end{cases} \tag{9.65}$$

For each value of κ, there is a unique solution of this equation that satisfies the correct boundary conditions.

Plugging the scaling form in Eq. (9.53) inside the differential equation for $P(\Delta, h)$, Eq. (9.49), is not sufficient to obtain closed equations for $p_-(t)$ and $p_+(t)$. These functions are non-universal and depend on the details of the problem – e.g., the value of $\widehat{\varphi}_g$ [89]. However, the differential equation for $P(\Delta, h)$, together with Eq. (9.61), gives a closed scaling equation for $p_0(t)$,

$$\begin{cases} \frac{a}{\kappa}p_0(t) + \frac{1}{2}tp_0'(t) = \frac{1}{2}p_0''(t) + \frac{\kappa-1}{\kappa}\left[p_0(t)\mathcal{M}(t)\right]', \\ p_0(t \to \infty) = t^{-2a/\kappa}, \qquad p_0(t \to -\infty) = |t|^{(1-\kappa+a)/(\kappa/2-1)}. \end{cases} \tag{9.66}$$

Universality then appears only in the matching regime of Eq. (9.53) for $P(\Delta, h)$. Note that Eq. (9.66) depends on both κ and a, but there is a unique value $a(\kappa)$ such that $p_0(t)$ satisfies the boundary conditions at $t \to \pm\infty$. Hence, for a given κ, Eqs. (9.65) and (9.66) fix the exponent $a(\kappa)$ and the scaling functions $\mathcal{M}(t)$, $p_0(t)$. Only the exponent κ thus remains undetermined at this stage.

The additional condition needed to determine the scaling solution of the fullRSB equations is provided by Eq. (6.23). Using Eq. (6.21), this equation can be written as

$$\frac{x(\Delta)}{\lambda(\Delta)} = \frac{1}{2}\frac{\int_{-\infty}^{\infty} dh\, P(\Delta, h)m''(\Delta, h)^2}{\int_{-\infty}^{\infty} dh\, P(\Delta, h)m'(\Delta, h)^2[1 + m'(\Delta, h)]}. \tag{9.67}$$

Plugging the scaling ansatz Eq. (9.53) and Eq. (9.61) into Eq. (9.67) gives

$$\frac{\kappa - 1}{\kappa} = \frac{1}{2}\frac{\int_{-\infty}^{\infty} dt\, p_0(t)\mathcal{M}''(t)^2}{\int_{-\infty}^{\infty} dt\, p_0(t)\mathcal{M}'(t)^2\left[1 - \mathcal{M}'(t)\right]}, \tag{9.68}$$

which provides a self-consistent condition for κ. One can then fix κ, solve Eqs. (9.65) and (9.66) to obtain $\mathcal{M}(t)$ and $p_0(t)$ and use Eq. (9.68) to obtain a new estimate of κ, repeating until convergence with arbitrary precision. This procedure gives [88]

$$\begin{aligned} a &= 0.29213\ldots, & \kappa &= 1.41574\ldots, \\ \gamma &= 0.41269\ldots, & \theta &= 0.42311\ldots. \end{aligned} \tag{9.69}$$

Note that within the reported numerical precision, the results of Eq. (9.69) satisfy the relation

$$a = 1 - \kappa/2. \tag{9.70}$$

Inserting Eq. (9.70) into Eq. (9.63) and Eq. (9.64), one obtains the scaling relations between κ, γ and θ given in Eq. (9.45) and Eq. (9.46). While it must be possible to prove this relation directly from the system of equations that define a and κ – i.e., Eqs. (9.65), (9.66) and Eq. (9.68) – such a proof has not yet been achieved. A different argument supporting the validity of Eq. (9.70) has, however, been obtained by investigating the jamming transition in the perceptron model [151].

We have thus far shown that the scaling of the fullRSB equations upon approaching jamming reproduces the observed scaling of the reduced pressure, $\widehat{p} \sim 1/(\widehat{\varphi_{\mathrm{j}}} - \widehat{\varphi})$, which also implies that the hard-sphere glass entropy diverges as $s \sim \log(\widehat{\varphi_{\mathrm{j}}} - \widehat{\varphi})$. The scaling of the mean square displacement, $\Delta_M \sim \widehat{p}^{-\kappa}$, is also reproduced, as shown in Figure 9.5, and the exact expression of κ is obtained analytically. Using Eq. (9.48), we can furthermore rewrite[12] the gap distribution function given by Eq. (6.19) in terms of $P(\Delta, h)$ and $f(\Delta, h)$,

$$\widehat{\varphi_g} e^h \overline{g}(h) = \theta(h) \gamma_{\Delta_M} \star \left[e^{-f(\Delta_M, h)} P(\Delta_M, h) \right] \xrightarrow[\Delta_M \to 0]{} \theta(h) \, p_+(h). \tag{9.71}$$

Therefore, Eq. (9.63) implies that $\overline{g}(h) \sim h^{-\gamma}$ for $h \to 0^+$, with the exponent γ given by Eq. (9.69), and the fullRSB scaling also reproduces the scaling of the gap distribution, providing an analytical expression of the associated exponent. To complete the discussion, the behaviour of the force distribution remains to be investigated. We refer the reader to [88] for the proof that $P(f) \sim f^\theta$ for hard spheres. In Section 9.4.2, we provide a much simpler derivation of this result using soft harmonic spheres.

9.4 The Overjammed Phase: Soft Harmonic Spheres

In this section, we consider the behaviour of soft harmonic spheres, with potential $\bar{v}(h) = (\varepsilon/2)h^2\theta(-h)$, in the overjammed phase. This analysis is interesting because it reproduces the scaling of pressure, energy and coordination discussed in Section 9.2.1. It is also a much easier model with which to prove isostaticity and compute $P(f)$ because the average coordination z is obtained by counting negative gaps, and, according to Eq. (9.22), forces are simply proportional to the modulus of negative gaps.

---

[12] To prove Eq. (9.71), it suffices to observe that for $h > 0$ and $\Delta_M \to 0$, one has $f(\Delta_M, h) \to 0$ according to Eq. (9.54) and $P(\Delta_M, h) \to p_+(h)$ according to Eq. (9.53). The convolution with γ_{Δ_M} then disappears because $\Delta_M \to 0$.

The scaled energy $\widehat{e}_{\text{liq}} = e_{\text{liq}}/d$ of a liquid of soft harmonic spheres in $d \to \infty$ is given by Eq. (2.78),

$$\widehat{e}_{\text{liq}}(\widehat{\varphi}, T) = \frac{\widehat{\varphi}}{2} \int_{-\infty}^{0} dh \, e^{h - \beta \frac{\varepsilon}{2} h^2} \frac{\varepsilon}{2} h^2, \tag{9.72}$$

which vanishes for $T \to 0$ at any finite density $\widehat{\varphi}$. The equilibrium zero-temperature liquid is therefore unjammed at all $\widehat{\varphi}$, and coincides with an equilibrium liquid of hard spheres[13] at the same $\widehat{\varphi}$. We can thus (1) prepare an initial equilibrium configuration of a soft-harmonic-sphere liquid at $(\widehat{\varphi}_g, T_g = 0)$, and (2) adiabatically follow the corresponding glass at state point $(\widehat{\varphi}, T)$, computing in particular its energy $\widehat{e}_g(\widehat{\varphi}, T | \widehat{\varphi}_g, 0)$ and reduced pressure $\widehat{p}_g(\widehat{\varphi}, T | \widehat{\varphi}_g, 0)$. According to the discussion of Section 9.2.1, as long as for $T \to 0$ the glass energy vanishes and the reduced pressure is finite, the glass is unjammed. The soft-sphere glass then coincides with the hard-sphere glass, and upon compression, the jamming point[14] is reached for $\widehat{\varphi} \to \widehat{\varphi}_j^-$, at which the glass reduced pressure $\widehat{p}_g(\widehat{\varphi}, 0 | \widehat{\varphi}_g, 0) \sim (\widehat{\varphi}_j - \widehat{\varphi})^{-1}$ diverges. The corresponding jamming scaling ansatz was studied in Section 9.3. Upon further compression, for $\widehat{\varphi} > \widehat{\varphi}_j$, the glass enters its overjammed phase, where energy and pressure are both positive, $\widehat{e}_g(\widehat{\varphi}, 0 | \widehat{\varphi}_g, 0) \sim (\widehat{\varphi} - \widehat{\varphi}_j)^2$ and $P_g(\widehat{\varphi}, 0 | \widehat{\varphi}_g, 0) \sim (\widehat{\varphi} - \widehat{\varphi}_j)$. In this section, we write the fullRSB equations that describe the overjammed phase and investigate their scaling behaviour when $\widehat{\varphi} \to \widehat{\varphi}_j^+$.

9.4.1 Zero-Temperature Limit of the fullRSB Equations

The overjammed phase can be studied by the formalism of Chapter 6, using the soft-harmonic-sphere potential and setting $T_g = 0$ and $T = 0$, with $\widehat{\varphi} > \widehat{\varphi}_j$. The first step consists in setting $T_g = 0$, which is easily done by replacing $e^{-\beta_g \bar{v}(h)} \to \theta(h)$ in Eq. (6.11). Using the modifications discussed in Section 9.3.1, one arrives at the same Eqs. (9.49) and (9.50), with the only difference that the initial condition for $f(\Delta, h)$ now reads

$$f(\Delta_M, h) = \log \gamma_{\Delta_M} \star e^{-\beta \frac{\varepsilon}{2} h^2 \theta(-h)}, \tag{9.73}$$

which corresponds to following the glass at state point $(\widehat{\varphi}, T)$. In the unjammed phase, $\widehat{\varphi} < \widehat{\varphi}_j$, we know from Section 9.3.1 that Δ_M remains finite when $T \to 0$. Hence, we can simply replace $e^{-\beta \frac{\varepsilon}{2} h^2 \theta(-h)} \to \theta(h)$ in Eq. (9.73) and the hard-sphere case is recovered. In the overjammed phase, $\widehat{\varphi} > \widehat{\varphi}_j$, by contrast, the mean square displacement vanishes – i.e., $\Delta_M \to 0$ when $T \to 0$ – because the system is in a

---

[13] Recall that a hard-sphere liquid exists up to the Kauzmann density $\widehat{\varphi}_K \sim \log d$, and a soft-harmonic-sphere liquid exists at densities below the Kauzmann transition line given by Eq. (7.9).

[14] Recall that the value of $\widehat{\varphi}_j$ depends on $\widehat{\varphi}_g$, but for notational convenience, we do not here indicate this dependence explicitly.

local minimum of the potential energy. The correct scaling solution is then obtained by assuming that

$$\Delta_M = \chi T, \qquad T \to 0 \text{ and } \widehat{\varphi} > \widehat{\varphi}_{\mathrm{j}}, \tag{9.74}$$

which can be justified by harmonic analysis, as discussed in Section 9.4.3.

Because one should take the limit $T \to 0$ and $\Delta_M = \chi T \to 0$ simultaneously, the function $f(\Delta_M, h)$ does not converge to the hard-sphere limit. Similarly to Eq. (9.51) for hard spheres, for overjammed soft harmonic spheres, it is convenient to define

$$y(\Delta) = x(\Delta)/\Delta_M,$$

$$\widehat{\lambda}(\Delta) = \lambda(\Delta)/\Delta_M = 1 + \int_{\Delta_M}^{\Delta} d\Delta' y(\Delta'),$$

$$\widehat{f}(\Delta, h) = \Delta_M f(\Delta, h), \tag{9.75}$$

$$m(\Delta, h) = \widehat{\lambda}(\Delta)\widehat{f}'(\Delta, h) = \lambda(\Delta)f'(\Delta, h).$$

In the zero temperature limit, $\Delta_M \to 0$ and

$$\widehat{f}(0, h) = \lim_{T \to 0} \widehat{f}(\Delta_M = \chi T, h)$$

$$= \lim_{T \to 0} \chi T \log \gamma_{\chi T} \star e^{-\beta \frac{\varepsilon}{2} h^2 \theta(-h)} = -\frac{\chi \varepsilon}{1 + \chi \varepsilon} \frac{h^2 \theta(-h)}{2}. \tag{9.76}$$

Note that the hard-sphere result, Eq. (9.54), is recovered for $\chi \to \infty$. We can then write Eq. (9.49) in terms of these scaled variables,

$$\begin{cases} \widehat{f}(0, h) & = -\frac{\chi \varepsilon}{1 + \chi \varepsilon} \frac{h^2 \theta(-h)}{2}, \\ \dot{\widehat{f}}(\Delta, h) & = \frac{1}{2}\left(\widehat{f}''(\Delta, h) + y(\Delta)\widehat{f}'(\Delta, h)^2\right), \\ P(\Delta_m, h) & = \widehat{\varphi}_g \gamma_{2\Delta_r - \Delta_m} \star e^{h + \eta - \Delta_r} \theta(h + \eta - \Delta_r), \\ \dot{P}(\Delta, h) & = -\frac{1}{2}\left[P''(\Delta, h) - 2y(\Delta)\left(P(\Delta, h)\widehat{f}'(\Delta, h)\right)'\right], \end{cases} \tag{9.77}$$

with $\Delta \in [0, \Delta_m]$. Eq. (9.50) also becomes

$$\frac{1}{\widehat{\lambda}(\Delta_m)} = -\frac{1}{2}\int_{-\infty}^{\infty} dh \, P(\Delta_m, h)[\widehat{f}''(\Delta_m, h) + \widehat{f}'(\Delta_m, h)],$$

$$\frac{2\Delta_r - \Delta_m}{\widehat{\lambda}(\Delta_m)^2} - \int_{\Delta_m}^{\Delta} d\Delta' \frac{1}{\widehat{\lambda}(\Delta')^2} = \frac{1}{2}\int_{-\infty}^{\infty} dh \, P(\Delta, h)\widehat{f}'(\Delta, h)^2, \tag{9.78}$$

which provides a system of equations that describe the zero-temperature over-jammed phase.

In the limit $\Delta \to 0$, the solution for $y(\Delta)$ is singular,

$$y(\Delta) \sim y_\chi \Delta^{-1/2}, \qquad \Delta \to 0, \tag{9.79}$$

where y_χ is a constant that depends on χ. We refer the reader to [151] for a detailed discussion. Similarly,

$$\widehat{\lambda}(\Delta) = 1 + \int_0^\Delta d\Delta' y(\Delta') \sim 1 + 2y_\chi \sqrt{\Delta}, \qquad \Delta \to 0. \tag{9.80}$$

Note that the same scaling is found in some spin glass models [254]. Correspondingly, $P(\Delta \to 0, h)$ is finite and smooth around $h = 0$ [151]. Finally, the parameter χ is determined by $P(0, h)$ through Eq. (6.21),

$$1 = \frac{1}{2} \int_{-\infty}^\infty dh \, P(\Delta, h) \widehat{f}''(\Delta, h)^2 \xrightarrow[\Delta \to 0]{} \frac{1}{2} \int_{-\infty}^0 dh \, P(0, h) \left(\frac{\chi \varepsilon}{1 + \chi \varepsilon} \right)^2. \tag{9.81}$$

9.4.2 Approaching Jamming from the Overjammed Phase

Taking the zero-temperature limit of Eq. (6.19) gives

$$\widehat{\varphi}_g e^h \overline{g}(h) = \begin{cases} P(0, h(1 + \chi\varepsilon))(1 + \chi\varepsilon) & \text{for } h < 0, \\ P(0, h) & \text{for } h > 0. \end{cases} \tag{9.82}$$

Similarly, from Eqs. (6.16) and (6.17), the scaled glass energy $\widehat{e}_g = e_g/d$ and pressure $\widehat{P}_g = P_g/(d\rho)$ become, for $T \to 0$,

$$
\begin{aligned}
\widehat{e}_g(\widehat{\varphi}) &= \frac{e_g(\widehat{\varphi}, 0 | \widehat{\varphi}_g, 0)}{d} = \frac{\varepsilon}{4(1 + \chi\varepsilon)^2} \int_{-\infty}^0 dh \, P(0, h) \, h^2, \\
\widehat{P}_g(\widehat{\varphi}) &= \frac{P_g(\widehat{\varphi}, 0 | \widehat{\varphi}_g, 0)}{\rho d} = \frac{\varepsilon}{2(1 + \chi\varepsilon)} \int_{-\infty}^0 dh \, P(0, h) \, |h|.
\end{aligned}
\tag{9.83}
$$

These expressions suggest that upon approaching the jamming transition from the overjammed side, $\widehat{\varphi} \to \widehat{\varphi}_j^+$, one has $\chi \to \infty$ because energy and pressure must then vanish. Consistently, the observation that Δ_M is finite for $\widehat{\varphi} < \widehat{\varphi}_j$ requires that χ diverges to match the hard-sphere and the soft-harmonic-sphere scalings.

A careful analysis [151] of the fullRSB equations derived in Section 9.4.1 shows that, for $\chi \to \infty$, one has

$$y(\Delta) \sim \begin{cases} y_\chi \Delta^{-1/2}, & \Delta \ll \Delta^*, \\ y_j \Delta^{-1/\kappa}, & \Delta \gg \Delta^*. \end{cases} \tag{9.84}$$

The crossover point Δ^* then scales as

$$\Delta^* \sim \chi^{-\frac{2\kappa}{\kappa - 2a}}, \qquad y_\chi \sim (\Delta^*)^{-1/2}, \qquad y(\Delta^*) \sim 1/\Delta^*. \tag{9.85}$$

In the jamming limit, $\Delta^* \to 0$, and, hence, we recover the jamming critical behaviour of hard spheres, $y(\Delta) \sim y_j \Delta^{-1/\kappa}$, over the full range $\Delta \in [0, \Delta_m]$. We refer the reader to [151] for a derivation of these results.

Using the scaling relation $a = 1 - \kappa/2$, Eq. (9.70), one obtains $\chi \sim (\Delta^*)^{(1-\kappa)/\kappa}$, and observing that in the jamming limit $\Delta^* \to 0$, one concludes that $P(0, h) \sim P(\Delta^*, h)$ follows the hard-sphere scaling in Eq. (9.53), because Δ^* is the boundary of the hard-sphere regime [151]. Hence,

$$P(0,h) \sim \begin{cases} \chi p_-(h\chi) & \text{for } h \sim -\chi^{-1}, \\ \chi^{-\frac{1-\kappa/2}{1-\kappa}} p_0\left(h\chi^{-\frac{\kappa}{2(1-\kappa)}}\right) & \text{for } |h| \sim \chi^{\frac{\kappa}{2(1-\kappa)}}, \\ p_+(h) & \text{for } |h| \gg \chi^{\frac{\kappa}{2(1-\kappa)}}. \end{cases} \qquad (9.86)$$

Eqs. (9.82) and (9.86) describe the critical behaviour of soft harmonic spheres for $\chi \to \infty$. The following properties are then indeed obtained.

1. Because $P(0, h) \sim \chi p_-(h\chi)$ for $h < 0$, Eq. (9.83) gives

$$\widehat{e}_g(\widehat{\varphi}) \sim \chi^{-4}, \qquad \widehat{P}_g(\widehat{\varphi}) \sim \chi^{-2}, \qquad (9.87)$$

which suggests that

$$\chi \sim (\widehat{\varphi} - \widehat{\varphi}_j)^{-1/2}, \qquad (9.88)$$

in order to reproduce the scaling of energy and pressure in Eqs. (9.8) and (9.9), respectively. Unfortunately, Eq. (9.88) cannot be proven from the scaling analysis. It must instead be obtained from a numerical solution of the fullRSB equations (similarly to the relation between reduced pressure and density in the case of hard spheres).

2. Because for $h < 0$ and $\chi \to \infty$, one has

$$\widehat{\varphi}_g e^h \overline{g}(h) = P(0, h(1 + \chi\varepsilon))(1 + \chi\varepsilon) = A_\chi p_-(A_\chi h), \qquad (9.89)$$

with $A_\chi = \chi(1 + \chi\varepsilon)$, the average contact number is

$$z = \rho\Omega_d \int_0^\ell dr\, g(r) = d\widehat{\varphi}_g \int_{-\infty}^0 dh\, e^h\, \overline{g}(h) = d \int_{-\infty}^0 dh\, p_-(h). \qquad (9.90)$$

In the limit $\chi \to \infty$, Eq. (9.81) becomes

$$2 = \int_{-\infty}^0 dh\, p_-(h), \qquad (9.91)$$

which implies $z = 2d$. The solution in the limit $d \to \infty$ thus predicts that jammed sphere packings are isostatic. It is also possible to show [151] that

$$z - 2d \sim 1/\chi \sim \sqrt{\widehat{P_g}} \sim \sqrt{\widehat{\varphi} - \widehat{\varphi}_j}, \tag{9.92}$$

which reproduces Eq. (9.17).

3. For $h > 0$,

$$\overline{g}(h) \sim P(0, h) \sim p_+(h) \underset{h \to 0^+}{\sim} h^{-\gamma}; \tag{9.93}$$

hence, the positive gaps are characterised by the same power-law divergence as those of hard spheres. The positive part of $\overline{g}(h)$ is indeed continuous at jamming.

4. For $h < 0$, Eq. (9.89) provides the scaling of the negative gaps, which are proportional to the contact forces. Because $f \propto |h|$, one has $P(f) = Bp_-(-Bf)/2$, which is correctly normalised as a consequence of Eq. (9.91). Normalising the average force to unity, as in Eq. (9.30), fixes the constant B and gives

$$P(f) = \frac{B}{2} p_-(-Bf), \qquad B = \frac{1}{2} \int_{-\infty}^{0} dh \, p_-(h)|h|. \tag{9.94}$$

From these results, it follows that $P(f) \sim f^\theta$ for small forces, with the exponent θ computed in Section 9.3.3 from the scaling of the fullRSB solution. Note that because at jamming the forces are entirely determined by the contact network, this result also holds upon approaching jamming from the hard-sphere side [88].

9.4.3 The Density of Harmonic Vibrations

In the overjammed phase at zero temperature, soft harmonic spheres are in a minimum of the potential energy, which makes a harmonic analysis possible. By definition, in an energy minimum, the eigenvalues λ_i of the Hessian matrix defined in Eq. (9.19) are all positive, and the frequency associated to each eigenvalue is $\omega_i = \sqrt{\lambda_i}$. Because each rattler gives d trivial zero eigenvalues, rattlers are once again removed from the analysis. One can then define the density of vibrational states as

$$D(\omega) = \frac{1}{dN_{cn}} \overline{\sum_{i=1}^{dN_{cn}} \delta(\omega - \omega_i)}, \tag{9.95}$$

where the average is over an ensemble of configurations generated by the same protocol. While an exact calculation of $D(\omega)$ has not yet been achieved for soft harmonic spheres in $d \to \infty$, this quantity has been studied analytically in simplified infinite-dimensional models such as the perceptron [150] and in overjammed sphere packings using an (approximate) effective medium theory [121]. This analysis shows that, if phonons are neglected,[15] $D(\omega)$ has the following general form

$$D(\omega) \sim \begin{cases} 0 & \omega \notin [\omega_0, \omega_{max}], \\ (\omega/\omega_*)^2 & \omega_0 \ll \omega \ll \omega_*, \\ \text{const.} & \omega_* \ll \omega \ll \omega_{max}, \end{cases} \tag{9.96}$$

which depends on three characteristic frequencies ω_0, ω_* and ω_{max}. The two frequencies ω_0 and ω_{max} are the two edges of the vibrational spectrum, while ω_* is an intermediate frequency scale. Within mean field theory [150], in the overjammed replica symmetric phase, $\omega_0 > 0$; hence, the density of states is gapped. No vibrational excitations of arbitrarily small frequency – i.e., soft modes – are then present. In the overjammed fullRSB phase, instead, $\omega_0 = 0$, and $D(\omega) \sim (\omega/\omega_*)^2$ down to zero frequency. Upon approaching the jamming point, the intermediate frequency $\omega_* \sim (\widehat{\varphi} - \widehat{\varphi}_j)^{1/2}$ also goes to zero. It follows that $D(\omega)$ remains constant down to zero frequency. The jamming point is then marginally stable also from the vibrational point of view, and a lot of vibrational modes with small frequency are observed. The scaling in Eq. (9.96) is also found in numerical simulations [92, 182, 228, 258, 277, 360], which additionally identified a class of localised vibrational modes that contribute a term $D_{loc}(\omega) \sim (\omega/\omega_*)^4$ to the density of states [228, 258]. These localised modes likely disappear upon increasing dimension, similarly to rattlers and bucklers. They are therefore unlikely to be present within mean field theory, but their disappearance has not yet been systematically studied.

To conclude, note that a single harmonic oscillator $x(t)$ of frequency ω has an equilibrium mean square displacement

$$\Delta = \lim_{t \to \infty} \langle (x(t) - x(0))^2 \rangle \propto T/\omega^2.$$

Harmonic analysis around a glass minimum provides dN_{cn} independent oscillators with frequencies distributed according to the density of states $D(\omega)$, hence [182]

$$\Delta = \chi T, \qquad \chi \propto \int_0^\infty d\omega\, D(\omega)/\omega^2, \tag{9.97}$$

which provides a physical argument in support of Eq. (9.74). Note that setting $\omega_0 = 0$ in Eq. (9.96) gives $D(\omega) = \mathcal{D}(\omega/\omega_*)$, with a scaling function $\mathcal{D}(x) \sim x^2$ for $x \ll 1$ and constant otherwise, which implies $\chi \sim 1/\omega_* \sim (\widehat{\varphi} - \widehat{\varphi}_j)^{-1/2}$, consistently with Eq. (9.88).

---

[15] Phonons provide a contribution that scales as ω^{d-1} and can therefore be neglected at low ω in large enough dimension.

9.5 Wrap-Up

9.5.1 Summary

In this chapter, we have seen that

- The jamming transition φ_j is the satisfiability transition of the packing problem. It separates unjammed configurations at $\varphi < \varphi_j$, which satisfy all the hard-core constraints, from overjammed configurations at $\varphi > \varphi_j$, which contain overlapping spheres (Section 9.1).
- The glass close packing density φ_{GCP} corresponds to the thermodynamic jamming transition; amorphous unjammed configurations do not exist for $\varphi > \varphi_{GCP}$. Local search algorithms that look for unjammed configurations, such as gradient descent from random initial conditions (Section 9.1.1), annealing soft harmonic spheres (Section 9.1.2) or compressing hard spheres (Section 9.1.3) generically get stuck at a lower density $\varphi_j < \varphi_{GCP}$. The jamming point φ_j thus strongly depends on the details of the preparation algorithm.
- Pressure, energy and coordination are always singular at φ_j (Section 9.2.1), but the structure of the Hessian matrix determines additional critical properties of jamming. If the pre-stress term is negative definite, then jamming is isostatic – e.g., for spherical particles – while otherwise, jamming is hypostatic – e.g., for ellipsoids (Section 9.2.2).
- Isostatic jammed sphere packings have universal, protocol-independent properties. They display a power-law distribution of small gaps, $g(r) \sim (r - \ell)^{-\gamma}$, and small forces, $P(f) \sim f^\theta$. Their mean square displacement vanishes as $\Delta_M \sim p^{-\kappa}$ in the unjammed phase, and their average coordination grows as $z - 2d \sim \sqrt{P}$ in the overjammed phase (Section 9.2.3). A mechanical marginal stability argument provides scaling relations between the exponents γ, θ, κ (Section 9.2.4).
- The fullRSB equations that describe adiabatic compression of a hard-sphere glass (Section 9.3.1) develop a scaling regime close to jamming, $\varphi \to \varphi_j^-$ (Section 9.3.2). An analysis of these equations leads to exact analytical results for the exponents γ, θ, κ (Section 9.3.3). Within this framework, isostaticity is a consequence of the marginality of the fullRSB solution, which also fixes the value of the exponents.
- The fullRSB equations that describe the overjammed phase of a soft-harmonic-sphere glass (Section 9.4.1) also develop a scaling regime close to jamming, $\varphi \to \varphi_j^+$ (Section 9.4.2). The analysis of soft harmonic spheres simplifies the derivation of isostaticity and of the force distribution, $P(f)$. Moreover, in the overjammed phase, the density of vibrational states $D(\omega)$ can be defined. It is

gapless in the fullRSB phase, where it behaves as $D(\omega) \sim (\omega/\omega_*)^2$ for $\omega \ll \omega_*$ and it is flat for $\omega \gg \omega_*$. At jamming, ω_* vanishes, and $D(\omega)$ is finite at zero frequency. An anomalous abundance of zero-frequency modes is then observed (Section 9.4.3).

9.5.2 Further Reading

We provide here a list of references that can be consulted to further explore the subjects discussed in this chapter, selected according to the criteria discussed in Section 1.6.2.

The jamming transition has been observed in a variety of materials, ranging from colloids to foams, granulars and emulsions. Complete reviews on the application of jamming ideas to soft condensed matter can be found in

- Van Hecke, *Jamming of soft particles: Geometry, mechanics, scaling and isostaticity* [348]
- Liu and Nagel, *The jamming transition and the marginally jammed solid* [234]
- Liu, Nagel, Van Saarloos, et al., *The jamming scenario: An introduction and outlook* [232]

A complete review of the marginal stability ideas described in Section 9.2.4 can be found in Müller and Wyart, *Marginal stability in structural, spin, and electron glasses* [269].

A general introduction to satisfiability problems can be found in

- Percus, Istrate and Moore, *Computational complexity and statistical physics* [294]
- Arora and Barak, *Computational complexity: A modern approach* [19]
- Biere, Heule and van Maaren (eds), *Handbook of satisfiability* [50]

In particular, the first reference [294] contains a contribution from Cocco, Monasson, Montanari et al., *Approximate analysis of search algorithms with 'physical' methods*, specifically focused on the analysis of search algorithms. The last reference [50] contains a contribution from Altarelli, Monasson, Semerjian et al., *A review of the statistical mechanics approach to random optimization problems*, which reviews advanced statistical mechanics methods to compute the satisfiability threshold in these problems. A discussion on the extension of these results to continuous satisfiability problems, in relation to jamming, can be found in Franz, Parisi, Sevelev et al., *Universality of the SAT-UNSAT (Jamming) Threshold in Non-convex continuous constraint satisfaction problems* [151].

A very important practical problem is that of understanding the role of friction in jamming. Friction indeed plays a very important role in selecting the jammed states of granular materials. Introductory reviews to this topic are

- Liu and Nagel, *Jamming and rheology: Constrained dynamics on microscopic and macroscopic scales* [233]
- Bi, Henkes, Daniels et al., *The statistical physics of athermal materials* [48]
- Behringer and Chakraborty, *The physics of jamming for granular materials: A review* [31]

The study of frictional jammed packings requires the introduction of new ideas, because the Franz–Parisi and Monasson constructions discussed in this book are based on a Gibbs–Boltzmann approach, which is not appropriate for dissipative systems. The out-of-equilibrium dynamical equations for infinite-dimensional spheres could possibly be generalised to frictional systems.

10

Rheology of the Glass

In Chapters 4, 6 and 9, we investigated the behaviour of glasses adiabatically followed under (de)compression, cooling or heating. This chapter extends the state following construction to the response of an amorphous solid to an applied shear strain, which leads to a mean field theory of the rheology of these materials. Several physical phenomena ranging from elasticity to dilatancy, yielding and jamming by shear are described by this approach.

An elastic response to small applied deformations is the hallmark of solidity. While for perfect crystals, perturbative calculations are possible, for amorphous solids, the theory is much less developed. We discuss here the static elastic response and how the solid breaks when subjected to larger deformations. Because a liquid can adapt to a change of shape of its container, a response to strain is only observed in the dynamically arrested phase unless dynamics is considered.

Because the topic of this chapter is rapidly evolving, we limit ourselves to a description of the formalism and of a few selected results. The first steps towards a theory of the rheology of amorphous solids using the techniques we have described in this book were taken in [366, 368], while the exact solution for particles in the limit $d \to \infty$ was derived in [298, 299, 369].

10.1 Perturbing the Glass by a Shear Strain

Within the state following approach described in Chapter 4, a reference replica $\underline{Y}$ is prepared in equilibrium at a state point (φ_g, T_g) in the dynamically arrested region, which selects a glass, and a second replica $\underline{X}$ in constrained equilibrium within this glass is compressed or cooled to a state point (φ, T); see Figure 4.2 for an illustration. In this section, we generalise this construction to the case in which a shear strain is also applied to $\underline{X}$. If the system is dynamically arrested, it behaves as an amorphous solid and thus reacts elastically to a small applied strain. The force (per unit surface) that the system exerts against the deformation is the shear stress

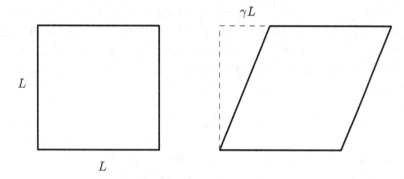

Figure 10.1 A two-dimensional illustration of a shear deformation, with shear strain γ.

(or simply the stress) [6, 175, 222]. One of the goals of this chapter is to compute the stress as a function of the strain; the resulting stress–strain curves describe the rheology of that solid.

10.1.1 State Following with Shear Strain

Following [366, 368], we consider particles confined in a cubic box of linear size L and volume L^d, and we denote as $\mathbf{x}' = \{x'_\mu\}_{\mu=1,\dots,d}$ a Cartesian system of coordinates in the laboratory frame, such that the box sides are parallel to the coordinate axes. In this reference frame, the box is subjected to a shear strain γ, as illustrated in Figure 10.1. Note that because the reference replica $\underline{Y}$ is used to select the glass in equilibrium, it should not be strained, while the constrained replica $\underline{X}$ should be strained. Hence, in the frame $\mathbf{x}'$, replica $\underline{Y}$ is in the original cubic box, while replica $\underline{X}$ is in the strained box. Because in the state following construction, it is not practical to compare replicas in different boxes, for convenience, we introduce a 'strained' frame with coordinates $\mathbf{x} = \{x_\mu\}_{\mu=1,\dots,d}$, to bring back the strained box to a cubic shape. If the box is strained along the first direction $\mu = 1$, then the transformation is

$$x'_1 = x_1 + \gamma x_2, \qquad x'_\mu = x_\mu, \quad \forall \mu = 2, \dots, d, \tag{10.1}$$

and its inverse is

$$x_1 = x'_1 - \gamma x'_2, \qquad x_\mu = x'_\mu, \quad \forall \mu = 2, \dots, d. \tag{10.2}$$

To limit the use of indices, we introduce a strain matrix $\mathcal{S}(\gamma) = \hat{1} + \gamma \hat{\mathbf{x}}_1 \hat{\mathbf{x}}_2^T$, where $\hat{1}$ is the $d \times d$ identity matrix and $\hat{\mathbf{x}}_\mu$ is the unit vector parallel to the coordinate axis μ, such that

$$\mathbf{x}' = \mathcal{S}(\gamma)\mathbf{x} = \mathbf{x} + \gamma \hat{\mathbf{x}}_1 x_2, \qquad \mathbf{x} = \mathcal{S}(-\gamma)\mathbf{x}'. \tag{10.3}$$

Note that $S(\gamma)S(-\gamma) = \hat{1}$, which simply means that reversing the applied strain brings the system back to the unstrained state. The particles of replica $\underline{X}$, in the laboratory frame $\mathbf{x}'$, interact via the normal interaction potential $v(|\mathbf{x}' - \mathbf{y}'|)$. Hence, in the strained frame, their pair interaction is $v(|S(\gamma)(\mathbf{x} - \mathbf{y})|)$ [366, 368], which results in a total potential energy for replica $\underline{X}$

$$V[\underline{X}, \ell, \gamma] = \sum_{i<j} v(|S(\gamma)(\mathbf{x}_i - \mathbf{x}_j)|). \tag{10.4}$$

The average free energy of a glass prepared in equilibrium at (φ_g, T_g) and adiabatically followed to (φ, T, γ) is then simply obtained from Eq. (4.9) by using Eq. (10.4) as potential for replica $\underline{X}$,

$$f_g(\varphi, T, \gamma; \varphi_g, T_g, \mathsf{D}_r) = -\frac{T}{N} \int \frac{d\underline{Y}}{Z[\varphi_g, \beta_g]} e^{-\beta_g V[\underline{Y}, \ell_g]} \log Z[\varphi, \beta, \gamma; \underline{Y}, \mathsf{D}_r],$$

$$Z[\varphi, \beta, \gamma; \underline{Y}, \mathsf{D}_r] = \int d\underline{X} e^{-\beta V[\underline{X}, \ell, \gamma]} \delta(\mathsf{D}_r - \mathsf{D}(\underline{X}, \underline{Y})). \tag{10.5}$$

The average of the logarithm can then be computed by introducing additional replicas, as in Section 4.1.3, where now replica 1 has no shear strain, $\gamma_1 = 0$, while replicas $a = 2, \cdots, s + 1$ have shear strain $\gamma_a = \gamma$. Note that because the mean square displacement $\mathsf{D}(\underline{X}, \underline{Y})$ is computed in the strained frame, the 'affine part' of the displacement – i.e., the linear part corresponding to the straining of the box – is removed, and $\mathsf{D}(\underline{X}, \underline{Y})$ only measures the 'non-affine' contribution to the mean square displacement. Note also that in full thermodynamic equilibrium, the free energy does not depend on the shape of the box, even for a solid [314]. Hence, the free energy in Eq. (10.5) only depends on γ because replica $\underline{X}$ is in a constrained, metastable equilibrium within the glass state selected by replica $\underline{Y}$.

10.1.2 Replicated Free Energy in Infinite Dimensions

In the limit $d \to \infty$, the derivation of Section 4.2 can be followed identically[1] up to Eq. (4.44), which is now replaced by[2]

$$f_{\mathrm{eff}}(\mathbf{R}) = e^{-\beta_g \bar{v}[d(R/\ell_g - 1)]} \langle e^{-\beta \sum_{a=2}^n \bar{v}[d(x_a/\ell - 1)]} \rangle - 1, \tag{10.6}$$

with

$$x_a = |S(\gamma_a)(\mathbf{R} + \mathbf{w}^a)|, \qquad \mathbf{w}^a = \mathbf{u}^a - \mathbf{v}^a. \tag{10.7}$$

---

[1] In Section 4.2, rotational invariance was used to obtain, for example, Eq. (4.30). In the presence of a shear strain, the coordinates $\mu = 1, 2$ are special, and rotational invariance only holds in the subspace of $d - 2$ coordinates $\mu = 3, \cdots, d$. When $d \to \infty$, however, one can show that this anisotropy can be neglected [59].

[2] By contrast to Section 4.2, we here explicitly take into account that replica 1 is at state point (φ_g, T_g), a priori different from the state point of the other replicas.

From this point on, however, the derivation of Section 4.2 should be adapted to take the shear strain into account [59]. For $d \to \infty$, one can show that the term $\mathcal{S}(\gamma_a)\mathbf{w}^a$ gives subleading contributions in $1/d$ and can be neglected. Then $x_a \sim |\mathbf{R} + \mathbf{w}^a + \gamma_a \hat{\mathbf{x}}_1 R_2|$, and one has

$$y_a = x_a^2 - R^2 \sim |\mathbf{w}^a| + 2\mathbf{R} \cdot \mathbf{w}^a + 2\gamma_a R_1 R_2 + \gamma_a^2 R_2^2, \tag{10.8}$$

where the term $2\gamma_a w_1^a R_2$ has been neglected because it also gives subleading contributions. The first two terms in Eq. (10.8) can be analysed as in Section 4.2.3, resulting in Eq. (4.49) being modified by the addition of the last two terms in Eq. (10.8),

$$y_a = 2\alpha_{aa}\ell^2/d + 2R\ell z_a/d + 2\gamma_a R_1 R_2 + \gamma_a^2 R_2^2. \tag{10.9}$$

Following the same reasoning as in Section 4.2.3, we expect that $y_a \propto 1/d$. Expanding at leading order in $1/d$, with $R = \ell_g(1 + h/d)$, one obtains a modification of Eq. (4.50),

$$d(x_a/\ell - 1) \sim h - \eta + \frac{d\,y_a}{2\ell^2} = h - \eta + \alpha_{aa} + z_a + d\gamma_a \frac{R_1 R_2}{R^2} + d\gamma_a^2 \frac{R_2^2}{R^2}. \tag{10.10}$$

Here, the variables z_a are distributed according to Eq. (4.48) and are independent of $\mathbf{R}$. Eq. (10.6) has to be integrated over $\mathbf{R}$ within the second virial coefficient, which amounts to integrating over $R = |\mathbf{R}|$ and over the unit vector $\hat{\mathbf{R}} = \mathbf{R}/R$. It can be shown that, when $d \to \infty$, the integration over the components of $\hat{\mathbf{R}}$ can be replaced by an average, $\hat{R}_\mu = R_\mu/R \to g_\mu/\sqrt{d}$, where g_μ are independent random Gaussian variables with zero mean and unit variance. Hence, Eq. (10.10) becomes

$$d(x_a/\ell - 1) \sim h - \eta + \alpha_{aa} + z_a + \gamma_a g_1 g_2 + \gamma_a^2 g_2^2. \tag{10.11}$$

Eq. (4.52) is then replaced by

$$\overline{f}_{\text{eff}}(h, g_1, g_2) = e^{-\beta_g \bar{v}(h)} e^{\sum_{a,b=2}^n \alpha_{ab} \frac{\partial^2}{\partial h_a \partial h_b} + \sum_{a=2}^n \left(\gamma_a g_1 g_2 + \frac{1}{2}\gamma_a^2 g_2^2\right)\frac{\partial}{\partial h_a}}$$
$$\times e^{-\sum_{a=2}^n \beta \bar{v}(h_a - \eta)}\Big|_{h_a = h} - 1, \tag{10.12}$$

and the excess free energy becomes

$$-\beta f^{\text{ex}} = \frac{d\widehat{\varphi}_g}{2} \int \mathcal{D}g_1 \mathcal{D}g_2 \int_{-\infty}^{\infty} dh\, e^h \overline{f}_{\text{eff}}(h, g_1, g_2), \tag{10.13}$$

where $\mathcal{D}g = dg\, e^{-g^2/2}/\sqrt{2\pi}$ denotes the integration over a Gaussian variable of zero mean and unit variance [59, 299]. By integrating by parts and expressing the free energy in terms of $\hat{\Delta}$, one of the two Gaussian integrations can be eliminated, and

$$-\beta f^{ex} = \frac{d\widehat{\varphi}_g}{2} \int \mathcal{D}g \int_{-\infty}^{\infty} dh \, e^h \left[f^d(\hat{\Delta}, h, g) - 1 \right], \tag{10.14}$$

$$f^d(\hat{\Delta}, h, g) = e^{-\frac{1}{2} \sum_{a,b=1}^{n} \left[\Delta_{ab} - \frac{g^2}{2}(\gamma_a - \gamma_b)^2 \right] \frac{\partial^2}{\partial h_a \partial h_b}} e^{-\sum_{a=1}^{n} \beta_a \bar{v}(h_a - \eta_a)} \Big|_{h_a = h},$$

where $\gamma_1 = 0$, $\eta_1 = 0$, and $\gamma_a = \gamma$, $\eta_a = \eta$, $\forall a \geq 2$, as in Chapter 4. The addition of a shear strain γ thus only modifies the interaction term by an additional Gaussian integration.

As discussed in Chapter 4, the matrix $\hat{\Delta}$ should be determined by extremising the free energy. Once the solution for $\hat{\Delta}$ is found, the physical observables can thus be obtained by taking the appropriate derivatives of the free energy. For example, the energy and pressure are still given by the derivatives with respect to temperature and density, respectively. The shear stress is defined through response theory [366, 368] as

$$\sigma(\varphi, T, \gamma; \varphi_g, T_g) = \frac{\partial f_g(\varphi, T, \gamma; \varphi_g, T_g)}{\partial \gamma} = d \, T \, \widehat{\sigma}(\varphi, T, \gamma; \varphi_g, T_g). \tag{10.15}$$

Because the glass free energy is proportional to d, so is the stress; hence, as for pressure, we define an adimensional scaled stress $\widehat{\sigma} = \beta\sigma/d$ that remains finite when $d \to \infty$. Moreover, because, for hard spheres, βf_g is independent of temperature, so is $\widehat{\sigma}$. Eq. (10.15) thus allows one to obtain the stress–strain curve of the glass within the state following approach.

Note that if all the replicas, including the reference replica $a = 1$, are subjected to the same shear strain, $\gamma_a = \gamma$, $\forall a$, then the free energy in Eq. (10.14) becomes independent of γ. This result is correct, because a molecular liquid is globally a liquid and its free energy should not depend on the shape of its container. Consistently, according to Eq. (10.15), the shear stress then vanishes – i.e., $\sigma = 0$. Recall that the state following construction keeps the reference replica unstrained, $\gamma_1 = 0$, and applies the strain on the constrained replicas, $\gamma_a \neq 0$. This produces a nontrivial result because replicas $a > 1$ are not liquid but, rather, constrained to remain within the glass state selected by the first replica. Note that within the Monasson construction discussed in Chapter 7, in which all replicas are equivalent, extracting the rheological properties of the glass requires a slightly different procedure, as discussed in [366, 368].

As in Chapter 4, $\hat{\Delta}$ is assumed to be a hierarchical matrix, the simplest choice being the replica symmetric ansatz. Its stability can be checked along the same lines as in Chapter 6 [59, 298, 299]. Using the replica symmetric ansatz in Eq. (4.62) for the matrix $\hat{\Delta}$, the RS free energy is

$$-\beta f_g = \frac{d}{2} \log\left(\frac{\pi e \Delta}{d^2}\right) + \frac{d}{2} \frac{2\Delta_r - \Delta}{\Delta}$$

$$+ \frac{d\widehat{\varphi}_g}{2} \int dh \, e^h \, q_\gamma(2\Delta_r - \Delta, \beta_g; h) \log q \, (\Delta, \beta, h - \eta),$$

(10.16)

where

$$q_\gamma(\Delta, \beta; h) = \int \mathcal{D}g \, q(\Delta + \gamma^2 g^2, \beta; h).$$ (10.17)

Note that for $\gamma = 0$, $q_\gamma(\Delta, \beta; h) = q(\Delta, \beta; h)$ and the replica symmetric expression without shear, Eq. (4.75), is recovered. The parameters Δ and Δ_r are fixed by extremising Eq. (10.16), and the resulting equations are identical to Eq. (4.76), with the replacement $q(2\Delta_r - \Delta, \beta_g; h) \rightarrow q_\gamma(2\Delta_r - \Delta, \beta_g; h)$. The solution to these equations can thus be found by starting from equilibrium at $\beta = \beta_g$ and $\eta = \gamma = 0$, where $\Delta = \Delta_r$, and then following the solution upon changing β, η and γ.

10.2 Linear Response

In the rest of this chapter, we discuss a selection of results for the hard-sphere potential. Before discussing the full stress–strain curves, we focus on the linear response regime in the small strain limit. We consider two observables: the shear modulus, which characterises the rigidity of glasses, and the dilatancy, which describes the propensity of glasses to expand upon constant-pressure straining.

10.2.1 Shear Modulus

Because the free energy, Eq. (10.14), is an even function of γ, the stress vanishes at zero strain. Physically, this is obvious because the glass is then in mechanical equilibrium. At very small strain, elasticity theory assumes that the stress increases linearly with γ, with a proportionality constant given by the shear modulus,

$$\mu = \left.\frac{d\sigma}{d\gamma}\right|_{\gamma=0} = dT\widehat{\mu}, \qquad \widehat{\mu} = \left.\frac{d\widehat{\sigma}}{d\gamma}\right|_{\gamma=0}.$$ (10.18)

The higher the shear modulus is, the more rigid the glass is, because its elastic response is stronger. Note that it is convenient to define a scaled modulus $\widehat{\mu}$ that remains finite for $d \rightarrow \infty$ and is independent of temperature for hard spheres. In the small strain limit, within the replica symmetric solution, one has [299, 369]

$$\widehat{\mu} = \frac{1}{\Delta},$$ (10.19)

where Δ is evaluated at $\gamma = 0$. The scaled shear modulus is then simply the inverse of the mean square displacement plateau. This linear response result connects the

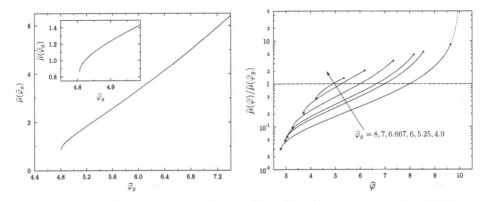

Figure 10.2 Shear modulus of the hard-sphere glass [299]. (Left) Shear modulus $\widehat{\mu}(\widehat{\varphi}_g)$ on the equilibrium line. For $\widehat{\varphi}_g > \widehat{\varphi}_d$, the shear modulus is finite. It displays a square root singularity before jumping to zero for $\widehat{\varphi} \to \widehat{\varphi}_d^+$, Eq. (10.20), as better shown in the inset. (Right) Shear modulus for glasses prepared at different $\widehat{\varphi}_g$, as a function of $\widehat{\varphi}$. Upon decompression, the shear modulus jumps to zero at melting (open squares, as in Figure 4.4), with a square root singularity inherited from the square root singularity of the mean square displacement. Upon compression, the RS solution becomes unstable at the Gardner transition (triangles, as in Figure 6.1), beyond which a fullRSB description is needed. The unstable continuation of the RS shear modulus diverges at jamming (dashed line, shown only for $\widehat{\varphi}_g = 8$).

linear rheology of the glass to its unstrained properties. It is consistent, moreover, with physical intuition: a glass with a smaller cage is more rigid.

We have seen in Chapter 4 that, at the dynamical transition, Δ jumps from infinity in the liquid phase to a finite value in the dynamically arrested phase. Particle cages are then infinitely long lived, and the system suddenly becomes rigid for an infinite time. Consequently, the shear modulus also jumps from zero (in the liquid) to a finite value at the dynamical transition, as shown in Figure 10.2. This observation has a precise dynamical interpretation. Upon approaching the dynamical glass transition from the liquid phase, $\widehat{\varphi} \to \widehat{\varphi}_d$, the time scale t_p – at which the dynamical mean square displacement shows a plateau before reaching the diffusive regime – diverges as a power law; see Section 3.4.2. On times $t \lesssim t_p$, the system is effectively frozen into an amorphous state; it is then able to sustain a shear strain on this time scale. In other words, if a shear strain is applied at $t = 0$, the shear stress $\sigma(t)$ would only decay to zero at times $t \gg t_p$. Only at the dynamical glass transition can the system sustain a finite stress for arbitrarily long times. While rigidity at infinite times emerges discontinuously, the time over which the system behaves as a solid diverges continuously.[3] Upon approaching

---

[3] We note that this effect is a $d \to \infty$ artefact. In any finite d, the dynamical glass transition becomes a smooth crossover, and a finite shear stress over an infinite time can only be sustained when the system is frozen in the ideal glass state – i.e., beyond the Kauzmann point. Yet, because experiments are always performed over finite time scales, the system becomes effectively solid when t_p is larger than the experimentally accessible time scales [80, 120].

the dynamical transition from the dynamically arrested phase, the shear modulus displays a square root singularity before jumping to zero,

$$\widehat{\mu} \sim \widehat{\mu}_d + C(\widehat{\varphi} - \widehat{\varphi}_d)^{1/2}, \qquad \widehat{\varphi} \to \widehat{\varphi}_d^+, \tag{10.20}$$

where $\widehat{\mu}_d = 1/\Delta(\widehat{\varphi}_d)$ is the shear modulus at the dynamical glass transition.

One can also consider the shear modulus of a glass prepared at $\widehat{\varphi}_g$ and followed adiabatically at $\widehat{\varphi}$. Upon decompression, the shear modulus decreases and also displays a square root singularity before jumping to zero at the melting spinodal of the glass. Upon compression, the shear modulus increases. It diverges at the jamming transition, where $\Delta \to 0$, because a hard-sphere system forms an infinitely rigid contact network and cannot be deformed anymore. The replica symmetric approximation gives $\Delta \sim \widehat{p}^{-1}$, and $\widehat{\mu} \sim \widehat{p}$ thus diverges upon approaching the jamming point, but we will see in Section 10.2.2 that this scaling is modified by fullRSB effects. Results for different $\widehat{\varphi}_g$ are illustrated in Figure 10.2.

10.2.2 Hierarchy of Shear Moduli in the Gardner Phase

We now discuss the behaviour of the shear modulus in the Gardner phase, in which replica symmetry is broken. Although this chapter only presents results for hard spheres, the discussion of this section applies to any interaction potential. As we discussed in Chapter 5, the phase space structure in the Gardner phase is ultrametrically organised, with individual glass states grouped in sub-basins, themselves grouped in bigger sub-basins, up to the largest metabasins, as described by a function $\Delta(x)$, or $x(\Delta)$. The linear response of the system in equilibrium can be probed by considering an initially equilibrated system, applying an infinitesimal shear strain γ at time $t = 0$ and monitoring the shear stress $\widehat{\sigma}(t)$. This perturbation takes a long time to relax because, in the Gardner phase, the ultrametric structure of glass states is associated to an infinite hierarchy of time scales [93, 112]. After a short transient, the system equilibrates in the individual glass state in which it was initially prepared, the mean square displacement approaches Δ_M and, correspondingly, the stress decays to a value $\widehat{\sigma}(t) \sim \widehat{\mu}(\Delta_M)\gamma$. The shear modulus $\widehat{\mu}(\Delta_M)$ corresponding to the intra-state relaxation is then given by

$$\widehat{\mu}(\Delta_M) = \frac{1}{\Delta_M} = \frac{1}{\lambda(\Delta_M)}, \tag{10.21}$$

which generalises Eq. (10.19) to the Gardner phase [369]. Waiting longer times (that diverge with the system size), the system is able to leave the initial glass state to find other states, thus exploring a wider portion of the ultrametric tree of states. Correspondingly, the mean square displacement increases to values $\Delta > \Delta_M$. The value of Δ and its associated index $x(\Delta)$ label the level of the hierarchy. One can associate to it a shear modulus $\widehat{\mu}(\Delta) = 1/\lambda(\Delta)$, where $\lambda(\Delta)$ is defined in Eq. (9.47) [369], which gives the stress response on the time scale over which values of $\Delta > \Delta_M$ are explored. A full exploration of the glass metabasin then gives a shear modulus $\widehat{\mu}(\Delta_m) = 1/\lambda(\Delta_m)$ [369]. Note that this idealised situation corresponds to an equilibrium exploration of the glass metabasin. In reality, the system explores the metabasin out of equilibrium, and establishing a correspondence between $\widehat{\mu}(\Delta)$ and the dynamical time scales is tricky [112, 192, 369]. Note also that this distribution of shear moduli provides a clear physical meaning to the function $\lambda(\Delta)$, similarly to the distribution of linear magnetic susceptibilities in spin glasses [254].

For hard spheres, upon approaching jamming one has

$$\Delta_M \sim \widehat{p}^{-\kappa} \qquad \Rightarrow \qquad \widehat{\mu}(\Delta_M) \sim \widehat{p}^{\kappa}, \tag{10.22}$$

as determined by the scaling solution discussed in Chapter 9. The divergence of the intra-state shear modulus is thus characterised by the exponent κ. Conversely, for any finite $\Delta > \Delta_M$, one has $\lambda(\Delta) = \widehat{\lambda}(\Delta)/\widehat{p}$, from Eq. (9.51), and the corresponding shear modulus $\widehat{\mu}(\Delta) = \widehat{p}/\widehat{\lambda}(\Delta)$ diverges proportionally to the reduced pressure $\widehat{p}$ [192, 369].

10.2.3 Dilatancy

Pressure also exhibits an interesting behaviour upon straining the system. Expanding the RS free energy of the strained glass for $\gamma \to 0$,

$$f_g(\eta, \gamma) \simeq f_g(\eta) + \frac{1}{2}\mu(\eta)\gamma^2 + O(\gamma^4), \tag{10.23}$$

and recalling that $\beta\mu(\eta)/d = 1/\Delta(\eta)$, Eq. (6.17) gives

$$p_g(\eta, \gamma) = \frac{\partial(\beta f_g(\eta, \gamma))}{\partial\eta} = p_g(\eta, \gamma = 0) + \frac{\beta R(\eta)}{\rho}\gamma^2 + O(\gamma^4), \tag{10.24}$$

where the coefficient

$$\frac{\beta R(\eta)}{\rho} = \frac{d}{2}\frac{d}{d\eta}\frac{1}{\Delta(\eta)} = -\frac{d}{2}\frac{1}{\Delta(\eta)^2}\frac{d\Delta}{d\eta} \tag{10.25}$$

is the dilatancy. A positive dilatancy indicates that the system, kept at fixed pressure, expands under strain. In hard-sphere glasses, $\Delta(\eta)$ decreases upon increasing η, as shown in Chapter 4; hence, the dilatancy is always positive, as shown in Figure 10.3. Note that the dilatancy diverges at melting because $\Delta(\eta)$ has a square root singularity, and at jamming because $\Delta(\eta) \to 0$ while $d\Delta/d\eta$ remains finite.

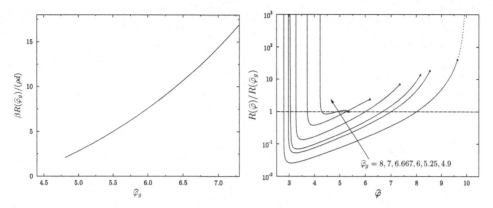

Figure 10.3 Dilatancy of the hard-sphere glass [299]. (Left) Dilatancy $R(\widehat{\varphi}_g)$ on the equilibrium line. For $\widehat{\varphi}_g > \widehat{\varphi}_d$, the dilatancy is finite. It scales linearly before jumping down to zero for $\widehat{\varphi} \to \widehat{\varphi}_d^+$. (Right) Dilatancy of glasses prepared at different $\widehat{\varphi}_g$, as a function of $\widehat{\varphi}$. Upon decompression, the dilatancy diverges at melting, because of the square root singularity of Δ. Upon compression, the RS solution becomes unstable at the Gardner transition (triangles, as in Figure 6.1), beyond which a fullRSB solution is needed. The unstable continuation of the RS dilatancy diverges proportionally to $1/\Delta^2$ at jamming (dashed line, shown only for $\widehat{\varphi}_g = 8$).

10.3 Stress–Strain Curves

Solving the equations for Δ and Δ_r as a function of γ provides the full stress–strain curve beyond the small strain, linear response regime. These curves correspond to an adiabatic (or 'quasi-static') following of the glass state under the applied strain. In this section, we discuss the behaviour of these curves for hard-sphere glasses in different regimes.

10.3.1 The Yielding Transition

Figure 10.4 shows the stress–strain curves of hard-sphere glasses prepared in equilibrium at $\widehat{\varphi} = \widehat{\varphi}_g$ and subjected to a shear strain γ at constant density. In agreement with Figure 10.2, more stable glasses at higher $\widehat{\varphi}_g$ have a larger shear modulus in the linear regime, at small γ. Upon increasing γ, all glasses undergo a Gardner transition, at which replica symmetry spontaneously breaks due to the applied strain. Recall that beyond the Gardner transition, the replica symmetric solution is only approximate, and a RSB calculation is needed. See [298] for a preliminary study. Here, we limit our discussion to the RS results. At high enough shear strain, after the stress reaches a maximum, a spinodal transition takes place, at which the solution for Δ_r and Δ is lost. The spinodal point $\gamma_Y(\widehat{\varphi}_g)$ then corresponds, within mean field theory, to the yielding point of the glass, i.e., where it breaks.

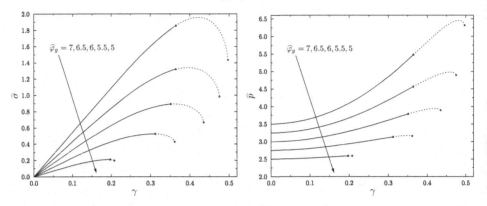

Figure 10.4 Stress–strain (left) and pressure–strain (right) curves for hard-sphere glasses prepared in equilibrium at $\widehat{\varphi}_g$ and strained at constant density. Both stress and pressure increase with strain, and a Gardner transition (triangle) is observed before the stress and pressure overshoot. At the yielding transition (diamond), the solution for Δ and Δ_r is lost via a spinodal mechanism in the Franz–Parisi potential, indicating the breakdown of the glass.

Note that upon increasing the preparation density $\widehat{\varphi}_g$, the yielding point $\gamma_Y(\widehat{\varphi}_g)$ increases and, before yielding, the stress–strain curves display a more pronounced stress overshoot. In Figure 10.4 the pressure as a function of the strain is also reported. As predicted by Eq. (10.25), pressure increases quadratically in γ, and the dilatancy is larger for more stable glasses.

At the yielding transition, both Δ and Δ_r display a square root singularity, indicating that the local minimum of the Franz–Parisi potential becomes an inflection point and then disappears, as it does upon approaching the dynamical transition. One can then consider the fluctuations of Δ_r, which define a susceptibility

$$\chi_r = N\left[\langle \Delta_r^2 \rangle - \langle \Delta_r \rangle^2\right]. \tag{10.26}$$

Because χ_r is related to the curvature at the local minimum of the Franz–Parisi potential, it diverges at the yielding point. The yielding transition in large dimension is thus a critical spinodal with disorder [293, 299].

10.3.2 Stability Map under Shear Strain and Compression

In Section 10.3.1, we discussed the behaviour of the stress–strain curves for glasses prepared at $\widehat{\varphi}_g$ and strained at constant density. More generally, one can consider following a glass state under a joint applied (de)compression (i.e., a compressive strain) and shear strain. In other words, one can prepare a glass at $(\widehat{\varphi}_g, \gamma = 0)$ and then follow it in the plane $(\widehat{\varphi}, \gamma)$. In the rest of this section, we consider $\widehat{\varphi}_g$ as fixed; hence, we do not indicate the dependence on $\widehat{\varphi}_g$ explicitly.

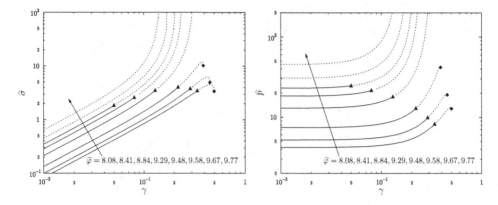

Figure 10.5 Stress–strain (left) and pressure–strain (right) curves for hard-sphere glasses prepared in equilibrium at $\widehat{\varphi}_g = 8$, compressed at $\widehat{\varphi} > \widehat{\varphi}_g$ and strained at constant density [346]. At low $\widehat{\varphi}$, both stress and pressure increase with strain, and a Gardner transition (triangle) is observed at $\gamma_G(\widehat{\varphi})$ before the stress and pressure overshoot, and the glass yields at $\gamma_Y(\widehat{\varphi})$ (diamond). At higher $\widehat{\varphi}$, a Gardner transition is observed, but instead of yielding, the stress and pressure diverge at a finite $\gamma_j(\widehat{\varphi})$, which indicates shear jamming. At even higher $\widehat{\varphi} > \widehat{\varphi}_G(\gamma = 0)$, the whole curve is unstable towards RSB.

In Figure 10.5, stress–strain and pressure–strain curves for a glass compressed at $\widehat{\varphi} > \widehat{\varphi}_g$, and then strained, are reported. Both the shear modulus and the dilatancy increase with $\widehat{\varphi}$, resulting in a steeper increase of stress and pressure with γ. The yielding strain $\gamma_Y(\widehat{\varphi})$, however, decreases upon increasing $\widehat{\varphi}$. For high enough $\widehat{\varphi}$, a qualitative difference is found in the stress–strain curves: the yielding instability disappears, and the stress and pressure diverge at a finite strain $\gamma_j(\widehat{\varphi})$, indicating that the system jams under strain.

The yielding instability at $\gamma_Y(\widehat{\varphi})$ and the jamming transition at $\gamma_j(\widehat{\varphi})$ thus delimit a region in which a stable glass exists under strain. Moreover, a Gardner transition line $\gamma_G(\widehat{\varphi})$ is also observed under strain. These three lines are plotted in Figure 10.6, resulting in a 'stability map' [8, 193, 346] of the glass in the $(\widehat{\varphi}, \gamma)$ plane. At zero strain, the stability region is delimited by the melting point in decompression and by the jamming point $\widehat{\varphi}_j$ in compression. At non-zero strain, the yielding line extends from the melting point at low $\widehat{\varphi}$, while the jamming line extends from $\widehat{\varphi}_j$ at high density. The shear jamming and shear yielding lines meet at a critical point $(\widehat{\varphi}_c, \gamma_c)$, at which the system yields at extremely high (divergent) stress.

Note that the Gardner transition line in Figure 10.6 extends from the zero-strain Gardner transition and crosses the yielding line at a lower density. Hence, the shear jamming line is fully contained within the RSB region. Shear jamming is thus described by the same fullRSB scaling solution described in Chapter 9. It is possible to show that the exponents γ, θ, κ are universal on the shear jamming

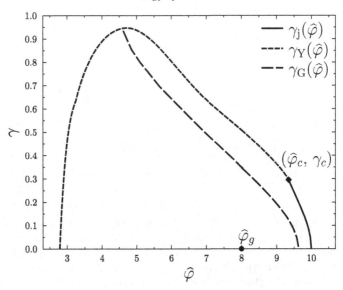

Figure 10.6 Stability map of a hard-sphere glass prepared at $\widehat{\varphi}_g = 8$, obtained by plotting in the $(\widehat{\varphi}, \gamma)$ plane the lines $\gamma_j(\widehat{\varphi})$, $\gamma_Y(\widehat{\varphi})$ and $\gamma_G(\widehat{\varphi})$ defined in Figure 10.5 [8, 346]. The preparation point $(\widehat{\varphi}_g, 0)$ is indicated by a dot. At $\gamma = 0$, the glass exists for densities larger than the melting point and smaller than the jamming point. At high density, close to the jamming transition $\widehat{\varphi}_j$, the glass jams under shear due to dilatancy; at low density, the glass yields under shear. The shear jamming line and part of the yielding line lie within the RSB phase, $\gamma > \gamma_G(\widehat{\varphi})$. Under the RS approximation, the shear jamming and shear yielding lines meet at a critical point $(\widehat{\varphi}_c, \gamma_c)$ (diamond) where the glass yields at divergent shear stress.

line [298]. The critical point $(\widehat{\varphi}_c, \gamma_c)$ also falls within the RSB region, and because the Gardner line crosses the yielding line, yielding can be either RS (at low $\widehat{\varphi}$) or RSB (at high $\widehat{\varphi}$). Different critical properties around the spinodal are expected in these two regions [298].

10.4 Wrap-Up

10.4.1 Summary

In this chapter, we have seen that

- The state following formalism can be extended to analyse the behaviour of glasses subjected to shear strain, leading to a theory of the elasticity and rheology of amorphous solids (Section 10.1).
- At small shear strains, the glass responds elastically, and the theory predicts the shear modulus and the dilatancy. These quantities depend on the degree of annealing of the glass and on the underlying phase space organisation of glass states. In particular, in the Gardner phase, spontaneous replica symmetry breaking gives rise to a hierarchy of shear moduli (Section 10.2).

- Beyond the linear response regime, the stress–strain curves can be computed up to the yielding point, where the glass breaks under the applied strain. Within mean field theory, yielding is a critical spinodal of the Franz–Parisi potential. Depending on the glass preparation and on the straining protocol, a Gardner transition can be present before yielding (Section 10.3.1).
- When a hard-sphere glass prepared at sufficiently high pressure is sheared, it can undergo a shear jamming transition that prevents yielding. The yielding and shear jamming lines delimit the stability region of the solid in the (φ, γ) phase diagram. These lines meet at a critical point, where yielding happens with a divergent stress (Section 10.3.2).

10.4.2 Further Reading

We provide here a list of references that can be consulted to further explore the subjects discussed in this chapter, selected according to the criteria discussed in Section 1.6.2. Because the literature on the rheology of glasses is extremely vast and scattered among several disciplines (physics, materials science, engineering), we only provide here a list of references that are closely related to the material discussed in this chapter.

General introductory reviews to the rheology of amorphous materials are

- Berthier, *Yield stress, heterogeneities and activated processes in soft glassy materials* [38]
- Rodney, Tanguy and Vandembroucq, *Modeling the mechanics of amorphous solids at different length scale and time scale* [305]
- Bonn, Denn, Berthier et al., *Yield stress materials in soft condensed matter* [62]
- Nicolas, Ferrero, Martens et al., *Deformation and flow of amorphous solids: a review of mesoscale elastoplastic models* [272]

The criticality of yielding has been studied in

- Lin, Lerner, Rosso et al., *Scaling description of the yielding transition in soft amorphous solids at zero temperature* [231]
- Parisi, Procaccia, Rainone et al., *Shear bands as manifestation of a criticality in yielding amorphous solids* [293]
- Ozawa, Berthier, Biroli et al., *Random critical point separates brittle and ductile yielding transitions in amorphous materials* [280]

Beyond yielding, soft glasses such as pastes and emulsions break and start to flow. A dynamical investigation is then required. Mean field dynamical equations for the description of this flow regime have been developed in

- Berthier, Barrat and Kurchan, *A two-time-scale, two-temperature scenario for nonlinear rheology* [36]

- Fuchs and Cates, *Theory of nonlinear rheology and yielding of dense colloidal suspensions* [156]
- Brader, Voigtmann, Fuchs et al., *Glass rheology: from mode-coupling theory to a dynamical yield criterion* [67]

The relation between the Gardner transition, plasticity and avalanches has been discussed in

- Biroli and Urbani, *Breakdown of elasticity in amorphous solids* [58]
- Nakayama, Yoshino and Zamponi, *Protocol-dependent shear modulus of amorphous solids* [270]
- Franz and Spigler, *Mean-field avalanches in jammed spheres* [148]
- Jin and Yoshino, *Exploring the complex free-energy landscape of the simplest glass by rheology* [192]
- Jin, Urbani, Zamponi et al., *A stability-reversibility map unifies elasticity, plasticity, yielding and jamming in hard sphere glasses* [193]

An extension of the theory presented here to attractive colloids can be found in Altieri, Urbani and Zamponi, *Microscopic theory of two-step yielding in attractive colloids* [7].

References

[1] Adam, G., and Gibbs, J. 1965. On the temperature dependence of cooperative relaxation properties in glass-forming liquids. *The Journal of Chemical Physics*, **43**, 139.

[2] Adda-Bedia, M., Katzav, E., and Vella, D. 2008. Solution of the Percus–Yevick equation for hard hyperspheres in even dimensions. *The Journal of Chemical Physics*, **129**, 144506.

[3] Agoritsas, E., Maimbourg, T., and Zamponi, F. 2018. Out-of-equilibrium dynamical equations of infinite-dimensional particle systems. I. The isotropic case. *Journal of Physics A: Mathematical and Theoretical*, **52**, 144002.

[4] Aktekin, N. 1997. Simulation of the eight-dimensional Ising model on the Creutz cellular automaton. *International Journal of Modern Physics C*, **8**, 287.

[5] Alder, B. J., and Wainwright, T. E. 1957. Phase transition for a hard sphere system. *The Journal of Chemical Physics*, **27**, 1208.

[6] Alexander, S. 1998. Amorphous solids: Their structure, lattice dynamics and elasticity. *Physics Reports*, **296**, 65.

[7] Altieri, A., Urbani, P., and Zamponi, F. 2018. Microscopic theory of two-step yielding in attractive colloids. *Physical Review Letters*, **121**, 185503.

[8] Altieri, A., and Zamponi, F. 2019. Mean-field stability map of hard-sphere glasses. *Physical Review E*, **100**, 032140.

[9] Aluffi-Pentini, F., Parisi, V., and Zirilli, F. 1988. A global optimization algorithm using stochastic differential equations. *ACM Transactions on Mathematical Software*, **14**, 345.

[10] Amit, D. J. 1992. *Modeling brain function: The world of attractor neural networks*. Cambridge University Press.

[11] Amit, D. J., and Martin-Mayor, V. 2005. *Field theory, the renormalization group, and critical phenomena*. 3rd edn. World Scientific.

[12] Andreanov, A., and Scardicchio, A. 2012. Random perfect lattices and the sphere packing problem. *Physical Review E*, **86**, 041117.

[13] Angelani, L., and Foffi, G. 2007. Configurational entropy of hard spheres. *Journal of Physics: Condensed Matter*, **19**, 256207.

[14] Angelini, M. C., and Biroli, G. 2015. Spin glass in a field: A new zero-temperature fixed point in finite dimensions. *Physical Review Letters*, **114**, 095701.

[15] Angelini, M. C., Parisi, G., and Ricci-Tersenghi, F. 2013. Ensemble renormalization group for disordered systems. *Physical Review B*, **87**, 134201.

[16] Angell, C. 1997. Entropy and fragility in supercooling liquids. *Journal of research of the National Institute of Standards and Technology*, **102**, 171.

[17] Apostolico, A., Comin, M., Dress, A., et al. 2013. Ultrametric networks: A new tool for phylogenetic analysis. *Algorithms for Molecular Biology*, **8**, 7.

[18] Aron, C., Biroli, G., and Cugliandolo, L. F. 2010. Symmetries of generating functionals of Langevin processes with colored multiplicative noise. *Journal of Statistical Mechanics: Theory and Experiment*, **2010**, P11018.

[19] Arora, S., and Barak, B. 2009. *Computational complexity: A modern approach*. Cambridge University Press.

[20] Asenjo, D., Paillusson, F., and Frenkel, D. 2014. Numerical calculation of granular entropy. *Physical Review Letters*, **112**, 098002.

[21] Ashcroft, N. W., and Mermin, N. D. 1976. *Solid state physics*. Thomson Learning.

[22] Aspelmeier, T., Bray, A., and Moore, M. 2004. Complexity of Ising spin glasses. *Physical Review Letters*, **92**, 087203.

[23] Baity-Jesi, M., Baños, R., Cruz, A., et al. 2013. Critical parameters of the three-dimensional Ising spin glass. *Physical Review B*, **88**, 224416.

[24] Baity-Jesi, M., Goodrich, C. P., Liu, A. J., et al. 2017. Emergent SO(3) symmetry of the frictionless shear jamming transition. *Journal of Statistical Physics*, **167**, 735.

[25] Bannerman, M. N., Lue, L., and Woodcock, L. V. 2010. Thermodynamic pressures for hard spheres and closed-virial equation-of-state. *The Journal of Chemical Physics*, **132**, 084507.

[26] Banos, R. A., Cruz, A., Fernandez, L., et al. 2010. Nature of the spin-glass phase at experimental length scales. *Journal of Statistical Mechanics: Theory and Experiment*, **2010**, P06026.

[27] Barrat, A., Burioni, R., and Mézard, M. 1996. Dynamics within metastable states in a mean-field spin glass. *Journal of Physics A: Mathematical and General*, **29**, L81.

[28] Barrat, A., Franz, S., and Parisi, G. 1997. Temperature evolution and bifurcations of metastable states in mean-field spin glasses, with connections with structural glasses. *Journal of Physics A: Mathematical and General*, **30**, 5593.

[29] Barrat, A., Kurchan, J., Loreto, V., et al. 2001. Edwards measures: A thermodynamic construction for dense granular media and glasses. *Physical Review E*, **63**, 051301.

[30] Baule, A., Morone, F., Herrmann, H. J., et al. 2018. Edwards statistical mechanics for jammed granular matter. *Reviews of Modern Physics*, **90**, 015006.

[31] Behringer, R. P., and Chakraborty, B. 2019. The physics of jamming for granular materials: A review. *Reports on Progress in Physics*, **82**, 012601.

[32] Belletti, F., Cruz, A., Fernandez, L., et al. 2009. An in-depth view of the microscopic dynamics of Ising spin glasses at fixed temperature. *Journal of Statistical Physics*, **135**, 1121.

[33] Bernal, J., and Mason, J. 1960. Packing of spheres: Co-ordination of randomly packed spheres. *Nature*, **188**, 910.

[34] Bernal, J., Mason, J., and Knight, K. 1962. Radial distribution of the random close packing of equal spheres. *Nature*, **194**, 957.

[35] Bernard, E. P., and Krauth, W. 2011. Two-step melting in two dimensions: First-order liquid-hexatic transition. *Physical Review Letters*, **107**, 155704.

[36] Berthier, L., Barrat, J., and Kurchan, J. 2000. A two-time-scale, two-temperature scenario for nonlinear rheology. *Physical Review E*, **61**, 5464.

[37] Berthier, L., Biroli, G., Bouchaud, J.-P., et al. 2011. *Dynamical heterogeneities and glasses*. Oxford University Press.

[38] Berthier, L. 2003. Yield stress, heterogeneities and activated processes in soft glassy materials. *Journal of Physics: Condensed Matter*, **15**, S933.

[39] Berthier, L. 2013. Overlap fluctuations in glass-forming liquids. *Physical Review E*, **88**, 022313.

[40] Berthier, L., and Biroli, G. 2011. Theoretical perspective on the glass transition and amorphous materials. *Reviews of Modern Physics*, **83**, 587.

[41] Berthier, L., and Kurchan, J. 2013. Non-equilibrium glass transitions in driven and active matter. *Nature Physics*, **9**, 310.

[42] Berthier, L., and Witten, T. A. 2009. Glass transition of dense fluids of hard and compressible spheres. *Physical Review E*, **80**, 021502.

[43] Berthier, L., Jacquin, H., and Zamponi, F. 2011. Microscopic theory of the jamming transition of harmonic spheres. *Physical Review E*, **84**, 051103.

[44] Berthier, L., Charbonneau, P., Jin, Y., et al. 2016. Growing timescales and length-scales characterizing vibrations of amorphous solids. *Proceedings of the National Academy of Sciences*, **113**, 8397.

[45] Berthier, L., Charbonneau, P., Coslovich, D., et al. 2017. Configurational entropy measurements in extremely supercooled liquids that break the glass ceiling. *Proceedings of the National Academy of Sciences*, **114**, 11356.

[46] Berthier, L., Charbonneau, P., Ninarello, A., et al. 2019. Zero-temperature glass transition in two dimensions. *Nature communications*, **10**, 1508.

[47] Bhattacharyya, S. M., Bagchi, B., and Wolynes, P. G. 2008. Facilitation, complexity growth, mode coupling, and activated dynamics in supercooled liquids. *Proceedings of the National Academy of Sciences*, **105**, 16077.

[48] Bi, D., Henkes, S., Daniels, K. E., et al. 2015. The statistical physics of athermal materials. *Annual Reviews of Condensed Matter Physics*, **6**, 63.

[49] Biazzo, I., Caltagirone, F., Parisi, G., et al. 2009. Theory of amorphous packings of binary mixtures of hard spheres. *Physical Review Letters*, **102**, 195701.

[50] Biere, A., Heule, M., and van Maaren, H. (eds). 2009. *Handbook of satisfiability*. Frontiers in Artificial Intelligence and Applications, vol. 185. IOS Press.

[51] Binder, K., and Young, A. 1986. Spin glasses: Experimental facts, theoretical concepts, and open questions. *Reviews of Modern Physics*, **58**, 801.

[52] Binder, K. 1987. Theory of first-order phase transitions. *Reports on Progress in Physics*, **50**, 783.

[53] Binder, K., and Kob, W. 2011. *Glassy materials and disordered solids: An introduction to their statistical mechanics*. World Scientific.

[54] Binder, K., and Luijten, E. 2001. Monte Carlo tests of renormalization-group predictions for critical phenomena in Ising models. *Physics Reports*, **344**, 179.

[55] Binder, K., Block, B. J., Virnau, P., et al. 2012. Beyond the van der Waals loop: What can be learned from simulating Lennard-Jones fluids inside the region of phase coexistence. *American Journal of Physics*, **80**, 1099.

[56] Biroli, G., Bouchaud, J., Miyazaki, K., et al. 2006. Inhomogeneous mode-coupling theory and growing dynamic length in supercooled liquids. *Physical Review Letters*, **97**, 195701.

[57] Biroli, G., Bouchaud, J., Cavagna, A., et al. 2008. Thermodynamic signature of growing amorphous order in glass-forming liquids. *Nature Physics*, **4**, 771.

[58] Biroli, G., and Urbani, P. 2016. Breakdown of elasticity in amorphous solids. *Nature Physics*, **12**, 1130.

[59] Biroli, G., and Urbani, P. 2018. Liu-Nagel phase diagrams in infinite dimension. *SciPost Physics*, **4**, 020.

[60] Blichfeldt, H. F. 1929. The minimum value of quadratic forms, and the closest packing of spheres. *Mathematische Annalen*, **101**, 605.

[61] Bolhuis, P. G., Frenkel, D., Mau, S.-C., et al. 1997. Entropy difference between crystal phases. *Nature*, **388**, 235.

[62] Bonn, D., Denn, M. M., Berthier, L., et al. 2017. Yield stress materials in soft condensed matter. *Reviews of Modern Physics*, **89**, 035005.

[63] Bouchaud, J.-P., and Biroli, G. 2004. On the Adam-Gibbs-Kirkpatrick-Thirumalai-Wolynes scenario for the viscosity increase in glasses. *The Journal of Chemical Physics*, **121**, 7347.

[64] Bouchaud, J.-P., and Biroli, G. 2005. Nonlinear susceptibility in glassy systems: A probe for cooperative dynamical length scales. *Physical Review B*, **72**, 064204.

[65] Bouchaud, J., Cugliandolo, L., Kurchan, J., et al. 1998. Out of equilibrium dynamics in spin-glasses and other glassy systems. In Young, A. (ed), *Spin glasses and random fields*. World Scientific.

[66] Bowles, R. K., and Ashwin, S. 2011. Edwards entropy and compactivity in a model of granular matter. *Physical Review E*, **83**, 031302.

[67] Brader, J. M., Voigtmann, T., Fuchs, M., et al. 2009. Glass rheology: From mode-coupling theory to a dynamical yield criterion. *Proceedings of the National Academy of Sciences*, **106**, 15186.

[68] Bray, A. J., and Moore, M. A. 1987. Chaotic nature of the spin-glass phase. *Physical Review Letters*, **58**, 57.

[69] Brézin, E. 2010. *Introduction to statistical field theory*. Cambridge University Press.

[70] Brito, C., Ikeda, H., Urbani, P., et al. 2018. Universality of jamming of nonspherical particles. *Proceedings of the National Academy of Sciences*, **115**, 11736.

[71] Butera, P., and Pernici, M. 2012. High-temperature expansions of the higher susceptibilities for the Ising model in general dimension d. *Physical Review E*, **86**, 011139.

[72] Caltagirone, F., Ferrari, U., Leuzzi, L., et al. 2012. Critical slowing down exponents of mode coupling theory. *Physical Review Letters*, **108**, 085702.

[73] Cammarota, C., Cavagna, A., Giardina, I., et al. 2010. Phase-separation perspective on dynamic heterogeneities in glass-forming liquids. *Physical Review Letters*, **105**, 055703.

[74] Campa, A., Dauxois, T., and Ruffo, S. 2009. Statistical mechanics and dynamics of solvable models with long-range interactions. *Physics Reports*, **480**, 57.

[75] Capaccioli, S., Ruocco, G., and Zamponi, F. 2008. Dynamically correlated regions and configurational entropy in supercooled liquids. *The Journal of Physical Chemistry B*, **112**, 10652.

[76] Cardenas, M., Franz, S., and Parisi, G. 1998. Glass transition and effective potential in the hypernetted chain approximation. *Journal of Physics A: Mathematical and General*, **31**, L163.

[77] Cardenas, M., Franz, S., and Parisi, G. 1999. Constrained Boltzmann–Gibbs measures and effective potential for glasses in hypernetted chain approximation and numerical simulations. *The Journal of Chemical Physics*, **110**, 1726.

[78] Castellana, M., and Parisi, G. 2015. Non-perturbative effects in spin glasses. *Scientific Reports*, **5**, 8697.

[79] Castellani, T., and Cavagna, A. 2005. Spin-glass theory for pedestrians. *Journal of Statistical Mechanics: Theory and Experiment*, **2005**, P05012.

[80] Cavagna, A. 2009. Supercooled liquids for pedestrians. *Physics Reports*, **476**, 51.

[81] Cavagna, A., Giardina, I., and Parisi, G. 1998. Stationary points of the Thouless-Anderson-Palmer free energy. *Physical Review B*, **57**, 11251.

[82] Chakraborty, B. 2010. Statistical ensemble approach to stress transmission in granular packings. *Soft Matter*, **6**, 2884.

[83] Charbonneau, B., Charbonneau, P., Jin, Y., et al. 2013. Dimensional dependence of the Stokes–Einstein relation and its violation. *The Journal of Chemical Physics*, **139**, 164502.

[84] Charbonneau, B., Charbonneau, P., and Szamel, G. 2018. A microscopic model of the Stokes–Einstein relation in arbitrary dimension. *The Journal of Chemical Physics*, **148**, 224503.

[85] Charbonneau, P., Ikeda, A., Parisi, G., et al. 2012. Dimensional study of the caging order parameter at the glass transition. *Proceedings of the National Academy of Sciences*, **109**, 13939.

[86] Charbonneau, P., Ikeda, A., Parisi, G., et al. 2011. Glass transition and random close packing above three dimensions. *Physical Review Letters*, **107**, 185702.

[87] Charbonneau, P., Corwin, E. I., Parisi, G., et al. 2012. Universal microstructure and mechanical stability of jammed packings. *Physical Review Letters*, **109**, 205501.

[88] Charbonneau, P., Kurchan, J., Parisi, G., et al. 2014. Exact theory of dense amorphous hard spheres in high dimension. III. The full replica symmetry breaking solution. *Journal of Statistical Mechanics: Theory and Experiment*, **2014**, P10009.

[89] Charbonneau, P., Kurchan, J., Parisi, G., et al. 2014. Fractal free energies in structural glasses. *Nature Communications*, **5**, 3725.

[90] Charbonneau, P., Jin, Y., Parisi, G., et al. 2014. Hopping and the Stokes–Einstein relation breakdown in simple glass formers. *Proceedings of the National Academy of Sciences*, **111**, 15025.

[91] Charbonneau, P., Corwin, E. I., Parisi, G., et al. 2015. Jamming criticality revealed by removing localized buckling excitations. *Physical Review Letters*, **114**, 125504.

[92] Charbonneau, P., Corwin, E. I., Parisi, G., et al. 2016. Universal non-Debye scaling in the density of states of amorphous solids. *Physical Review Letters*, **117**, 045503.

[93] Charbonneau, P., Kurchan, J., Parisi, G., et al. 2017. Glass and jamming transitions: From exact results to finite-dimensional descriptions. *Annual Review of Condensed Matter Physics*, **8**, 265.

[94] Charbonneau, P., Corwin, E. I., Fu, L., et al. 2019. Glassy, Gardner-like phenomenology in minimally polydisperse crystalline systems. *Physical Review E*, **99**, 020901(R).

[95] Charbonneau, P., Hu, Y., Raju, A., et al. 2019. Morphology of renormalization-group flow for the de Almeida-Thouless-Gardner universality class. *Physical Review E*, **99**, 022132.

[96] Clisby, N., and McCoy, B. M. 2006. Ninth and tenth order virial coefficients for hard spheres in D dimensions. *Journal of Statistical Physics*, **122**, 15.

[97] Cohn, H. 2002. New upper bounds on sphere packings II. *Geometry & Topology*, **6**, 329.

[98] Cohn, H. 2016. Packing, coding, and ground states. *arXiv:1603.05202*.

[99] Cohn, H. 2017. A conceptual breakthrough in sphere packing. *Notices of the American Mathematical Society*, **64**, 102.

[100] Cohn, H., and Elkies, N. 2003. New upper bounds on sphere packings I. *Annals of Mathematics*, **157**, 689.

[101] Cohn, H., Kumar, A., Miller, S. D., et al. 2017. The sphere packing problem in dimension 24. *Annals of Mathematics*, **185**, 1017.

[102] Coluzzi, B., Mézard, M., Parisi, G., et al. 1999. Thermodynamics of binary mixture glasses. *The Journal of Chemical Physics*, **111**, 9039.

[103] Conway, J. H., and Sloane, N. J. A. 1993. *Sphere packings, lattices and groups*. Spriger-Verlag.

[104] Conway, J. B. 1990. *A course in functional analysis*. Springer-Verlag.

[105] Costigliola, L., Schroder, T. B., and Dyre, J. C. 2016. Studies of the Lennard–Jones fluid in 2, 3, and 4 dimensions highlight the need for a liquid-state $1/d$ expansion. *The Journal of Chemical Physics*, **144**, 231101.

[106] Crisanti, A., and De Dominicis, C. 2015. Replica Fourier transform: Properties and applications. *Nuclear Physics B*, **891**, 73.

[107] Crisanti, A., and Rizzo, T. 2002. Analysis of the ∞-replica symmetry breaking solution of the Sherrington-Kirkpatrick model. *Physical Review E*, **65**, 046137.

[108] Crisanti, A., and Leuzzi, L. 2006. Spherical $2 + p$ spin-glass model: An analytically solvable model with a glass-to-glass transition. *Physical Review B*, **73**, 014412.

[109] Cugliandolo, L. F., and Kurchan, J. 1993. Analytical solution of the off-equilibrium dynamics of a long-range spin-glass model. *Physical Review Letters*, **71**, 173.

[110] Cugliandolo, L. F., Kurchan, J., and Peliti, L. 1997. Energy flow, partial equilibration, and effective temperatures in systems with slow dynamics. *Physical Review E*, **55**, 3898.

[111] Cugliandolo, L. F. 2003. Dynamics of glassy systems. In Barrat, J., Feigelman, M., Kurchan, J., et al. (eds), *Slow relaxations and nonequilibrium dynamics in condensed matter*. Springer-Verlag.

[112] Cugliandolo, L. F., and Kurchan, J. 1994. On the out-of-equilibrium relaxation of the Sherrington-Kirkpatrick model. *Journal of Physics A: Mathematical and General*, **27**, 5749.

[113] de Almeida, J., and Thouless, D. 1978. Stability of the Sherrington-Kirkpatrick solution of a spin glass model. *Journal of Physics A: Mathematical and General*, **11**, 983.

[114] De Dominicis, C. 1978. Dynamics as a substitute for replicas in systems with quenched random impurities. *Physical Review B*, **18**, 4913.

[115] De Dominicis, C., and Giardina, I. 2006. *Random fields and spin glasses: A field theory approach*. Cambridge University Press.

[116] De Dominicis, C., Carlucci, D., and Temesvari, T. 1997. Replica Fourier tansforms on ultrametric trees, and block-diagonalizing multi-replica matrices. *Journal de Physique I*, **7**, 105.

[117] De Dominicis, C., Temesvari, T., and Kondor, I. 1998. On Ward-Takahashi identities for the Parisi spin glass. *Journal de Physique IV*, **8**, Pr6.13.

[118] De Dominicis, C. 1962. Variational formulations of equilibrium statistical mechanics. *Journal of Mathematical Physics*, **3**, 983.

[119] Debenedetti, P. G., and Stillinger, F. H. 2001. Supercooled liquids and the glass transition. *Nature*, **410**, 259.

[120] Debenedetti, P. 1996. *Metastable liquids: Concepts and principles*. Princeton University Press.

[121] DeGiuli, E., Laversanne-Finot, A., Düring, G., et al. 2014. Effects of coordination and pressure on sound attenuation, boson peak and elasticity in amorphous solids. *Soft Matter*, **10**, 5628.

[122] DeGiuli, E., Lerner, E., Brito, C., et al. 2014. Force distribution affects vibrational properties in hard-sphere glasses. *Proceedings of the National Academy of Sciences*, **111**, 17054.

[123] DeGiuli, E., Lerner, E., and Wyart, M. 2015. Theory of the jamming transition at finite temperature. *The Journal of Chemical Physics*, **142**, 164503.

[124] Delamotte, B. 2012. An introduction to the nonperturbative renormalization group. In Polonyi, J., and Schwenk, A. (eds), *Renormalization group and effective field theory approaches to many-body systems*. Springer-Verlag.

[125] Derrida, B. 1981. Random-energy model: An exactly solvable model of disordered systems. *Physical Review B*, **24**, 2613.

[126] DiVincenzo, D. P., and Steinhardt, P. J. 1999. *Quasicrystals: The state of the art.* World Scientific.

[127] Donev, A., Cisse, I., Sachs, D., et al. 2004. Improving the density of jammed disordered packings using ellipsoids. *Science*, **303**, 990.

[128] Donev, A., Torquato, S., and Stillinger, F. H. 2005. Pair correlation function characteristics of nearly jammed disordered and ordered hard-sphere packings. *Physical Review E*, **71**, 011105.

[129] Donev, A., Connelly, R., Stillinger, F. H., et al. 2007. Underconstrained jammed packings of nonspherical hard particles: Ellipses and ellipsoids. *Physical Review E*, **75**, 051304.

[130] Donth, E. 2001. *The glass transition: Relaxation dynamics in liquids and disordered materials.* Springer-Verlag.

[131] Dotsenko, V., Franz, S., and Mézard, M. 1994. Partial annealing and overfrustration in disordered systems. *Journal of Physics A: Mathematical and General*, **27**, 2351.

[132] Drouffe, J.-M., Parisi, G., and Sourlas, N. 1979. Strong coupling phase in lattice gauge theories at large dimension. *Nuclear Physics B*, **161**, 397.

[133] Duplantier, B. 1981. Comment on Parisi's equation for the SK model for spin glasses. *Journal of Physics A: Mathematical and General*, **14**, 283.

[134] Dyre, J. C. 2016. Simple liquids quasiuniversality and the hard-sphere paradigm. *Journal of Physics: Condensed Matter*, **28**, 323001.

[135] Ediger, M. D. 2000. Spatially heterogeneous dynamics in supercooled liquids. *Annual Review of Physical Chemistry*, **51**, 99.

[136] Edwards, S. F., and Anderson, P. W. 1975. Theory of spin glasses. *Journal of Physics F: Metal Physics*, **5**, 965.

[137] El-Showk, S., Paulos, M. F., Poland, D., et al. 2014. Solving the 3D Ising model with the conformal bootstrap II. c-minimization and precise critical exponents. *Journal of Statistical Physics*, **157**, 869.

[138] Engel, A., and Van den Broeck, C. 2001. *Statistical mechanics of learning.* Cambridge University Press.

[139] Fernández, L., Martin-Mayor, V., Seoane, B., et al. 2012. Equilibrium fluid-solid coexistence of hard spheres. *Physical Review Letters*, **108**, 165701.

[140] Fischer, K., and Hertz, J. 1991. *Spin glasses.* Cambridge University Press.

[141] Fisher, D. S., and Huse, D. A. 1988. Equilibrium behavior of the spin-glass ordered phase. *Physical Review B*, **38**, 386.

[142] Fisher, M. E., and Gaunt, D. S. 1964. Ising model and self-avoiding walks on hypercubical lattices and 'high-density' expansions. *Physical Review*, **133**, A224.

[143] Franz, S., and Montanari, A. 2007. Analytic determination of dynamical and mosaic length scales in a Kac glass model. *Journal of Physics A: Mathematical and Theoretical*, **40**, F251.

[144] Franz, S., and Parisi, G. 1995. Recipes for metastable states in spin glasses. *Journal de Physique I*, **5**, 1401.

[145] Franz, S., Parisi, G., Ricci-Tersenghi, F., et al. 2011. Field theory of fluctuations in glasses. *The European Physical Journal E*, **34**, 1.

[146] Franz, S., and Parisi, G. 2000. On non-linear susceptibility in supercooled liquids. *Journal of Physics: Condensed Matter*, **12**, 6335.

[147] Franz, S., and Parisi, G. 2016. The simplest model of jamming. *Journal of Physics A: Mathematical and Theoretical*, **49**, 145001.

[148] Franz, S., and Spigler, S. 2017. Mean-field avalanches in jammed spheres. *Physical Review E*, **95**, 022139.

[149] Franz, S., Jacquin, H., Parisi, G., et al. 2012. Quantitative field theory of the glass transition. *Proceedings of the National Academy of Sciences*, **109**, 18725.

[150] Franz, S., Parisi, G., Urbani, P., et al. 2015. Universal spectrum of normal modes in low-temperature glasses. *Proceedings of the National Academy of Sciences*, **112**, 14539.

[151] Franz, S., Parisi, G., Sevelev, M., et al. 2017. Universality of the SAT-UNSAT (jamming) threshold in non-convex continuous constraint satisfaction problems. *SciPost Physics*, **2**, 019.

[152] Frenkel, D., and Smit, B. 2001. *Understanding molecular simulation: From algorithms to applications*. Elsevier.

[153] Frisch, H. L., and Percus, J. K. 1999. High dimensionality as an organizing device for classical fluids. *Physical Review E*, **60**, 2942.

[154] Frisch, H. L., Rivier, N., and Wyler, D. 1985. Classical hard-sphere fluid in infinitely many dimensions. *Physical Review Letters*, **54**, 2061.

[155] Frisch, H., and Percus, J. 1987. Nonuniform classical fluid at high dimensionality. *Physical Review A*, **35**, 4696.

[156] Fuchs, M., and Cates, M. E. 2002. Theory of nonlinear rheology and yielding of dense colloidal suspensions. *Physical Review Letters*, **89**, 248304.

[157] Fullerton, C. J., and Berthier, L. 2017. Density controls the kinetic stability of ultrastable glasses. *Europhysics Letters*, **119**, 36003.

[158] Gallavotti, G. 2000. *Statistical mechanics. A short treatise*. Springer-Verlag.

[159] Gardiner, C. W. 1985. *Handbook of stochastic methods for physics, chemistry and natural sciences*. Springer-Verlag.

[160] Gardner, E. 1985. Spin glasses with p-spin interactions. *Nuclear Physics B*, **257**, 747.

[161] Gardner, E., and Derrida, B. 1988. Optimal storage properties of neural network models. *Journal of Physics A: Mathematical and General*, **21**, 271.

[162] Geirhos, K., Lunkenheimer, P., and Loidl, A. 2018. Johari-Goldstein relaxation far below T_g: Experimental evidence for the Gardner transition in structural glasses? *Physical Review Letters*, **120**, 085705.

[163] Georges, A., and Yedidia, J. S. 1991. How to expand around mean-field theory using high-temperature expansions. *Journal of Physics A: Mathematical and General*, **24**, 2173.

[164] Georges, A., Kotliar, G., Krauth, W., et al. 1996. Dynamical mean-field theory of strongly correlated fermion systems and the limit of infinite dimensions. *Reviews of Modern Physics*, **68**, 13.

[165] Gofman, M., Adler, J., Aharony, A., et al. 1993. Series and Monte Carlo study of high-dimensional Ising models. *Journal of Statistical Physics*, **71**, 1221.

[166] Goldstein, M. 1969. Viscous liquids and the glass transition: A potential energy barrier picture. *The Journal of Chemical Physics*, **51**, 3728.

[167] Goodrich, C. P., Liu, A. J., and Sethna, J. P. 2016. Scaling ansatz for the jamming transition. *Proceedings of the National Academy of Sciences*, **113**, 9745.

[168] Götze, W. 2008. *Complex dynamics of glass-forming liquids: A mode-coupling theory*. Oxford University Press.

[169] Götze, W. 1999. Recent tests of the mode-coupling theory for glassy dynamics. *Journal of Physics: Condensed Matter*, **11**, A1.

[170] Gross, D. J., and Mézard, M. 1984. The simplest spin glass. *Nuclear Physics B*, **240**, 431.

[171] Gross, D., Kanter, I., and Sompolinsky, H. 1985. Mean-field theory of the Potts glass. *Physical Review Letters*, **55**, 304.

[172] Haji-Akbari, A., Engel, M., Keys, A. S., et al. 2009. Disordered, quasicrystalline and crystalline phases of densely packed tetrahedra. *Nature*, **462**, 773.

[173] Hales, T., Adams, M., Bauer, G., et al. 2017. A formal proof of the Kepler conjecture. In *Forum of Mathematics, Pi*, vol. 5. Cambridge University Press.

[174] Hales, T. C. 2005. A proof of the Kepler conjecture. *Annals of Mathematics*, **162**, 1065.

[175] Hansen, J.-P., and McDonald, I. R. 1986. *Theory of simple liquids* (3rd edition). Academic Press.

[176] Henkel, M., Pleimling, M., and Sanctuary, R. (eds). 2007. *Ageing and the glass transition*. Springer.

[177] Heuer, A. 2008. Exploring the potential energy landscape of glass-forming systems: From inherent structures via metabasins to macroscopic transport. *Journal of Physics: Condensed Matter*, **20**, 373101.

[178] Hicks, C. L., Wheatley, M. J., Godfrey, M. J., et al. 2018. Gardner transition in physical dimensions. *Physical Review Letters*, **120**, 225501.

[179] Hopkins, A. B., Stillinger, F. H., and Torquato, S. 2013. Disordered strictly jammed binary sphere packings attain an anomalously large range of densities. *Physical Review E*, **88**, 022205.

[180] Hull, D., and Bacon, D. J. 2011. *Introduction to dislocations*. Elsevier.

[181] Hunter, G. L., and Weeks, E. R. 2012. The physics of the colloidal glass transition. *Reports on Progress in Physics*, **75**, 066501.

[182] Ikeda, A., Berthier, L., and Biroli, G. 2013. Dynamic criticality at the jamming transition. *The Journal of Chemical Physics*, **138**, 12A507.

[183] Ikeda, A., and Miyazaki, K. 2010. Mode-coupling theory as a mean-field description of the glass transition. *Physical Review Letters*, **104**, 255704.

[184] Ikeda, H., Miyazaki, K., Yoshino, H., et al. 2017. Decoupling phenomena and replica symmetry breaking of binary mixtures. *arXiv:1710.08373*.

[185] Ikeda, H., Zamponi, F., and Ikeda, A. 2017. Mean field theory of the swap Monte Carlo algorithm. *The Journal of Chemical Physics*, **147**, 234506.

[186] Irving, J. H., and Kirkwood, J. G. 1950. The statistical mechanical theory of transport processes. IV. The equations of hydrodynamics. *The Journal of Chemical Physics*, **18**, 817.

[187] Janssen, H.-K. 1976. On a Lagrangean for classical field dynamics and renormalization group calculations of dynamical critical properties. *Zeitschrift für Physik B Condensed Matter*, **23**, 377.

[188] Janssen, L. M. C. 2018. Mode-coupling theory of the glass transition: A primer. *Frontiers in Physics*, **6**, 97.

[189] Janssen, L. M. C., and Reichman, D. R. 2015. Microscopic dynamics of supercooled liquids from first principles. *Physical Review Letters*, **115**, 205701.

[190] Jaric, M. V. 2012. *Introduction to the mathematics of quasicrystals*. Elsevier.

[191] Jenssen, M., Joos, F., and Perkins, W. 2019. On the hard sphere model and sphere packings in high dimensions. *Forum of Mathematics, Sigma*, **7**, E1.

[192] Jin, Y., and Yoshino, H. 2017. Exploring the complex free-energy landscape of the simplest glass by rheology. *Nature Communications*, **8**, 14935.

[193] Jin, Y., Urbani, P., Zamponi, F., et al. 2018. A stability-reversibility map unifies elasticity, plasticity, yielding and jamming in hard sphere glasses. *Science Advances*, **4**, eaat6387.

[194] Johari, G. 2000. A resolution for the enigma of a liquids configurational entropy-molecular kinetics relation. *The Journal of Chemical Physics*, **112**, 8958.

[195] Johnson, K. L. 1987. *Contact mechanics*. Cambridge University Press.

[196] Joslin, C. 1982. Third and fourth virial coefficients of hard hyperspheres of arbitrary dimensionality. *The Journal of Chemical Physics*, **77**, 2701.

[197] Kabatiansky, G. A., and Levensthein, V. I. 1978. Bounds for packings on a sphere and in space. *Problems on Information Transmission*, **14**, 1.

[198] Kallus, Y. 2013. Statistical mechanics of the lattice sphere packing problem. *Physical Review E*, **87**, 063307.

[199] Kallus, Y., Marcotte, E., and Torquato, S. 2013. Jammed lattice sphere packings. *Physical Review E*, **88**, 062151.

[200] Kamenev, A. 2009. *Field theory of non-equilibrium systems*. Cambridge University Press.

[201] Kauzmann, W. 1948. The glassy state and the behaviour of liquids at low temperature. *Chemical Reviews*, **43**, 219.

[202] Kirkpatrick, S., Gelatt, C. D., and Vecchi, M. P. 1983. Optimization by simulated annealing. *Science*, **220**, 671.

[203] Kirkpatrick, T. R., and Thirumalai, D. 1987. Dynamics of the structural glass transition and the p-spin-interaction spin-glass model. *Physical Review Letters*, **58**, 2091.

[204] Kirkpatrick, T. R., and Thirumalai, D. 1988. Comparison between dynamical theories and metastable states in regular and glassy mean-field spin models with underlying first-order-like phase transitions. *Physical Review A*, **37**, 4439.

[205] Kirkpatrick, T. R., and Thirumalai, D. 1989. Random solutions from a regular density functional Hamiltonian: A static and dynamical theory for the structural glass transition. *Journal of Physics A: Mathematical and General*, **22**, L149.

[206] Kirkpatrick, T. R., and Wolynes, P. G. 1987. Connections between some kinetic and equilibrium theories of the glass transition. *Physical Review A*, **35**, 3072.

[207] Kirkpatrick, T. R., and Wolynes, P. G. 1987. Stable and metastable states in mean-field Potts and structural glasses. *Physical Review B*, **36**, 8552.

[208] Kirkpatrick, T. R., Thirumalai, D., and Wolynes, P. G. 1989. Scaling concepts for the dynamics of viscous liquids near an ideal glassy state. *Physical Review A*, **40**, 1045.

[209] Kirkwood, J. G., and Monroe, E. 1940. On the theory of fusion. *The Journal of Chemical Physics*, **8**, 845.

[210] Klein, W., and Frisch, H. 1986. Instability in the infinite dimensional hard sphere fluid. *The Journal of Chemical Physics*, **84**, 968.

[211] Koch, H., Radin, C., and Sadun, L. 2005. Most stable structure for hard spheres. *Physical Review E*, **72**, 016708.

[212] Kossevich, A. M. 1999. *The crystal lattice: Phonons, solitons, dislocations*. Wiley Online Library.

[213] Kraichnan, R. H. 1962. Stochastic models for many-body systems. I. Infinite systems in thermal equilibrium. *Journal of Mathematical Physics*, **3**, 475.

[214] Krzakala, F., and Zdeborová, L. 2010. Following Gibbs states adiabatically–The energy landscape of mean-field glassy systems. *Europhysics Letters*, **90**, 66002.

[215] Krzakala, F., and Kurchan, J. 2007. Landscape analysis of constraint satisfaction problems. *Physical Review E*, **76**, 021122.

[216] Krzakala, F., and Zdeborová, L. 2013. Performance of simulated annealing in p-spin glasses. *Journal of Physics: Conference Series*, **473**, 12022.

[217] Krzakala, F., Montanari, A., Ricci-Tersenghi, F., et al. 2007. Gibbs states and the set of solutions of random constraint satisfaction problems. *Proceedings of the National Academy of Sciences*, **104**, 10318.

[218] Kurchan, J. 2003. Supersymmetry, replica and dynamic treatments of disordered systems: A parallel presentation. *Markov Processes and Related Fields*, **9**, 243.

[219] Kurchan, J., and Levine, D. 2010. Order in glassy systems. *Journal of Physics A: Mathematical and Theoretical*, **44**, 035001.

[220] Kurchan, J., Parisi, G., and Zamponi, F. 2012. Exact theory of dense amorphous hard spheres in high dimension. I. The free energy. *Journal of Statistical Mechanics: Theory and Experiment*, **2012**, P10012.

[221] Kurchan, J., Parisi, G., Urbani, P., et al. 2013. Exact theory of dense amorphous hard spheres in high dimension. II. The high density regime and the Gardner transition. *The Journal of Physical Chemistry B*, **117**, 12979.

[222] Landau, L. D., and Lifshitz, E. M. 1986. *Course of theoretical physics, vol.7: Theory of elasticity*. Butterworth-Heinemann.

[223] Landau, L. D., and Lifshitz, E. M. 1980. *Course of theoretical physics, vol.5: Statistical physics*. Butterworth-Heinemann.

[224] Larson, D., Katzgraber, H. G., Moore, M., et al. 2013. Spin glasses in a field: Three and four dimensions as seen from one space dimension. *Physical Review B*, **87**, 024414.

[225] Le Doussal, P., Müller, M., and Wiese, K. J. 2012. Equilibrium avalanches in spin glasses. *Physical Review B*, **85**, 214402.

[226] Lebowitz, J., and Penrose, O. 1964. Convergence of virial expansions. *Journal of Mathematical Physics*, **5**, 841.

[227] Lerner, E., During, G., and Wyart, M. 2013. Low-energy non-linear excitations in sphere packings. *Soft Matter*, **9**, 8252.

[228] Lerner, E., Düring, G., and Bouchbinder, E. 2016. Statistics and properties of low-frequency vibrational modes in structural glasses. *Physical Review Letters*, **117**, 035501.

[229] Leuzzi, L., Parisi, G., Ricci-Tersenghi, F., et al. 2008. Dilute one-dimensional spin glasses with power law decaying interactions. *Physical Review Letters*, **101**, 107203.

[230] Liao, Q., and Berthier, L. 2019. Hierarchical landscape of hard disk glasses. *Physical Review X*, **9**, 011049.

[231] Lin, J., Lerner, E., Rosso, A., et al. 2014. Scaling description of the yielding transition in soft amorphous solids at zero temperature. *Proceedings of the National Academy of Sciences*, **111**, 14382.

[232] Liu, A., Nagel, S., Van Saarloos, W., et al. 2011. The jamming scenario – an introduction and outlook. In Berthier, L., Biroli, G., Bouchaud, J.-P., et al. (eds), *Dynamical Heterogeneities and Glasses*. Oxford University Press.

[233] Liu, A. J., and Nagel, S. R. 2001. *Jamming and rheology: Constrained dynamics on microscopic and macroscopic scales*. CRC Press.

[234] Liu, A. J., and Nagel, S. R. 2010. The jamming transition and the marginally jammed solid. *Annual Review of Condensed Matter Physics*, **1**, 347.

[235] Lubachevsky, B. D., and Stillinger, F. H. 1990. Geometric properties of random disk packings. *Journal of Statistical Physics*, **60**, 561.

[236] Luban, M., and Baram, A. 1982. Third and fourth virial coefficients of hard hyperspheres of arbitrary dimensionality. *The Journal of Chemical Physics*, **76**, 3233.

[237] Lundow, P. H., and Markström, K. 2009. Critical behavior of the Ising model on the four-dimensional cubic lattice. *Physical Review E*, **80**, 031104.

[238] Maimbourg, T., and Kurchan, J. 2016. Approximate scale invariance in particle systems: A large-dimensional justification. *Europhysics Letters*, **114**, 60002.

[239] Maimbourg, T., Kurchan, J., and Zamponi, F. 2016. Solution of the dynamics of liquids in the large-dimensional limit. *Physical Review Letters*, **116**, 015902.

[240] Maimbourg, T., Sellitto, M., Semerjian, G., et al. 2018. Generating dense packings of hard spheres by soft interaction design. *SciPost Physics*, **4**, 1.

[241] Mangeat, M., and Zamponi, F. 2016. Quantitative approximation schemes for glasses. *Physical Review E*, **93**, 012609.

[242] Mari, R., and Kurchan, J. 2011. Dynamical transition of glasses: From exact to approximate. *The Journal of Chemical Physics*, **135**, 124504.

[243] Mari, R., Krzakala, F., and Kurchan, J. 2009. Jamming versus glass transitions. *Physical Review Letters*, **103**, 025701.

[244] Marinari, E., Parisi, G., Ricci-Tersenghi, F., et al. 2000. Replica symmetry breaking in short-range spin glasses: Theoretical foundations and numerical evidences. *Journal of Statistical Physics*, **98**, 973.

[245] Martin, P. C., Siggia, E. D., and Rose, H. A. 1973. Statistical dynamics of classical systems. *Physical Review A*, **8**, 423.

[246] Martinelli, F. 1999. Lectures on Glauber dynamics for discrete spin models. In Bernard, P. (ed), *Lectures on probability theory and statistics*. Springer-Verlag.

[247] Martiniani, S., Schrenk, K. J., Stevenson, J. D., et al. 2016. Turning intractable counting into sampling: Computing the configurational entropy of three-dimensional jammed packings. *Physical Review E*, **93**, 012906.

[248] Maxwell, J. C. 1864. On the calculation of the equilibrium and stiffness of frames. *The London, Edinburgh, and Dublin Philosophical Magazine and Journal of Science*, **27**, 294.

[249] Mehta, A. (ed). 1994. *Granular matter: An interdisciplinary approach*. Springer-Verlag.

[250] Mézard, M. 1999. How to compute the thermodynamics of a glass using a cloned liquid. *Physica A*, **265**, 352.

[251] Mézard, M., and Montanari, A. 2009. *Information, physics and computation*. Oxford University Press.

[252] Mézard, M., and Parisi, G. 1996. A tentative replica study of the glass transition. *Journal of Physics A: Mathematical and General*, **29**, 6515.

[253] Mézard, M., and Parisi, G. 2012. Glasses and replicas. In Wolynes, P. G., and Lubchenko, V. (eds), *Structural glasses and supercooled liquids: Theory, experiment and applications*. Wiley & Sons.

[254] Mézard, M., Parisi, G., and Virasoro, M. A. 1987. *Spin glass theory and beyond*. World Scientific.

[255] Mézard, M., and Parisi, G. 1991. Replica field theory for random manifolds. *Journal de Physique I*, **1**, 809.

[256] Mézard, M., and Parisi, G. 1999. A first-principle computation of the thermodynamics of glasses. *The Journal of Chemical Physics*, **111**, 1076.

[257] Mézard, M., and Parisi, G. 2000. Statistical physics of structural glasses. *Journal of Physics: Condensed Matter*, **12**, 6655.

[258] Mizuno, H., Shiba, H., and Ikeda, A. 2017. Continuum limit of the vibrational properties of amorphous solids. *Proceedings of the National Academy of Sciences*, **114**, E9767.

[259] Mon, K. K., and Percus, J. K. 1999. Virial expansion and liquid–vapor critical points of high dimension classical fluids. *The Journal of Chemical Physics*, **110**, 2734.

[260] Monasson, R. 1995. Structural glass transition and the entropy of the metastable states. *Physical Review Letters*, **75**, 2847.

[261] Montanari, A., and Semerjian, G. 2006. Rigorous inequalities between length and time scales in glassy systems. *Journal of Statistical Physics*, **125**, 23.

[262] Montanari, A., and Ricci-Tersenghi, F. 2003. On the nature of the low-temperature phase in discontinuous mean-field spin glasses. *The European Physical Journal B*, **33**, 339.

[263] Montanari, A., and Ricci-Tersenghi, F. 2004. Cooling-schedule dependence of the dynamics of mean-field glasses. *Physical Review B*, **70**, 134406.

[264] Moore, M., and Bray, A. J. 2011. Disappearance of the de Almeida–Thouless line in six dimensions. *Physical Review B*, **83**, 224408.

[265] Mora, T., Walczak, A. M., and Zamponi, F. 2012. Transition path sampling algorithm for discrete many-body systems. *Physical Review E*, **85**, 036710.

[266] Morita, T., and Hiroike, K. 1961. A new approach to the theory of classical fluids. III: General treatment of classical systems. *Progress of Theoretical Physics*, **25**, 537.

[267] Morse, P. K., and Corwin, E. I. 2014. Geometric signatures of jamming in the mechanical vacuum. *Physical Review Letters*, **112**, 115701.

[268] Moustrou, P. 2017. On the density of cyclotomic lattices constructed from codes. *International Journal of Number Theory*, **13**, 1261.

[269] Müller, M., and Wyart, M. 2015. Marginal stability in structural, spin, and electron glasses. *Annual Review of Condensed Matter Physics*, **6**, 177.

[270] Nakayama, D., Yoshino, H., and Zamponi, F. 2016. Protocol-dependent shear modulus of amorphous solids. *Journal of Statistical Mechanics: Theory and Experiment*, **2016**, 104001.

[271] Nebe, G., and Sloane, N. J. A. 2015. *Table of densest packings presently known*. www.math.rwth-aachen.de/~Gabriele.Nebe/LATTICES/density.html. Accessed: 2018-05-15.

[272] Nicolas, A., Ferrero, E. E., Martens, K., et al. 2018. Deformation and flow of amorphous solids: Insights from elastoplastic models. *Reviews of Modern Physics*, **90**, 045006.

[273] Ninarello, A., Berthier, L., and Coslovich, D. 2017. Models and algorithms for the next generation of glass transition studies. *Physical Review X*, **7**, 021039.

[274] Nishimori, H. 2001. *Statistical physics of spin glasses and information processing: An introduction*. Clarendon Press.

[275] Noya, E. G., and Almarza, N. G. 2015. Entropy of hard spheres in the close-packing limit. *Molecular Physics*, **113**, 1061.

[276] O'Hern, C. S., Langer, S. A., Liu, A. J., et al. 2002. Random packings of frictionless particles. *Physical Review Letters*, **88**, 075507.

[277] O'Hern, C. S., Silbert, L. E., Liu, A. J., et al. 2003. Jamming at zero temperature and zero applied stress: The epitome of disorder. *Physical Review E*, **68**, 011306.

[278] Ozawa, M., Kuroiwa, T., Ikeda, A., et al. 2012. Jamming transition and inherent structures of hard spheres and disks. *Physical Review Letters*, **109**, 205701.

[279] Ozawa, M., Berthier, L., and Coslovich, D. 2017. Exploring the jamming transition over a wide range of critical densities. *SciPost Physics*, **3**, 027.

[280] Ozawa, M., Berthier, L., Biroli, G., et al. 2018. Random critical point separates brittle and ductile yielding transitions in amorphous materials. *Proceedings of the National Academy of Sciences*, **115**, 6656.

[281] Panchenko, D. 2013. *The Sherrington-Kirkpatrick model*. Springer Science & Business Media.

[282] Parisi, G. 1988. *Statistical field theory*. Addison-Wesley.

[283] Parisi, G. 2008. On the most compact regular lattices in large dimensions: A statistical mechanical approach. *Journal of Statistical Physics*, **132**, 207.

[284] Parisi, G., and Slanina, F. 2000. Toy model for the mean-field theory of hard-sphere liquids. *Physical Review E*, **62**, 6554.

[285] Parisi, G. 1980. A sequence of approximated solutions to the SK model for spin glasses. *Journal of Physics A: Mathematical and General*, **13**, L115.

[286] Parisi, G. 2003. Glasses, replicas and all that. In Barrat, J.-L., Feigelman, M., Kurchan, J., et al. (eds), *Slow relaxations and nonequilibrium dynamics in condensed matter*. Springer-Verlag.

[287] Parisi, G., and Potters, M. 1995. On the number of metastable states in spin glasses. *Europhysics Letters*, **32**, 13.

[288] Parisi, G., and Rizzo, T. 2013. Critical dynamics in glassy systems. *Physical Review E*, **87**, 012101.

[289] Parisi, G., and Sourlas, N. 2000. P-adic numbers and replica symmetry breaking. *The European Physical Journal B*, **14**, 535.

[290] Parisi, G., and Temesvári, T. 2012. Replica symmetry breaking in and around six dimensions. *Nuclear Physics B*, **858**, 293.

[291] Parisi, G., and Zamponi, F. 2006. Amorphous packings of hard spheres for large space dimension. *Journal of Statistical Mechanics: Theory and Experiment*, **2006**, P03017.

[292] Parisi, G., and Zamponi, F. 2010. Mean-field theory of hard sphere glasses and jamming. *Reviews of Modern Physics*, **82**, 789.

[293] Parisi, G., Procaccia, I., Rainone, C., et al. 2017. Shear bands as manifestation of a criticality in yielding amorphous solids. *Proceedings of the National Academy of Sciences*, **114**, 5577.

[294] Percus, A., Istrate, G., and Moore, C. (eds). 2006. *Computational complexity and statistical physics*. Oxford University Press.

[295] Pinson, D., Zou, R. P., Yu, A. B., et al. 1998. Coordination number of binary mixtures of spheres. *Journal of Physics D: Applied Physics*, **31**, 457.

[296] Plefka, T. 1982. Convergence condition of the TAP equation for the infinite-ranged Ising spin glass model. *Journal of Physics A: Mathematical and General*, **15**, 1971.

[297] Presutti, E. 2008. *Scaling limits in statistical mechanics and microstructures in continuum mechanics*. Springer Science & Business Media.

[298] Rainone, C., and Urbani, P. 2016. Following the evolution of glassy states under external perturbations: The full replica symmetry breaking solution. *Journal of Statistical Mechanics: Theory and Experiment*, **2016**, P053302.

[299] Rainone, C., Urbani, P., Yoshino, H., et al. 2015. Following the evolution of hard sphere glasses in infinite dimensions under external perturbations: Compression and shear strain. *Physical Review Letters*, **114**, 015701.

[300] Rammal, R., Toulouse, G., and Virasoro, M. A. 1986. Ultrametricity for physicists. *Reviews of Modern Physics*, **58**, 765.

[301] Reichman, D. R., and Charbonneau, P. 2005. Mode-coupling theory. *Journal of Statistical Mechanics: Theory and Experiment*, **2005**, P05013.

[302] Richert, R., and Angell, C. 1998. Dynamics of glass-forming liquids. V. On the link between molecular dynamics and configurational entropy. *The Journal of Chemical Physics*, **108**, 9016.

[303] Rizzo, T. 2013. Replica-symmetry-breaking transitions and off-equilibrium dynamics. *Physical Review E*, **88**, 032135.

[304] Rizzo, T., and Voigtmann, T. 2015. Qualitative features at the glass crossover. *Europhysics Letters*, **111**, 56008.

[305] Rodney, D., Tanguy, A., and Vandembroucq, D. 2011. Modeling the mechanics of amorphous solids at different length scale and time scale. *Modelling and Simulation in Materials Science and Engineering*, **19**, 083001.

[306] Rogers, C. A. 1964. *Packing and covering*. Cambridge University Press.

[307] Rogers, C. A. 1947. Existence theorems in the geometry of numbers. *Annals of Mathematics*, **48**, 994.

[308] Rohrmann, R. D., Robles, M., de Haro, M. L., et al. 2008. Virial series for fluids of hard hyperspheres in odd dimensions. *The Journal of Chemical Physics*, **129**, 014510.

[309] Ruelle, D. 1982. Do turbulent crystals exist? *Physica A*, **113**, 619.

[310] Ruelle, D. 1999. *Statistical mechanics: Rigorous results*. World Scientific.

[311] Ruijgrok, T. W., and Tjon, J. 1973. Critical slowing down and nonlinear response in an exactly solvable stochastic model. *Physica*, **65**, 539.

[312] Russo, L. 2004. *The forgotten revolution: How science was born in 300 BC and why it had to be reborn*. Springer-Verlag.

[313] Sastry, S. 2000. Evaluation of the configurational entropy of a model liquid from computer simulations. *Journal of Physics: Condensed Matter*, **12**, 6515.

[314] Sausset, F., Biroli, G., and Kurchan, J. 2010. Do solids flow? *Journal of Statistical Physics*, **140**, 718.

[315] Scalliet, C., Berthier, L., and Zamponi, F. 2017. Absence of marginal stability in a structural glass. *Physical Review Letters*, **119**, 205501.

[316] Scalliet, C., Berthier, L., and Zamponi, F. 2019. Marginally stable phases in mean-field structural glasses. *Physical Review E*, **99**, 012107.

[317] Schmid, B., and Schilling, R. 2010. Glass transition of hard spheres in high dimensions. *Physical Review E*, **81**, 041502.

[318] Schreck, C. F., Mailman, M., Chakraborty, B., et al. 2012. Constraints and vibrations in static packings of ellipsoidal particles. *Physical Review E*, **85**, 061305.

[319] Schultz, A. J., and Kofke, D. A. 2014. Fifth to eleventh virial coefficients of hard spheres. *Physical Review E*, **90**, 023301.

[320] Schweizer, K. S., and Saltzman, E. J. 2003. Entropic barriers, activated hopping, and the glass transition in colloidal suspensions. *The Journal of Chemical Physics*, **119**, 1181.

[321] Sciortino, F. 2002. One liquid, two glasses. *Nature Materials*, **1**, 145.

[322] Sciortino, F., Kob, W., and Tartaglia, P. 1999. Inherent structure entropy of super-cooled liquids. *Physical Review Letters*, **83**, 3214.

[323] Sciortino, F., and Tartaglia, P. 2005. Glassy colloidal systems. *Advances in Physics*, **54**, 471.

[324] Seguin, A., and Dauchot, O. 2016. Experimental evidence of the Gardner phase in a granular glass. *Physical Review Letters*, **117**, 228001.

[325] Sellitto, M., and Zamponi, F. 2013. A thermodynamic description of colloidal glasses. *Europhysics Letters*, **103**, 46005.

[326] Seoane, B., and Zamponi, F. 2018. Spin-glass-like aging in colloidal and granular glasses. *Soft Matter*, **14**, 5222.

[327] Seoane, B., Reid, D. R., de Pablo, J. J., et al. 2018. Low-temperature anomalies of a vapor deposited glass. *Physical Review Materials*, **2**, 015602.

[328] Simon, B. 2014. *The statistical mechanics of lattice gases*. Princeton University Press.

[329] Singh, Y., Stoessel, J. P., and Wolynes, P. G. 1985. Hard-sphere glass and the density-functional theory of aperiodic crystals. *Physical Review Letters*, **54**, 1059.

[330] Skoge, M., Donev, A., Stillinger, F. H., et al. 2006. Packing hyperspheres in high-dimensional Euclidean spaces. *Physical Review E*, **74**, 041127.

[331] Sommers, H.-J. 1983. Properties of Sompolinsky's mean field theory of spin glasses. *Journal of Physics A: Mathematical and General*, **16**, 447.

[332] Sommers, H.-J. 1985. Parisi function $q(x)$ for spin glasses near T_c. *Journal de Physique Lettres*, **46**, 779.

[333] Sommers, H.-J., and Dupont, W. 1984. Distribution of frozen fields in the mean-field theory of spin glasses. *Journal of Physics C: Solid State Physics*, **17**, 5785.

[334] Stillinger, F. H. 1988. Supercooled liquids, glass transitions, and the Kauzmann paradox. *The Journal of Chemical Physics*, **88**, 7818.

[335] Stillinger, F. H., and Weber, T. A. 1983. Dynamics of structural transitions in liquids. *Physical Review A*, **28**, 2408.

[336] Stillinger, F. H., Debenedetti, P. G., and Truskett, T. M. 2001. The Kauzmann paradox revisited. *The Journal of Physical Chemistry B*, **105**, 11809.

[337] Svidzinsky, A., Scully, M., and Herschbach, D. 2014. Bohrs molecular model, a century later. *Physics Today*, **67**, 33.

[338] Swallen, S. F., Kearns, K. L., Mapes, M. K., et al. 2007. Organic glasses with exceptional thermodynamic and kinetic stability. *Science*, **315**, 353.

[339] Szamel, G. 2017. Simple theory for the dynamics of mean-field-like models of glass-forming fluids. *Physical Review Letters*, **119**, 155502.

[340] Talagrand, M. 2003. *Spin glasses: A challenge for mathematicians: Cavity and mean field models*. Springer Science & Business Media.

[341] Talagrand, M. 2010. *Mean field models for spin glasses: Volume I: Basic examples*. Springer Science & Business Media.

[342] Thouless, D., Anderson, P., and Palmer, R. 1977. Solution of 'Solvable model of a spin glass'. *Philosophical Magazine*, **35**, 593.

[343] Toda, M., Kubo, R., and Saito, N. 1992. *Statistical physics I: Equilibrium statistical mechanics*.

[344] Torquato, S., and Stillinger, F. H. 2010. Jammed hard-particle packings: From Kepler to Bernal and beyond. *Reviews of Modern Physics*, **82**, 2633.

[345] Torquato, S., and Jiao, Y. 2009. Dense packings of the Platonic and Archimedean solids. *Nature*, **460**, 876.

[346] Urbani, P., and Zamponi, F. 2017. Shear yielding and shear jamming of dense hard sphere glasses. *Physical Review Letters*, **118**, 038001.

[347] Vagberg, D., Olsson, P., and Teitel, S. 2011. Glassiness, rigidity, and jamming of frictionless soft core disks. *Physical Review E*, **83**, 031307.

[348] Van Hecke, M. 2010. Jamming of soft particles: Geometry, mechanics, scaling and isostaticity. *Journal of Physics: Condensed Matter*, **22**, 033101.

[349] Van Kampen, N. G. 1992. *Stochastic processes in physics and chemistry*. Elsevier.

[350] van Meel, J. A., Charbonneau, B., Fortini, A., et al. 2009. Hard-sphere crystallization gets rarer with increasing dimension. *Physical Review E*, **80**, 061110.

[351] Vance, S. 2011. Improved sphere packing lower bounds from Hurwitz lattices. *Advances in Mathematics*, **227**, 2144.

[352] Venkatesh, A. 2013. A note on sphere packings in high dimension. *International Mathematics Research Notices*, **2013**, 1628.

[353] Viazovska, M. S. 2017. The sphere packing problem in dimension 8. *Annals of Mathematics*, **185**, 991.

[354] Wales, D. 2003. *Energy landscapes: Applications to clusters, biomolecules and glasses*. Cambridge University Press.

[355] Wang, W., Machta, J., Munoz-Bauza, H., et al. 2017. Number of thermodynamic states in the three-dimensional Edwards-Anderson spin glass. *Physical Review B*, **96**, 184417.

[356] Witten, E. 1980. Quarks, atoms, and the $1/N$ expansion. *Physics Today*, **33**, 38.

[357] Wolynes, P., and Lubchenko, V. (eds). 2012. *Structural glasses and supercooled liquids: Theory, experiment, and applications.* Wiley & Sons.

[358] Woodcock, L. 1997. Entropy difference between the face-centred cubic and hexagonal close-packed crystal structures. *Nature*, **385**, 141.

[359] Wyart, M., Silbert, L., Nagel, S., et al. 2005. Effects of compression on the vibrational modes of marginally jammed solids. *Physical Review E*, **72**, 051306.

[360] Wyart, M., Nagel, S., and Witten, T. 2005. Geometric origin of excess low-frequency vibrational modes in weakly connected amorphous solids. *Europhysics Letters*, **72**, 486.

[361] Wyart, M. 2012. Marginal stability constrains force and pair distributions at random close packing. *Physical Review Letters*, **109**, 125502.

[362] Wyler, D., Rivier, N., and Frisch, H. L. 1987. Hard-sphere fluid in infinite dimensions. *Physical Review A*, **36**, 2422.

[363] Yaida, S., Berthier, L., Charbonneau, P., et al. 2016. Point-to-set lengths, local structure, and glassiness. *Physical Review E*, **94**, 032605.

[364] Yeo, J., and Moore, M. A. 2012. Renormalization group analysis of the M-p-spin glass model with $p = 3$ and $M = 3$. *Physical Review B*, **85**, 100405.

[365] Yerazunis, S., Cornell, S., and Wintner, B. 1965. Dense random packing of binary mixtures of spheres. *Nature*, **207**, 835.

[366] Yoshino, H. 2012. Replica theory of the rigidity of structural glasses. *The Journal of Chemical Physics*, **136**, 214108.

[367] Yoshino, H. 2018. Translational and orientational glass transitions in the large-dimensional limit: A generalized replicated liquid theory and an application to patchy colloids. *arXiv:1807.04095.*

[368] Yoshino, H., and Mézard, M. 2010. Emergence of rigidity at the structural glass transition: A first-principles computation. *Physical Review Letters*, **105**, 015504.

[369] Yoshino, H., and Zamponi, F. 2014. Shear modulus of glasses: Results from the full replica-symmetry-breaking solution. *Physical Review E*, **90**, 022302.

[370] Zdeborová, L., and Krzakala, F. 2007. Phase transitions in the coloring of random graphs. *Physical Review E*, **76**, 031131.

[371] Zdeborová, L., and Krzakala, F. 2010. Generalization of the cavity method for adiabatic evolution of Gibbs states. *Physical Review B*, **81**, 224205.

[372] Zdeborová, L., and Krzakala, F. 2016. Statistical physics of inference: Thresholds and algorithms. *Advances in Physics*, **65**, 453.

[373] Zhang, C., and Pettitt, B. M. 2014. Computation of high-order virial coefficients in high-dimensional hard-sphere fluids by Mayer sampling. *Molecular Physics*, **112**, 1427.

Index

Printed in the United States
by Baker & Taylor Publisher Services